NOVEL PROCESSING METHODS FOR PLANT-BASED HEALTH FOODS

Extraction, Encapsulation, and Health Benefits of Bioactive Compounds

Innovations in Agricultural and Biological Engineering

NOVEL PROCESSING METHODS FOR PLANT-BASED HEALTH FOODS

Extraction, Encapsulation, and Health Benefits of Bioactive Compounds

Edited by

Megh R. Goyal, PhD, PE

N. Veena, PhD

Ritesh B. Watharkar, PhD

AAP APPLE ACADEMIC PRESS

First edition published 2023

Apple Academic Press Inc.
1265 Goldenrod Circle, NE,
Palm Bay, FL 32905 USA

760 Laurentian Drive, Unit 19,
Burlington, ON L7N 0A4, CANADA

CRC Press
6000 Broken Sound Parkway NW,
Suite 300, Boca Raton, FL 33487-2742 USA

4 Park Square, Milton Park,
Abingdon, Oxon, OX14 4RN UK

Library and Archives Canada Cataloguing in Publication

Title: Novel processing methods for plant-based health foods : extraction, encapsulation, and health benefits of bioactive compounds / edited by Megh R. Goyal, PhD, PE, Ritesh B. Watharkar, PhD, N. Veena, PhD.
Names: Goyal, Megh R., editor. | Watharkar, Ritesh B., editor. | Veena, N. (Nagaraj), editor.
Series: Innovations in agricultural and biological engineering.
Description: First edition. | Series statement: Innovations in agricultural and biological engineering | Includes bibliographical references and index.
Identifiers: Canadiana (print) 2022042263X | Canadiana (ebook) 20220422656 | ISBN 9781774910740 (hardcover) | ISBN 9781774910757 (softcover) | ISBN 9781003284109 (ebook)
Subjects: LCSH: Food industry and trade. | LCSH: Bioactive compounds. | LCSH: Natural foods—Processing.
Classification: LCC TP248.65.F66 N68 2023 | DDC 664/.02—dc23

Library of Congress Cataloging-in-Publication Data

CIP data on file with US Library of Congress

ISBN: 978-1-77491-074-0 (hbk)
ISBN: 978-1-77491-075-7 (pbk)
ISBN: 978-1-00328-410-9 (ebk)

ABOUT THE BOOK SERIES: INNOVATIONS IN AGRICULTURAL AND BIOLOGICAL ENGINEERING

Under this book series, Apple Academic Press Inc. is publishing book volumes over a span of 8–10 years in the specialty areas defined by the American Society of Agricultural and Biological Engineers (<asabe.org>). Apple Academic Press Inc. aims to be a principal source of books in agricultural and biological engineering. We welcome book proposals from readers in areas of their expertise.

The mission of this series is to provide knowledge and techniques for agricultural and biological engineers (ABEs). The book series offers high-quality reference and academic content on agricultural and biological engineering (ABE) that is accessible to academicians, researchers, scientists, university faculty and university-level students, and professionals around the world.

Agricultural and biological engineers ensure that the world has the necessities of life, including safe and plentiful food, clean air and water, renewable fuel and energy, safe working conditions, and a healthy environment by employing knowledge and expertise of the sciences, both pure and applied, and engineering principles. Biological engineering applies engineering practices to problems and opportunities presented by living things and the natural environment in agriculture.

ABE embraces a variety of the following specialty areas (<asabe.org>): aquaculture engineering, biological engineering, energy, farm machinery and power engineering, food, and process engineering, forest engineering, information, and electrical technologies, soil, and water conservation engineering, natural resources engineering, nursery, and greenhouse engineering, safety, and health, and structures and environment.

For this book series, we welcome chapters on the following specialty areas (but not limited to):

- Academia to industry to end-user loop in agricultural engineering;
- Agricultural mechanization;
- Aquaculture engineering;
- Biological engineering in agriculture;

- Biotechnology applications in agricultural engineering;
- Energy source engineering;
- Farm to fork technologies in agriculture;
- Food and bioprocess engineering;
- Forest engineering;
- GPS and remote sensing potential in agricultural engineering;
- Hill land agriculture;
- Human factors in engineering;
- Impact of global warming and climatic change on agriculture economy;
- Information and electrical technologies;
- Irrigation and drainage engineering;
- Nanotechnology applications in agricultural engineering;
- Natural resources engineering;
- Nursery and greenhouse engineering;
- Potential of phytochemicals from agricultural and wild plants for human health;
- Power systems and machinery design;
- Robot engineering and drones in agriculture;
- Rural electrification;
- Sanitary engineering;
- Simulation and computer modeling;
- Smart engineering applications in agriculture;
- Soil and water engineering;
- Micro-irrigation engineering;
- Structures and environment engineering;
- Waste management and recycling;
- Any other focus areas.

For more information on this series, readers may contact:

Megh R. Goyal, PhD, PE
Book Series Senior Editor-in-Chief: Innovations in Agricultural and Biological Engineering
E-mail: goyalmegh@gmail.com

OTHER BOOKS ON AGRICULTURAL AND BIOLOGICAL ENGINEERING BY APPLE ACADEMIC PRESS, INC.

Management of Drip/Trickle or Micro Irrigation
Megh R. Goyal, PhD, PE, Senior Editor-in-Chief
Evapotranspiration: Principles and Applications for Water Management
Megh R. Goyal, PhD, PE and Eric W. Harmsen, PhD Editors

Book Series: Research Advances in Sustainable Micro Irrigation
Senior Editor-in-Chief: Megh R. Goyal, PhD, PE

- **Volume 1:** Sustainable Micro Irrigation: Principles and Practices
- **Volume 2:** Sustainable Practices in Surface and Subsurface Micro Irrigation
- **Volume 3:** Sustainable Micro Irrigation Management for Trees and Vines
- **Volume 4:** Management, Performance, and Applications of Micro Irrigation Systems
- **Volume 5:** Applications of Furrow and Micro Irrigation in Arid and Semi-Arid Regions
- **Volume 6:** Best Management Practices for Drip Irrigated Crops
- **Volume 7:** Closed Circuit Micro Irrigation Design: Theory and Applications
- **Volume 8:** Wastewater Management for Irrigation: Principles and Practices
- **Volume 9:** Water and Fertigation Management in Micro Irrigation
- **Volume 10:** Innovation in Micro Irrigation Technology

Book Series: Innovations and Challenges in Micro Irrigation
Senior Editor-in-Chief: Megh R. Goyal, PhD, PE

Volume 1: Management of Drip/Trickle or Micro Irrigation
Volume 2: Sustainable Micro Irrigation Design Systems for Agricultural Crops
Volume 3: Principles and Management of Clogging in Micro Irrigation
Volume 4: Performance Evaluation of Micro Irrigation Management

Volume 5: Potential Use of Solar Energy and Emerging Technologies in Micro Irrigation
Volume 6: Micro Irrigation Management: Technological Advances and Their Applications
Volume 7: Micro Irrigation Engineering for Horticultural Crops
Volume 8: Micro Irrigation Scheduling and Practices
Volume 9: Engineering Interventions in Sustainable Trickle Irrigation
Volume 10: Management Strategies for Water Use Efficiency and Micro Irrigated Crops
Volume 11: Fertigation Technologies in Micro Irrigation

Book Series: Innovations in Agricultural and Biological Engineering
Senior Editor-in-Chief: Megh R. Goyal, PhD, PE

- Advances in Food Process Engineering: Novel Processing, Preservation and Decontamination of Foods
- Advanced Research Methods in Food Processing Technologies
- Advances in Sustainable Food Packaging Technology
- Analytical Methods for Milk and Milk Products, 2-volume set:
    - Volume 1: Sampling Methods, Chemical and Compositional Analysis
    - Volume 2: Physicochemical Analysis of Concentrated, Coagulated, and Fermented Products
- Advances in Green and Sustainable Nanomaterials: Applications in Energy, Biomedicine, Agriculture, and Environmental Science
- Biological and Chemical Hazards in Food and Food Products: Prevention, Practices, and Management
- Bioremediation and Phytoremediation Technologies in Sustainable Soil Management
    - Volume 1: Fundamental Aspects and Contaminated Sites
    - Volume 2: Microbial Approaches and Recent Trends
- The Chemistry of Milk and Milk Products: Physicochemical Properties, Therapeutic Characteristics, and Processing Methods
- Dairy Engineering: Advanced Technologies and Their Applications
- Developing Technologies in Food Science: Status, Applications, and Challenges
- Emerging Technologies in Agricultural Engineering
- Engineering Interventions in Agricultural Processing
- Engineering Interventions in Foods and Plants
- Engineering Practices for Agricultural Production and Water Conservation: An Interdisciplinary Approach

- Engineering Practices for Management of Soil Salinity: Agricultural, Physiological, and Adaptive Approaches
- Engineering Practices for Milk Products: Dairyceuticals, Novel Technologies, and Quality
- Enzyme Inactivation in Food Processing: Technologies, Materials, and Applications
- Field Practices for Wastewater Use in Agriculture: Future Trends and Use of Biological Systems
- Flood Assessment: Modeling and Parameterization
- Food Engineering: Emerging Issues, Modeling, and Applications
- Food Process Engineering: Emerging Trends in Research and Their Applications
- Food Processing and Preservation Technology: Advances, Methods, and Applications
- Food Technology: Applied Research and Production Techniques
- Functional Dairy Ingredients and Nutraceuticals: Physicochemical, Technological, and Therapeutic Aspects
- Handbook of Research on Food Processing and Preservation Technologies:
  - Volume 1: Nonthermal and Innovative Food Processing Methods
  - Volume 2: Nonthermal Food Preservation and Novel Processing Strategies
  - Volume 3: Computer-Aided Food Processing and Quality Evaluation Techniques
  - Volume 4: Design and Development of Specific Foods, Packaging Systems, and Food Safety
  - Volume 5: Emerging Techniques for Food Processing, Quality, and Safety Assurance
- Modeling Methods and Practices in Soil and Water Engineering
- Nanotechnology Horizons in Food Process Engineering, 3-volume set:
  - Volume 1: Food Preservation, Food Packaging and Sustainable Agriculture
  - Volume 2: Scope, Biomaterials, and Human Health
  - Volume 3: Trends, Nanomaterials, and Food Delivery
- Nanotechnology and Nanomaterial Applications in Food, Health, and Biomedical Sciences
- Nanotechnology Applications in Agricultural and Bioprocess Engineering: Farm to Table

- Nanotechnology Applications in Dairy Science: Packaging, Processing, and Preservation
- Novel and Alternate Methods in Food Processing: Biotechnological, Physicochemical, and Mathematical Approaches
- Novel Dairy Processing Technologies: Techniques, Management, and Energy Conservation
- Novel Processing Methods for Plant-Based Health Foods: Extraction, Encapsulation and Health Benefits of Bioactive Compounds
- Novel Strategies to Improve Shelf-Life and Quality of Foods: Quality, Safety, and Health Aspects
- Processing of Fruits and Vegetables: From Farm to Fork
- Processing Technologies for Milk and Milk Products: Methods, Applications, and Energy Usage
- Quality Control in Fruit and Vegetable Processing: Methods and Strategies
- Scientific and Technical Terms in Bioengineering and Biological Engineering
- Soil and Water Engineering: Principles and Applications of Modeling
- Soil Salinity Management in Agriculture: Technological Advances and Applications
- State-of-the-Art Technologies in Food Science: Human Health, Emerging Issues and Specialty Topics
- Sustainable Biological Systems for Agriculture: Emerging Issues in Nanotechnology, Biofertilizers, Wastewater, and Farm Machines
- Sustainable Nanomaterials for Biomedical Engineering: Impacts, Challenges, and Future Prospects
- Sustainable Nanomaterials for Biosystems Engineering: Trends in Renewable Energy, Environment, and Agriculture
- Technological Interventions in Dairy Science: Innovative Approaches in Processing, Preservation, and Analysis of Milk Products
- Technological Interventions in Management of Irrigated Agriculture
- Technological Interventions in the Processing of Fruits and Vegetables
- Technological Processes for Marine Foods, from Water to Fork, Bioactive Compounds, Industrial Applications, and Genomics

ABOUT SENIOR-EDITOR-IN-CHIEF

Megh R. Goyal, PhD, PE

Megh R. Goyal, PhD, PE, is, currently a retired professor of agricultural and biomedical engineering from the General Engineering Department at the College of Engineering at the University of Puerto Rico–Mayaguez Campus (UPRM); and Senior Acquisitions Editor and Senior Technical Editor-in-Chief for Agricultural and Biomedical Engineering for Apple Academic Press Inc.

During his long career, he has worked as a Soil Conservation Inspector; Research Assistant at Haryana Agricultural University and Ohio State University; Research Agricultural Engineer/Professor at the Department of Agricultural Engineering of UPRM; and Professor of Agricultural and Biomedical Engineering in the General Engineering Department of UPRM. He spent a one-year sabbatical leave in 2002–2003 at the Biomedical Engineering Department of Florida International University, Miami, USA.

Dr. Goyal was the first agricultural engineer to receive the professional license in agricultural engineering from the College of Engineers and Surveyors of Puerto Rico. In 2005, he was proclaimed the "Father of Irrigation Engineering in Puerto Rico for the Twentieth Century" by the American Society of Agricultural and Biological Engineers, Puerto Rico Section, for his pioneering work on micro irrigation, evapotranspiration, agroclimatology, and soil and water engineering.

During his professional career of 52 years, he has received many awards, including Scientist of the Year, Membership Grand Prize for the American Society of Agricultural Engineers Campaign, Felix Castro Rodriguez Academic Excellence Award, Man of Drip Irrigation by the Mayor of Municipalities of Mayaguez/Caguas/Ponce and Senate/Secretary of Agriculture of ELA, Puerto Rico, and many others.

He has been recognized as one of the experts "who rendered meritorious service for the development of [the] irrigation sector in India" by the Water Technology Centre of Tamil Nadu Agricultural University in Coimbatore, India, and ASABE who bestowed on him the 2018 Netafim Microirrigation Award.

VDGOOD Professional Association of India awarded Lifetime Achievement Award at 12th Annual Meeting on Engineering, Science and Medicine that was held on 20-21 of November of 2020 in Visakhapatnam, India.

Dr. Goyal has authored more than 200 journal articles and edited more than 100 books. AAP has published many of his books, including *Management of Drip/Trickle or Micro Irrigation; Evapotranspiration: Principles and Applications for Water Management;* ten-volume set on *Research Advances in Sustainable Micro Irrigation.* Dr. Goyal has also developed several book series with AAP, including Innovations in Agricultural & Biological Engineering (with over 60 titles in the series to date), Innovations and Challenges in Micro Irrigation; and Innovations in Plant Science for Better Health: From Soil to Fork.

Dr. Goyal received his BSc degree in Engineering from Punjab Agricultural University, Ludhiana, India, and his MSc and PhD degrees from the Ohio State University, Columbus, Ohio, USA. He also earned a Master of Divinity degree from the Puerto Rico Evangelical Seminary, Hato Rey, Puerto Rico, USA.

Readers may contact him at goyalmegh@gmail.com.

ABOUT THE EDITORS

N. Veena, PhD
Assistant Professor (Dairy Chemistry),
Dairy Chemistry Department,
College of Dairy Science and Technology,
Guru Angad Dev Veterinary and Animal Sciences
University (GADVASU), Ludhiana, 141004,
Punjab, India, Mobile: +91-9855886831,
E-mail: veena.ndri@gmail.com

N. Veena, PhD, is working as an Assistant Professor in the Department of Dairy Chemistry, College of Dairy Science and Technology, Guru Angad Dev Veterinary and Animal Sciences University, Ludhiana, Punjab, India.

She received her BTech degree in Dairy Technology from Karnataka Veterinary, Animal, and Fisheries Sciences University, Bidar, Karnataka, India, and her MTech and PhD degrees in Dairy Chemistry from ICAR-National Dairy Research Institute, Karnal, Haryana, India.

Her area of expertise includes fortification of milk, development and characterization of functional dairy foods, and nanoencapsulation of nutraceuticals and their delivery. She is the recipient of junior research and institutional fellowships during her master's and doctoral programs from the ICAR-National Dairy Research Institute, Karnal, Haryana, India. She has been involved in different institutional and externally funded research projects such as Milkfed (Punjab), ICAR, DST, and RKVY as PI and Co-PI.

She has published 17 research papers in peer-reviewed journals, one edited book, 11 book chapters, 15 popular articles, and seven practical manuals. She has developed the e-learning content for post-graduate dairy chemistry courses under an NAHEP project. She has presented her research work at various national and international seminars and conferences and received several awards for best oral and poster presentations. She holds membership in various professional bodies such the Indian Dairy Association, Association of Food Scientists and Technologists (India) and Dairy Technology Society of India, etc.

Readers may contact her at: veena.ndri@gmail.com.

Ritesh B. Watharkar, PhD
Assistant Professor, College of Agriculture and Food Business Management,
MIT-ADT University, Rajbaugh,
Solapur–Pune Hwy, Near Bharat Petrol Pump,
Loni Kalbhor 412201, Maharashtra–India;
Mobile: +91-9370614686,
E-mail: watharkarritesh2019@gmail.com

Ritesh B. Watharkar, PhD, is currently working as an Assistant Professor in the College of Agriculture and Food Business Management at MIT-ADT University, Maharashtra, India.

He received his BTech degree in Food Technology from Dr. Balasaheb Sawant Konkan Agriculture University, Dapoli, Maharashtra, India; his MTech in Food Processing and Engineering degree is from the Karunya University, India; his PhD in Food Engineering and Technology degree from Tezpur University (A Central University), Assam, India.

He was formerly a Senior Research Fellow under a funded project by the Ministry of Food Processing Industries, New Delhi (2013–2015), and Assistant Professor at K.K. Wagh Food Tech College (2011–2013); at Shramshakti College of Food Technology, affiliated by Mahatma Phule Agricultural University (2019–2020).

During his professional career, he received awards such as senior research fellow under the UNESCO/China Great Wall Fellowship Program and worked for one year in the School of Biosystem Engineering and Food Science at Zhejiang University, Hangzhou, China. He was also awarded for best popular article from *Krishidut Magazine,* Maharashtra, India. Currently, he is a life member of the Association of Food Scientists and Technologists, India. He has published more than 10 journal articles and attended more than 10 national and international conferences. He also published more than 10 popular articles in the *Marathi Literacy Magazine*.

Readers may contact him at: watharkarritesh2019@gmail.com.

CONTENTS

CONTRIBUTORS

Sadaf Ahmad
Assistant Professor, Civil Engineering Section, University Polytechnic, Aligarh Muslim University, Aligarh–202002, Uttar Pradesh, India, Mobile: +91-9634105536, E-mail: itzsadaf.ahmad@gmail.com

Varisha Anjum
Assistant Professor, School of Pharmacy, Lingayas Vidyapeeth, Faridabad–121002, Haryana, India, Mobile: +91-9667048242, E-mail: varishaanjum786@gmail.com

U. S. Annapure
Professor and Head, Department of Food Engineering and Technology, Institute of Chemical Technology (ICT), Mumbai–400019, Maharashtra, India, Mobile: +91-9819682821, E-mail: udayannapure@gmail.com

Z. R. A. A. Azad
Associate Professor, Department of Postharvest Engineering and Technology, Faculty of Agricultural Sciences, Aligarh Muslim University, Aligarh–202002, Uttar Pradesh, India, Mobile: +91-9810810816, E-mail: zrazad@gmail.com

Fungja Basumatary
BTech, Technician, Department of Food Processing and Technology, Central Institute of Technology, Kokrajhar–783370, Assam, India, Mobile: +91-7002814796, E-mail: xengsi95@gmail.com

B. B. Borse
Senior Principal Scientist and Deputy Director, Department of Spices and Flavor Science, CSIR–Central Food Technological Research Institute, KRS Road, Mysore–570020, Karnataka, India, Mobile: +91-7259376170, E-mails: borsebb@cftri.res.in; bbborase@yahoo.com

Ashok Daimary
Technician, Department of Food Processing and Technology, Central Institute of Technology, Kokrajhar–783370, Assam, India, Mobile: +91-6002462872, E-mail: ashokdaimarycit@gmail.com

Munir A. Dandago
Associate Professor, Department of Food Science and Technology, Kano University of Science and Technology, Wudil–3244, Kano State, Nigeria, Mobile: +234-8161733527, E-mail: dandago223@gmail.com

Pradip B. Dhamole
Assistant Professor, Department of Chemical Engineering, Visvesvaraya National Institute of Technology, South Ambazari Road, Nagpur–440022, Maharashtra, India, Mobile: +91-9922458202, E-mail: pradipdhamole@gmail.com

B. L. Dinesha
Center for Nanotechnology, Department of Processing and Food Engineering, College of Agricultural Engineering, University of Agricultural Sciences, Raichur–584104, Karnataka, India, Mobile: +91-9900960821, E-mail: dinirbdgtc@gmail.com

Ahmad Gambo
PhD Research Scholar, Department of Food Science and Technology, Federal University of Technology, Akure, Nigeria; Lecturer, Department of Food Science and Technology, Bayero University, Kano, Nigeria, Mobile: +234-7063856163, E-mail: agambo.fst@buk.edu.ng

Namrata A. Giri
Scientist, Division of Crop Utilization, ICAR–Central Tuber Crops Research Institute, Thiruvananthapuram–695017, Kerala, India, Mobile: +91-8975963929, E-mail: namrata_cft@yahoo.in

Megh R. Goyal
Retired Faculty in Agricultural and Biomedical Engineering from College of Engineering at University of Puerto Rico–Mayaguez Campus; and Senior Editor-in-Chief in Agricultural and Biomedical Engineering for Apple Academic Press Inc.; PO Box 86, Rincon–PR–00677-0086, USA, E-mail: goyalmegh@gmail.com

Kuldeep Gupta
Research Associate, Department of Molecular Biology and Biotechnology, Tezpur University, Tezpur–784028, Assam, India, Mobile: +91-8486743328, E-mail: kuldeep.mbbt19@gmail.com

Anuck Islary
Assistant Professor, Department of Food Engineering and Technology, Central Institute of Technology, Kokrajhar–783370, Assam, India, Mobile: +91-9854926163, E-mail: a.islary@cit.ac.in

Balarabe B. Ismail
PhD Research Scholar, College of Biosystem Engineering and Food Science, Zhejiang University, Hangzhou–310058, Peoples Republic of China; and Lecturer, Department of Food Science and Technology, Bayero University Kano, Nigeria, Mobile: +861-7816871004, E-mail: bbismail@zju.edu.cn

Sweta Joshi
Assistant Professor, Department of Food Technology, School of Interdisciplinary Sciences, Jamia Hamdard–110062, New Delhi, India, Mobile: +91-8368054899, E-mail: swetajoshi@jamiahamdard.ac.in

Bhupinder Kaur
PhD Research Scholar, Agricultural and Food Engineering Department, Indian Institute of Technology, Kharagpur–721302, West Bengal, India, Mobile: +91-9547484984, E-mail: bhupisangra@gmail.com

Anjineyulu Kothakota
Scientist, Agro-Processing and Technology Division, CSIR–National Institute for Interdisciplinary Science and Technology (NIIST), Trivandrum–695019, Kerala, India, Mobile: +91-9719427663, E-mail: anjineyuluk@niist.res.in

Sheetal Kothawale
MTech Student, Department of Chemical Engineering, Visvesvaraya National Institute of Technology, South Ambazari Road, Nagpur–440022, Maharashtra, India, Mobile: +91-9975563853, E-mail: ssckothawale@gmail.com

S. S. Lele
Director, Institute of Chemical Technology, Marathwada Campus, Jalna, BT-6/7, Biotechnology Park, Additional MIDC Area, Aurangabad Road, Jalna–431203, Maharashtra, India, Mobile; +91-9619423666, E-mail: Smita.Lele@gmail.com

Nikhil Kumar Mahnot
Assistant Professor, Department of Bioengineering and Technology, Gauhati University Institute of Science and Technology (GUIST), Gauhati–781014, Assam, India, Mobile: +91-7896434467, E-mail: nikhil.mahnot@gmail.com

Santosh Mandal
BTech, Technician, Department of Food Processing and Technology, Central Institute of Technology, Kokrajhar–783370, Assam, India, Mobile: +91-9706968853, E-mail: santoshmandal13@gmail.com

Sangita Muchahary
PhD Research Scholar, Department of Food Engineering and Technology, Tezpur University, Tezpur–784028, Assam, India, Mobile: +91-7002731924, E-mail: sangita.muchahary@gmail.com

Sanusi S. Nassarawa
PhD Research Scholar, College of Biosystem Engineering and Food Science, Zhejiang University, Hangzhou–310058, Peoples Republic of China; and Lecturer, Department of Food Science and Technology, Bayero University, Kano, Nigeria, Mobile: +861-7816871004,
E-mails: shamsudeensunusi2015@gmail.com; snsanusi.fst@buk.edu.ng

R. Pandiselvam
Scientist, Biochemistry and Post-Harvest Technology Division, ICAR–Central Plantation Crops Research Institute, Kasaragod–671124, Kerala, India, Mobile: +91-4994-232232,
E-mail: anbupandi1989@yahoo.co.in

Debraj Pradhan
PhD Research Scholar, Department of Food Science, Gauhati University, Guwahati –781014, Assam, India, Mobile: +91-7002167421, E-mail: debrajpradhan1@gmail.com

Madhulekha Rakshit
Assistant Professor, Department of Food Technology, Faculty of Engineering and Technology, JAIN (Deemed-to-be-University), Jain Global Campus, Jakkasandra, Kanakapura, Ramanagara–562112, Karnataka, India; Mobile: +91-9967360209; Email: madhulekha91@gmail.com

R. C. Ranveer
Associate Professor, Department of Post-Harvest Management of Meat, Poultry, and Fish, Post Graduate Institute of Post-Harvest Management, Dapoli, Killa-Roha–402116, Maharashtra, India, Mobile: +91-9967309295, E-mail: rahul.ranveer@gmail.com

B. Renuka
Former DST-Women Scientist Fellow, CSIR–Central Food Technological Research Institute, KRS Road, Mysore–570020, Karnataka, India, Mobile: +91-9035376157, E-mail: renukabs@gmail.com

Sangeeta Saikia
Visiting Faculty, Department of Bioengineering and Technology, Gauhati University Institute of Science and Technology (GUIST), Gauhati–781014, Assam, India, Mobile: +91-7086608604,
E-mail: sangeetasaikia15@gmail.com

Bhagwan K. Sakhale
Associate Professor, Department of Chemical Technology, Dr. Babasaheb Ambedkar Marathwada University (BAMU), Aurangabad–431004, Maharashtra, India, Mobile: +91-9422701632,
E-mail: bksakhale@gmail.com

Vatsala Sharma
PhD Research Scholar, Amity Institute of Food Technology, Amity University Uttar Pradesh, Noida–201303, Uttar Pradesh, India, Mobile: +91-9818635929, E-mail: vsharma6@amity.edu

Ashmita Singh
PhD Research Scholar, Amity Institute of Food Technology, Amity University Uttar Pradesh, Noida–201303, Uttar Pradesh, India, Mobile: +91-8077094179, E-mail: ashmita0908@gmail.com

Salam Maheshkumar Singh
PhD Research Scholar, Center for Rural Development and Innovative Sustainable Technology, Indian Institute of Technology, Kharagpur–721302, West Bengal, India, Mobile: +91-9089519974,
E-mail: salammaheshkumar@gmail.com

Prem P. Srivastav
Associate Professor, Agricultural and Food Engineering Department, Indian Institute of Technology, Kharagpur–721302, West Bengal, India, Mobile: +91-9434043426, E-mail: pps@agfe.iitkgp.ernet.in

Nazia Tabassum
Assistant Professor, Department of Post-Harvest Engineering and Technology,
Faculty of Agricultural Sciences, Aligarh Muslim University, Aligarh–202002, Uttar Pradesh, India,
Mobile: +91-8266051107, E-mail: nazia.tabassum@gmail.com

Monika Thakur
Assistant Professor, Amity Institute of Food Technology, Amity University Uttar Pradesh,
Noida–201303, Uttar Pradesh, India, Mobile: +91-9810495426,
E-mails: mthakur1@amity.edu; monika.harsh05@gmail.com

Soubhagya Tripathy
PhD Research Scholar, Agricultural and Food Engineering Department,
Indian Institute of Technology, Kharagpur–721302, West Bengal, India,
Mobile: +91-8457829200, E-mail: mr.soubhagyatripathy@gmail.com

N. Veena
Assistant Professor, Department of Dairy Chemistry, College of Dairy Science and Technology,
Guru Angad Dev Veterinary and Animal Sciences University, Ludhiana–141004, Punjab, India,
Mobile: +91-9855886831, E-mail: veena.ndri@gmail.com

Ritesh B. Watharkar
Assistant Professor, College of Agriculture and Food Business Management, MIT-ADT University,
Rajbaugh, Solapur–Pune Hwy, Near Bharat Petrol Pump, Loni Kalbhor–412201, Maharashtra, India,
Mobile: +91-9370614686, E-mail: watharkarritesh2019@gmail.com

ABBREVIATIONS

Å	Angstrom
AA	amino acid
AAFCO	Association of American Feed Control Officials
AB	*Agaricus bisporus*
AC	alternating current
ACE	angiotensin-converting enzyme
ACN	acetonitrile
ADI	acceptable daily intake
AEAC	ascorbic acid equivalent antioxidant capacity
ALA	α-Linolenic acid
$AlCl_3$	aluminum chloride
ANZTPA	Australia New Zealand therapeutic products agency
Aq	aqueous
ASE	accelerated solvent extraction
ATPS	aqueous two-phase system
a_w	water activity
BAFS	biologically active food supplements
BI	*Boletus impolitus*
BP	bioactive peptides
BSA	bovine serum albumin
$C_2H_9NaO_5$	sodium acetate trihydrate
Ca	calcium
CAGR	compound annual growth rate
CE	conventional extraction
CEM	conventional extraction methods
CH_3COOH	acetic acid
CHD	coronary heart diseases
CIT	Central Institute of Technology
Cl	chlorine
CMC	carboxymethylcellulose
Co	cobalt
CO_2	carbon dioxide
COVID-19	coronavirus disease
CPE	cloud point extraction

CPE	cytopathic effect
Cr	chromium
Cu	copper
$CuSO_4$	copper sulfate
CVDs	cardiovascular diseases
DAD	diode array detector
DC	direct current
DE	dextrose equivalent
D_e	diffusivity
DFs	dietary fibers
DHA	docosahexaenoic acid
DNA	deoxyribonucleic acid
DPPH	2,2-diphenyl-1-picrylhydrazyl
DPP-IV	dipeptidyl peptidase-IV
DSM	Dutch state mines
E_a	activation energy
EAA	essential amino acids
EAE	enzyme-assisted extraction
EE	encapsulation efficiency
EFSA	European Food Safety Authority
EGCG	epigallocatechin gallate
EPA	eicosapentaenoic acid
EPS	exo-polysaccharides
ER	estrogen receptor
EY	encapsulation yield
FAO	Food and Agriculture Organization
FDA	Food and Drug Administration
Fe	iron
FFA	free fatty acid
FFQ	food frequency questionnaire
FoSHU	food for specified health use
FRAP	ferric reducing antioxidant power
FRS	free radical scavenging
FTIR	Fourier transform infrared spectroscopy
GABA	gamma-aminobutyric acid
GAE	gallic acid equivalents
GHz	gigahertz
GI	gastrointestinal
GLA	gamma-linolenic acid

GRAS	generally regarded as safe
GUA	α-L-guluronic acid
H_2SO_4	sulfuric acid
HACCP	hazard analysis and critical control point
HCl	hydrochloric acid
HFC	hydrofluorocarbon
HFC-134a	hydrofluorocarbon-134a
HFCA	health food control act
HFF	health/functional food
HILF	high-intensity low-frequency
HPE	high pressure extraction
HP-β-CD	hydroxypropyl-β-cyclodextrin
I	iodine
IBD	inflammatory bowel disease
IDF	insoluble dietary fibers
IgA	immunoglobulin A
IL-6	interleukin-6
ILs	ionic liquids
iNOS	inducible nitric oxide synthase
ISO	International Organization for Standardization
K	potassium
kHz	kilohertz
kJ/mol	kilojoule/mole
KOH	potassium hydroxide
LA	lactic acid
LAB	lactic acid bacteria
LDL	low density lipoprotein
LIHF	low-intensity high-frequency
LLE	liquid-liquid extraction
LPS	lipopolysaccharide
M/G	ratio of mannuronate residues to guluronate residues
MAATPE	microwave-assisted aqueous two-phase extraction
MAE	microwave-assisted extraction
MDCK	Madin-Darby canine kidney
Meth	Methanolic
Mg	magnesium
mg/g	milligram per gram
MLV	multilamellar vesicle
Mn	manganese

Mo	molybdenum
MP	*Macrolepiota procera*
MPa	megapascal
MRL	maximum residual levels
MRSA	methicillin resistant *Staphylococcus aureus*
MSR	mushroom
MUA	mannuronic acid
MUAAEE	microwave-ultrasonic assisted aqueous enzymatic extraction
MUFA	monounsaturated fatty acids
Na	sodium
Na_2CO_3	sodium carbonate
$NaNO_2$	sodium nitrate
NaOH	sodium hydroxide
NDC	non-digestible carbohydrates
NFkB	nuclear factor-kB
NHANES	National Health and Nutrition Examination Survey
NLCs	nanostructured lipid carriers
nm	nanometer
NO	nitric oxide
NPC1L1	Niemann-pick C1-like 1
O/W	oil-in-water
O_2	oxygen
OSA	octenyl succinic anhydride
OVX	ovariectomized
P	phosphorus
PCL	polycaprolactone
PEG	poly-ethyl glycol
PFA	prevention of food adulteration
PIE	polyol-induced extraction
pKa	acid dissociation constant
PLE	pressurized liquid extraction
PMTDI	provisional maximum tolerable daily intake
PO	*Pleurotus ostreatus*
psi	pound per square inch
PTI	provisional tolerable intake
PTMI	provisional tolerable monthly intake
PTWI	provisional tolerable weekly intake
PUFA	polyunsaturated fatty acids
QE	quercetin equivalent

ROS	reactive oxygen species
rpm	rotation per minute
RSLE	rapid solid-liquid extraction
S	sulfur
SAS	supercritical anti-solvent
SDF	soluble dietary fibers
Se	selenium
SE	Soxhlet extraction
SEM	scanning electron microscope
SF	supercritical fluid
SFDA	State Food and Drug Administration
SFE	supercritical fluid extraction
SLNs	solid lipid nanoparticles
SMP	skimmed milk powder
SNEDDS	self-generating nano-emulsifying drug delivery system
TCM	traditional Chinese medicine
T_{eu}	eutectic temperature
T_g	glass transition temperature
TOA	Tennessee orthopedic alliance
TPC	total phenolic content
TPTZ	2,4,6-tripyridyl-s-triazine
UAE	ultrasonic-assisted extraction
UAE-TICPE	ultrasonic-assisted extraction integrated temperature-induced cloud point extraction
UAME	ultrasound-assisted microwave extraction
ULV	unilamellar vesicle
UTI	urinary tract infections
UV	ultraviolet
W	Watt
W/cm^2	Watt per square centimeter
W/mL	Watt per milliliter
W/O/W	water-in-oil-in-water
WHO	World Health Organization
WPC	whey protein concentrate
Zn	zinc
ZnO_2	zinc peroxide
β-CD	β-cyclodextrin
μm	micrometer

PREFACE

In the present era, the increase in health awareness among consumers caused an enhanced demand for healthy and nutritious food options. Recently, coupled with consumers' interest, plant-derived bioactive compounds have emerged potential and alternative therapeutic candidates that substitute synthetic compounds. Many functional foods contain bioactive compounds that can be derived from medicinal plants, fruits, vegetables, wastes, and byproducts. Currently, there is an increased demand for food from plant materials and plant-based bioactive compounds that are considered as fresh, natural, safe, and with high nutritive value while produced in sustainable ways.

Bioactive compounds derived from natural food sources have ample scope to be used as nutraceuticals as therapeutics for chronic metabolic disorders viz. diabetes, cancer, hypertension, neurodegenerative diseases, cardiovascular diseases (CVDs), etc. Researchers in recent times have identified and determined the therapeutic roles of many of these bioactive compounds. The major hindrance in the use of bioactive compounds as nutraceuticals is their limited bioavailability and absorption in the body. To overcome this limitation, researchers are on the pursuit to design delivery matrixes and systems for nutraceuticals. Specially designed delivery systems have for nutraceuticals has shown a substantial potential to increase the bioavailability and bioaccessibility of varied bioactives.

This book illustrates various applications of novel food processing extraction and encapsulation techniques and the health and safety aspects of plant-derived bioactive compounds and functional foods. The book consists of three main sections. The first part, Novel Extraction Methods of Bioactive Compounds, explores the principle and application of various advanced extraction techniques (like ultrasonic-assisted, microwave-assisted, rapid solid-liquid, supercritical fluid extraction [SFE], and other hybrid technologies, etc.) for obtaining the valuable bioactive compounds from various foods or food processing waste/agriculture biomass/wastes and fermentation broths for industrial food applications.

The second part, Encapsulation Methods of Bioactive Compounds, addresses the recent advancements in the various encapsulation technologies (such as spray drying, lyophilization, spray cooling, coacervation, liposomal formation, ionic gelation, emulsion, and molecular inclusion complexation) to entrap various plant-derived bioactive compounds; their role and application in protection, and stabilization; and as targeted delivery system for enhanced nutritional health benefits.

The third part, Health Promoting Activities of Bioactive Compounds, provides an overview of the health-promoting activities of various plant-based bioactive compounds, safety aspects, and their uses in the development of health foods has been deliberated.

This book volume, *Novel Processing Methods for Plant-Based Health Foods: Extraction, Encapsulation, and Health Benefits of Bioactive Compounds*, is a bouquet of the novel methods for processing of food and will be quite relevant for the food industry and academic professionals. It is a treasure chest of information and excellent reference source for researchers, scientists, students, growers, traders, processors, industries, and others for emerging food processing approaches for extraction and encapsulation of plant-based bioactive compounds and health-promoting properties of plant-derived nutraceuticals and safety aspects in production of functional foods.

This book has exceeded our anticipation due to the support of all contributing authors to this book, who have been most valuable in this compilation. Their names are mentioned in each chapter and in the list of contributors. We are pleased and obliged to all authors for their proficiency, pledge, and perseverance. We confident that this volume will provide valuable information on different areas of food industry including food processing, preservation, health-promoting properties, safety, and quality evaluation of plant-based foods.

We would like to thank the editorial staff at Apple Academic Press, Inc. for their valuable assistance and great support.

Also, we also wish to thank our friends and families for their unlimited support, encouragement, love, and affection during the course of editing this book volume. Finally, and most importantly we would like to commend our spouses Subhadra, Mohan Kumar and Aishwarya, for their understanding and patience throughout this project.

We appeal to the reader to suggest your feedback that may benefit to improve the subsequent edition of this book.

—Editors

PART I

NOVEL EXTRACTION METHODS FOR BIOACTIVE COMPOUNDS

CHAPTER 1

PRINCIPLES AND APPLICATIONS OF EXTRACTION TECHNOLOGIES IN THE FOOD INDUSTRY

R. PANDISELVAM, B. L. DINESHA, and ANJINEYULU KOTHAKOTA

ABSTRACT

One of the important unit operations followed in the food industry is extraction. It is used for the extraction of valuable ingredients from natural foods. Beneficial active components from the sample matrix were recovered using polar as well as nonpolar solvents. Conventional extraction methods (CEM) like squeezing, maceration, decoction, percolation, Soxhlet, and steam distillation have been used for several decades for oils, food colors, flavors, and essence extraction. These techniques have many disadvantages, such as laborious processes, toxic solvent residues, and thermal degradation of valuable substances. To overcome these disadvantages, novel techniques like microwave-assisted extraction (MAE), ultrasonic-assisted extraction (UAE), supercritical fluid extraction (SFE), accelerated solvent extraction (ASE), and rapid solid-liquid extraction (RSLE) have been used in food industries in recent years. These are highly sustainable and green extraction techniques in terms of the precision and accuracy of analytical results. Both these conventional and novel methods were used for the extraction of essential oils, oleoresins, juices, bioactive compounds, food essences, pigments, pectins, vitamins, carbohydrates, proteins, etc.

Novel Processing Methods for Plant-Based Health Foods: Extraction, Encapsulation, and Health Benefits of Bioactive Compounds. Megh R. Goyal, N. Veena & Ritesh B. Watharkar (Eds.)

1.1 INTRODUCTION

Extraction is one of the important unit operations which is followed in the food industry. Food products are found in our prehistoric civilization as a key source of additives for nutritional, artistic, and religious applications [18]. Various food industries are adopted the extraction process, and it is a prime unit operation in the processing line. It is actually the process of separating one or more analytes of interest from the sample matrix to another phase [6]. More attention is to create in the food processing industry is extraction and separation process process. The solvent is added originally to extract compounds present in the materials, and separation is used to separate compounds dissolved in the solvent. Extraction can be achieved through two phases such as solid-liquid and solid-gas. Solid-liquid extraction technique has been used in the vegetable oil industry for a long time, and it is commonly called leaching [5].

Mechanical pressed extraction of oil from maize, rice bran, and soybean is not suitable because of rancidity and poor stability. Hence, the solvent extraction method is most preferred in commercial industries. Olive oil is commonly extracted by mechanical pressing as a first operation. The final recovery of virgin olive oil can be obtained by repressing the residue after the first pass by using solvent extraction. Peanut oil is extracted by mechanical extraction, and residual oil can be removed through solvent extraction [33].

The quality of oilseed meals obtained by mechanical pressing is good in terms of residual protein and is suitable for animal feed. Essential oils and natural aromatic flavors extraction practices are followed in the commercial industries. The extraction of functional food additives from food and agricultural waste is creating more opportunities [17]. In the sugar cane processing industry, extraction is used to separate the juice from sugar beets by using a multistage mechanical expeller. In this technique, water may be added in between the extraction process. Hence it can also be contemplated as a type of extraction [5].

Roller mills are used for the mechanical expression of cane juice. This process is a capital-intensive, time-consuming, and energy thorough operation. Hence novel cane sugar industries adopted the modern cane juice expellers. Including the mechanical extraction method, several conventional solvent and novel extraction methods are being used in food industries for several applications such as extraction of important bioactive compounds, food essences, colorants, flavors, oils, oleoresins, etc. [11].

This chapter explores the potential applications of traditional and advanced extraction techniques in the food industry.

1.2 CLASSIFICATION: CONVENTIONAL AND ADVANCED EXTRACTION TECHNIQUES

Traditional extraction methods, viz. maceration, squeezing, percolation, and steam distillation have been used for many years in the food processing industries [16]. These traditional extraction techniques are very tedious and consume a higher amount of solvent. The solvents costs are high in the extraction process, and also it will create environmental problems. Consequently, there is a growing demand for novel extraction techniques that have several advantages like reduced extraction time, low amount of solvent consumption, less possibility of cross-contamination, and environmental pollution [1]. Recently several novel extraction methods have been developed by many researchers. These methods have major disadvantages, such as the cost of operation [39].

1.2.1 CONVENTIONAL EXTRACTION METHODS (CEM)

Conventional extraction methods (CEM) have been known for several decades back and are extensively used for the extraction of essential oils, food essences, flours, and colors. Further experiments are required to understand the mechanism of the extraction process [4]. It is feasible to differentiate the conventional liquid-solid extraction techniques. CEM like squeezing, distillation, counter-current extraction, maceration, and Soxhlet play a major role in liquid-solid extraction methods. Major disadvantages of these extraction techniques are the large quantity of solvent consumption, higher solvent evaporation rate and extraction time, fewer chances to select the proper solvent for the extraction process, and degradation of an important bioactive compound at higher temperature [22].

To control all these disadvantages, promising and new liquid-solid extraction methods like microwave assisted [12], ultrasound-assisted based [10], supercritical fluid [25], accelerated solvent [7], solid-phase microextraction [35], enzyme-based [26] and rapid liquid-solid extraction methods [29] could be used. These novel extraction methods were used in the commercial industries and scientific community to study the sustainable development of natural products. Common objectives of all these techniques are extraction of bioactive compounds from the sample matrix and bio-waste materials, detection, separation, and isolation of selective compounds. The economy of the extraction process depends on the viability, extraction yields, and economy of the process [10].

1.2.1.1 SQUEEZING

Squeezing is a classical method of extraction to get essential food constituents such as oils, oleoresins, food essence, flours, and colors [3]. This extraction process is simple, and it is worked based on the principle of high impact force acting on sample mass by using screws, mortars, mullers, etc. Sample matrixes are made in the form of exudates and then fed into the mechanical screw press for valuable components extraction [2].

This solid-liquid extraction technique is abnormal since, here on, solvents were used for molecular extraction of inner components in the solid materials. Extraction of essential oils and oleoresins from agricultural biomass and waste materials is found potential applications in the food industry. Thermal degradation of important bioactive compounds might be due to the induced higher temperature gradient and occurrence of peroxidation in the extracted oils and oleoresins.

Crushing solid materials is a crucial problem in the release of many active elements. Consequently, some of the crave compounds get contaminated by unwanted compounds. Hence, the products obtained by the squeezing process are rarely used, and it is applied for the extraction of important compounds. In most cases, in order to isolate the specific compounds, it is crucial to resort to more sophisticated separation processes. For this reason, a limited number of applications were present in the squeezing-like ancient technique. Priority of discovering a new extraction technique, it is required a good yield of essential oils. Vongsak et al. [38] reported on the squeezing method of extraction for phytochemical constituents from *M. oleifera* leaves.

1.2.1.2 MACERATION

Maceration is an easy and cheaper extraction technique, and it could be done in small and large capacity steel containers in the form of the pilot to commercial-scale level. This is a kind of liquid-solid-based method of extraction work based on the process of osmosis and diffusion. Hence it is considered a control technique for the extraction of many components from biomaterials [27].

Solids are placed into the inner container with solvent. To obtain maximum recovery, the hermetically sealed containers were used to agitate and in order to make the diffusion of extracted compounds in a batch process. The extraction process is influenced by several factors such as temperature, type of solvent used for maceration, and contact time of solvent [36]. The kinetic energy of molecules within the solids was increased with the increasing temperature by assisting the microwaves or ultrasounds. Extraction of thermolabile and soluble compounds on therapeutic interest, the maceration technique is generally recommended. Particularly in this process, small quantities of solvent in several cycles were used for extraction [11].

The solvent absorption capacity depends on the sample matrix and the type of solvent in which important compounds are dissolved. Cujic et al. [13] used the maceration process for the extraction of important compounds from chokeberry fruit. In the case of the maceration technique, it is very difficult to maintain room temperature during the extraction process because extraction takes place in the aqueous phase. To overcome this problem, the infusion process can be done in 1–2 min. In this process, extraction is very fast, and simultaneously thermal degradation also takes place. Comparatively, maceration is an excellent extraction method in the area of liquid-solid extraction. It is a simple extraction process, and the loss of contents in the sample matrix is also pertinent [11].

1.2.1.3 DECOCTION

The decoction is an extraction process in which the sample matrix was boiled with solvent for 30 min at boiling temperature. This technique is specially used for compact materials such as thermoresistive compounds. Fotakis et al. [21] reported on 10 herbal preparations and its metabolic activities. The results suggested that the extractability of phenolic compounds was positively affected by infusion compared to the decoction process. The shelf life of the decoction-based extracts is very short; hence extracts may be used immediately after the extraction.

Maceration is one of the alternative methods for digestion, which involves heating the sample matrix from 35 to 60°C with solvent. Moderate heating is allowed to obtain a higher extraction yield. The solvent power depends on the rate of change in temperature during the extraction process. Solvents used in the decoction process are highly volatile; hence refluxing condenser system is required to use and recover solvent. Manousi et al. [28] reported that extracts obtained from digestion yield more compared to the maceration process. Consequently, the decoction is a sound method for the extraction of biological compounds from aromatic and medicinal plants, but it is not suitable for thermo-degradable compounds [9].

1.2.1.4 PERCOLATION

A simple percolator contains a cylindrical type container filled with a sample matrix (Figure 1.1). A pump is used for circulating the extracting liquid. The nature of the extract depends on the sample matrix, and the percolator is made up of glass, iron, and steel materials. The sample requires good grinding, and the pulverization rate greatly depends on the effectiveness and the time of extraction [27]. The process of humidification is requisite for swelling of the particulate matrix within the solvent contact. The percolator was filled with a layer of sand and cotton. Interstitial granular spaces prevent the liquid flow without operation. Generally, the percolator was to fill with cotton and sand at the bottom in order to chunk the sample matrix and filtration element. In order to minimize the humidification of contents, the sample matrix is then added to the chromatographic column in a compact and uniform manner [35]. To soften the tissues, a preventive maceration process should be taken place. The extraction solvent is added at the head of the percolator, and it was in contact with the sample matrix to solubilization of liquid extract at dynamic action. In this method, osmosis and diffusion principles are used in like maceration process. The major material in the simple percolation process is extractant liquid which is continuously moved through the solid materials; this acts as a driving force in the percolator [3].

In the continuous percolation process, the sample matrix was fed continuously into the series of percolators, and extraction took place in counter-current motion by leachates. Extractant liquid obtained by successive percolators was less in extracted substances. The solvent passing from the subsequent diffuser and extractable components increased the solutes concentration. In each diffuser, the concentration gradient goes increases hence optimization of distribution ratio, and soaking also increases [23]. The operation of the percolator takes place in 5 to 10 series of diffusers.

Chanda et al. [9] conducted a comparative study on different extraction methods for antioxidants extraction from leaves of *Syzygium cumini* L. The results suggested that the best method for the extraction of antioxidants from *Syzygium cumini* L. leaves is the cold percolation method. This process does not require any trained person to operate the percolator. To increase the extraction efficiency of the percolator, heat, microwaves, and ultrasounds can be used to accelerate the extraction process. However, briefly, percolation is a very fast process for the extraction of essential oils and other important products from plant materials.

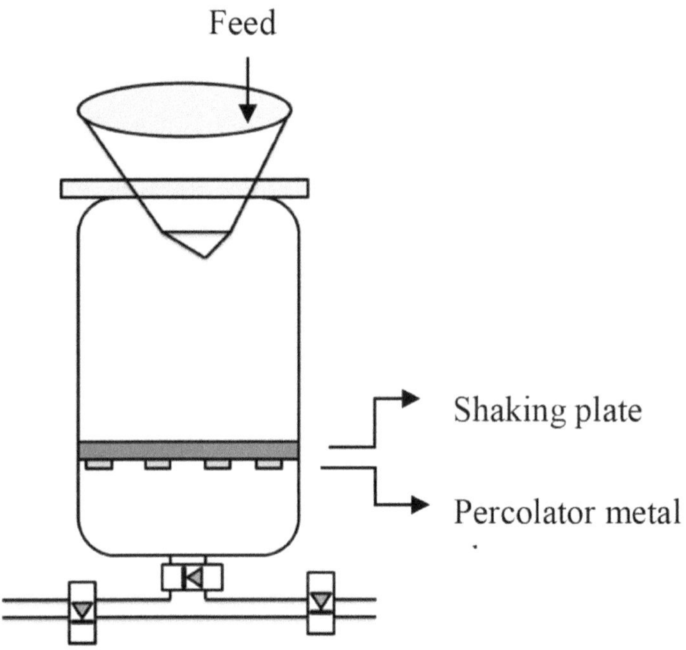

FIGURE 1.1 A commercial-scale percolator.

1.2.1.5 SOXHLET EXTRACTION (SE)

Soxhlet is a CEM. A large quantity of fresh sample matrix was added into the less solid-liquid ratio of extractant [27]. In the Soxhlet apparatus, the liquid used for extraction is in contact with the fresh sample matrix, and the liquid will get start boiling in the flask. This is a special extraction process used only for compounds of high thermal stability. Major advantages of this process are the use of less quantity of extractive solvent, continuous distillation, and purification of the extract is possible [15].

The Soxhlet operator consists of a heating base, sample flask, cellulose thimble, condenser, inlet, and outlet water connections for the condenser (Figure 1.2). The material is placed in the cellulose-based thimble that is to be placed on the heating coil inside the chamber. A distillation flask could be used to heat the solvent and sample matrix. As the liquid gets starts boiling, vapors stand up in the condenser mounted on the extractor. The vapors get condensate by using condensation liquid in the condenser and condensate liquid drops into the porous thimble in the extraction flask. This flask is inter-connected with the lateral siphon elbow, through which liquid extract gets separated [19].

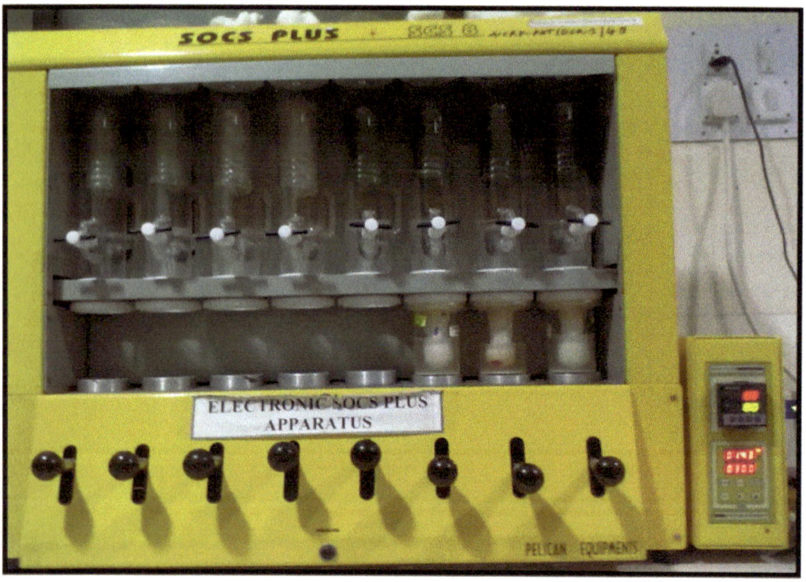

FIGURE 1.2 Laboratory scale Soxhlet extractor.

The repetitive vaporization and condensation take place in the distillation process. This repetitive solvent distillation and condensation process will take place until the extraction process completes. The extraction process can be done with nonpolar solvents such as n-hexane, ethanol, methanol, acetone, etc. Generally, the process of extraction takes place at 65 to 70°C with an extraction time of 2 to 3 h [20].

Major advantages of the Soxhlet extraction (SE) process are higher extraction yield and efficiency. Hence, it can be used for model extraction techniques for other novel extraction methods. The residual solvent present in the liquid extractant can be easily removed by applying desolventization up to some extent. Some of the disadvantages also exist in this extraction process, such as maximum solvent consumption, and more extraction time, and this process is not suitable for heat-sensitive compounds. To overcome these disadvantages, semi-automatic and fully automatic temperature controller-based, SE units have been available in recent days [28].

1.2.1.6 STEAM DISTILLATION

Steam distillation is a large-scale conventional extraction (CE) process in which essential oils, food colors, essence, and flavors can be extracted. The distillation unit consists of a sample flask or column in which the sample matrix is placed. Steam is passed through the column, and then volatile compounds rise up in the distillation flask or condenser (Figure 1.3). A simple distillation unit works based on the principle of vapor pressure difference in volatile substances. Hence it is called has solid-liquid extraction technique. Vaporized substances in the distillation were separated based on the volatile substances vapor pressure gradient. However, this method is used in the final effect extraction process. Distillation is simple and fast, and it will yield more compared to the other extraction methods.

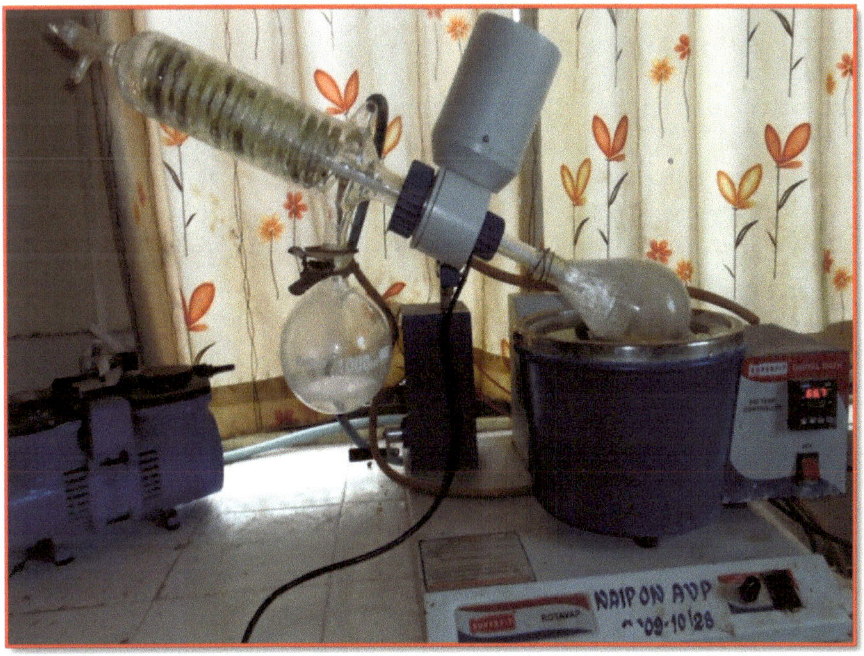

FIGURE 1.3 Rotary vacuum-based steam distillation unit.

Generally, water is used as the main fluid in the configuration of steam. This generated steam that has a higher latent heat value; hence it is especially acceptable for the extraction of oils and resins from plant materials. Wei et al. [40] have extracted bioactive compounds from the *Flaveria bidentis* (L.), and the study showed a higher extraction yield and a maximum of 96.8% purity of the extracted compounds. Nevertheless, in spite of its several advantageous applications, steam used in the extraction process will affect the critical molecules in extractant liquid [14].

1.2.2 NOVEL ADVANCED TECHNOLOGIES

1.2.2.1 MICROWAVE-ASSISTED EXTRACTION (MAE)

MAE is the new method of extraction, and it will use the microwave to heat the specimen with the solvent mixture in place to make faster extraction of the analyte (Figure 1.4). Compared to the conventional sources of heat, microwaves transfer the heat from the inner layer to the surface of the sample matrix. Transfer of heat source basically acts on two processes such as conduction and convection. In this process, the microwaves cover the entire homogeneous volume, polar molecules of the product, and localized heating centers. Presently, MAE is using in commercial industries for the extraction and isolation of very important constituents from bio-waste materials [37]. This process is advantageous in terms of reduction in extraction process time and consumption of solvent volume. Kaderides et al. [24] extracted the phenolic-based bioactive compounds by using MAE, and it was compared with ultrasound extraction. The results suggested that thermostable compounds were obtained after the filtration phase with high antioxidant activity. MAE can also be used as a pretreatment for the maceration process in the liquid-solid phase extraction process. The extraction process is accelerated by enhanced temperature, but at the same time, microwave energy is too high, and it is susceptible to solid matrix damage [8].

1.2.2.2 ULTRASONIC ASSISTED EXTRACTION (UAE)

Ultrasonic assisted extraction (UAE) is one of the important novel extraction processes (Figure 1.5) in which pulses of high-intensity ultrasound waves pass through a liquid medium containing the immersed titanium probes. Incompressibility can be achieved due to particle implosion. The cavitation was generated by a high-pressure vibration impulse. The cavitation phenomenon consists of an impulse pressure wave generated through a lack of minute bubbles formation at the instant phase of pessimistic pressure, and it will burst in the succeeding phases of compression [31].

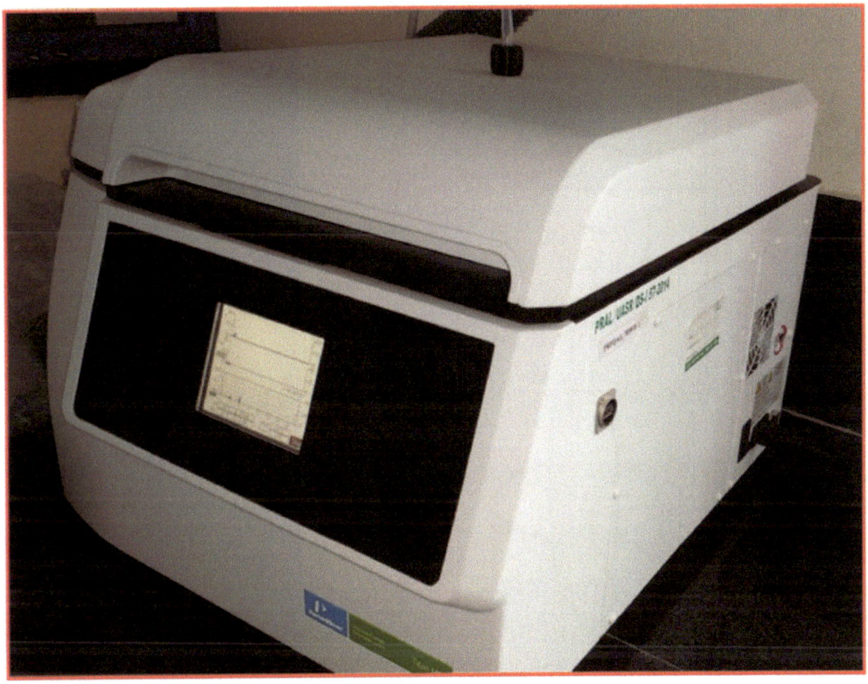

FIGURE 1.4 Laboratory scale microwave-assisted extraction unit.

Due to change in pressure and temperature within the system collapse of each bubble take place. These bubbles collapsed near the solid-liquid interface; it is very significant in a homogeneous cavitation phase. As a matter of fact, the expansion of ultrasound cycles passes through liquid molecules and travels apart from each other. The cavity will be generated by liquid molecules of acceptable intense negative pressure better than the tensile strength [28].

Cavitation bubbles are formed by liquid and inside the solid surfaces in the pre-existing weak points. These points were top-up together with gas in suspended solids and powdered materials. Proceeding to cavitation, micro-bubbles were swung in the irradiated liquid. Accordingly, the cellular structure was destructed by a chemical reaction. Inertial force re-compresses the build-up half-cycle and rapidly develops the high power. The bubble will grown up in each successive cycle, causing the increases in size, and it will repeat even during the critical size. Thermal energy affects the bubble formation in terms of collapsing the bubble. The bubble size reaches the resonant dimension, the cavity will develop to a critical dimension, and it will absorb the energy from ultrasonic radiations [31].

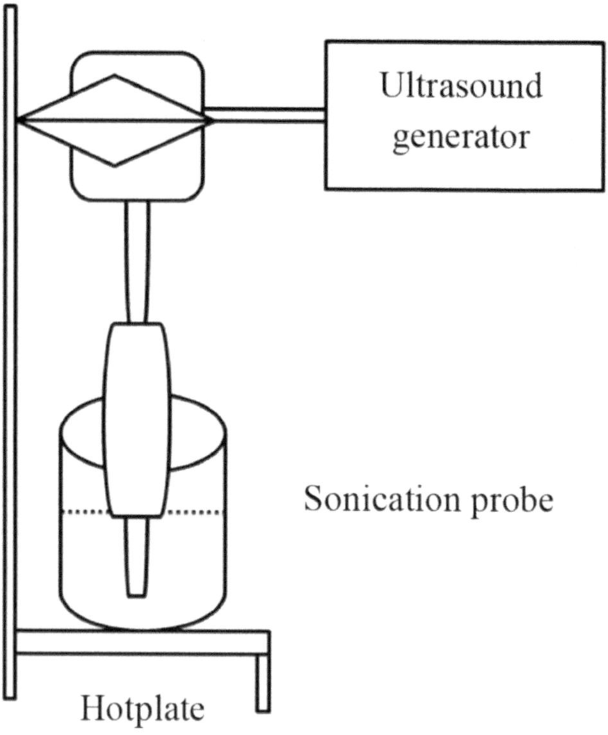

FIGURE 1.5 Ultrasonication-assisted extraction unit.

Tiwari [37] conducted an experiment on the extraction of total carotene from pomegranate wastes by using a solvent and ultrasound-based extraction process. The results suggested that a good remedy to avoid diffusion of solvent during the extraction process is the assistance of ultrasound in the extraction procedure. For extraction of vegetable oils, vegetables are filled in an extraction vessel with extractant liquid, and breaking of cellular structure was done by applying the squeezing technique. Extracted compounds were lost their beneficial activity due to ultrasound transformations generated. In ultrasound-based extraction, there are many factors to be considered in order to get a higher yield. Hence, optimization of the various parameters increases the experimental time. This technique is used for extraction of active compounds from plants [28]. Squeezing is generally used for comparative extraction method for extraction of active ingredients from plants [31].

1.2.2.3 SUPERCRITICAL FLUID EXTRACTION (SFE)

Supercritical fluid extraction (SFE) is a novel and simple liquid-solid extraction method. The process of SFE depends on the solvent used in the extraction system (Figure 1.6). Here carbon dioxide (CO_2) acts as a main solvent and ethanol, methanol, etc., are the co-solvents used in the extraction process [36]. The properties of CO_2 are intermediate in between gases and liquids; hence it is called supercritical fluid. The properties of CO_2 at the supercritical state vary with the changing temperature and pressure. At the critical state, the control phase behavior in the extraction and separation process is very much important. In practical, the solubility of supercritical fluid (SF) is regulated continuously by small change in isobaric temperature and isothermal pressure. The main feature of the SFE systems is based on the ability to regulate the solvent power at a supercritical state [25]. Supercritical fluids have special characters such as diffusivity, viscosity, and density that are present in intermediate solid and liquid states. The major advantages of this technique are the extraction will take place with CO_2 at atmospheric temperature and pressure [17]. SFE process is also called the green technique in solid-liquid extraction. Another major advantage of the SFE process is extraction speed and also density of SF is almost close to the conventional liquid phases [30].

It can be observed from the conventional liquids, the magnitude of viscosity may present in many orders. Diffusion coefficient larger than the typical ones can be observed in conventional fluids. CO_2 acts as a main solvent in SFE process [16]. SFE offers several advantages such as the solvent expands through sample matrix relatively larger molecular size, faster, and cheaper extraction process, and higher diffusion coefficient. Supercritical CO_2 have lower viscosity compared to the other solvents, good permeability, odorless, non-toxic, and eco-friendly [29].

The working temperature of supercritical CO_2 extraction process take place at normal atmospheric temperature hence, it is used for extraction of heat-sensitive materials. Single technique including extraction and separation are significantly increases the processing time. The operating conditions such as temperature and pressure depend on variable solvent power.

De-Silva et al. [14] reviewed on applications of SFs in the extraction of bioactive compounds. SFs technology offers many advantages compared to the traditional extraction techniques. Nevertheless, the major disadvantages of this technique are lower dissolution power led to the water-solubilizing capacity for water-soluble compounds. Generally, solvents such as ethanol, methanol, and acetone could be used in the novel extraction process. Putnik et al. [33] published the review article on extraction technologies on high-added value compounds from plant materials.

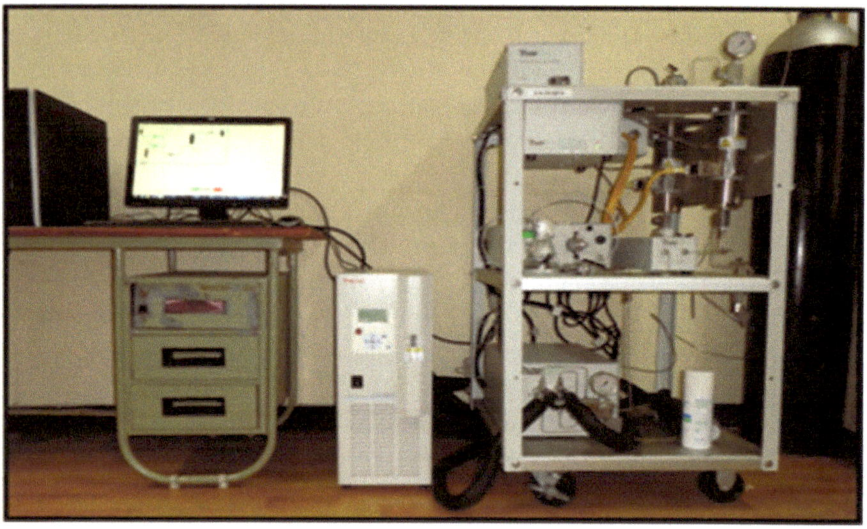

FIGURE 1.6 Laboratory scale supercritical fluid extraction unit.

1.2.2.4 ACCELERATED SOLVENT EXTRACTION (ASE)

ASE is mostly used for important compounds extraction in phase of liquid at high pressure and above the boiling temperature. The high pressures created by an increase in temperature beyond solvent boiling point. Desorption of analytes from the sample it might be due to an increase in temperature at accelerated condition. As per Fick's law of diffusion, extraction of constituents from the plant materials and their solubility enhanced with elevating the temperature at lower time of extraction [31].

At higher pressure and temperature, the use of solvents is viable to affect the extraction procedure by changing the physicochemical properties of the sample-solvent system. Effect of pressure at significant level is responsible to puncturing of sample and makes them in pores structure. The solvent in the liquid state is a critical function of the ASE process [32]. Major advantages of supercritical fluids are highest solvent strength, extraction process take place at atmospheric pressure, higher extraction yield and efficiency. Further superiority of this process is the use of pure solvents and no changes in the phase of atmospheric conditions. Hence, packed reactors and traps are not required for extraction and recovery of analytes from the sample matrix system [31].

Comparison of nutritional constituents in sweet potatoes using conventional and novel extraction techniques is reported by Cai et al. [7]. The negative decisions were obtained for the separation of phenolics/flavonoids by adopting three novel extracting methods. These negative results might be due to static extraction system and feasible deterioration bioactive compounds at the specific operating conditions. Even though, ASE is an excellent method for extraction of heat-abiding compounds at higher temperatures above the boiling point temperature. Sample matrix system in contact with the solvent at high temperature remains solid entire experimental period. Accelerated solvent-based extraction system generates the high pressure, hence, the sample used in this process is prepared at laboratory level [32].

1.2.2.5 RAPID SOLID-LIQUID EXTRACTION (RSLE)

Naviglio et al. [29] introduced the RSLE through the utilization of Naviglio extractor. Naviglio is an extraction process in which extractor constituted a reasonable alternative solution in solid-liquid extraction technique and escort notable advantage in obtaining high quality extracts. In the first step of the extraction process heat the system by the mechanical menace. Conventional methods of extraction like steam distillation, Soxhlet, percolation, and ultrasound methods are based on principles of osmosis and diffusion. These principles increase the extraction yield and efficiency with increasing the temperature. In the event of an increase in temperature put up the thermally degradable compounds.

RSLE is necessary for extraction cycles of 30 numbers, to complete the vegetable oil extraction process. Compared to the conventional maceration process, the RSLE has approved and comprehensive method. Furthermore, water is used as an extracting medium in case of RSLE. It is used in many applications such as oil plant materials because of lower extraction time. The major advantage of this technology is cheaper and low energy when compared to the other novel extraction technique. Posadino et al. [32] reported that polyphenolic antioxidants extraction by using Cagnulari grape. RSLE is a green extraction technique that can be used to extract antioxidants, and those can be used for the food and nutraceuticals applications.

1.2.3 COMPARISON OF DIFFERENT EXTRACTION TECHNIQUES BASED ON YIELD

Among the several extraction techniques explained above, the novel extraction techniques give a higher extraction yield. In general, 20 to 40% more yield can be obtained by using advanced extraction techniques. Optimized extraction temperature in both conventional and novel extraction methods ranges between 50°C and 60°C. Therefore, the same yield can be expected from all the extraction methods. In the SFE, supercritical CO_2 has been used instead of organic solvents, and these were called eco-friendly processes. Extraction process yield is more at minimum extraction time compared to the other conventional methods [24].

1.2.4 COMPARISON OF DIFFERENT EXTRACTION TECHNIQUES BASED ON TIME AND SOLVENTS CONSUMPTION

Generally, CEMs were required longer extraction time, i.e., 500 to 700 min and novel extraction methods like SFE, UAE, and MAE were required quiet shorter extraction time, i.e., 1 to 30 min. There is a 10 times reduction in extraction time in case of novel methods. Organic solvents consumption in CEM is about 100 to 200 ml, where as in the case of novel extraction methods will consume 50 ml. Hence, there will be 4 times decrease in solvent consumption in case of novel methods in comparison with the conventional methods [34].

1.2.5 DIFFERENT EXTRACTION METHODS: ADVANTAGES AND DISADVANTAGES

The extraction process depends on solid matrix compositions and surface complications. The alternative procedure and techniques connected to an extraction procedure is based on contact of liquid-solid phase and it is a very complicated process. Complexity of solid matrix composition majorly depends on the solid-liquid phase. Hence, it is very important to select a suitable extraction method for solid-liquid extraction method. Types of solvents and operating conditions are responsible for selecting suitable extraction techniques. Most important chemical compounds and their behavior were used at different sample–solvent ratio. Large extent operating conditions for extraction of vegetable oils depends on different geometry of solids, solvents used for extraction, temperature, and ratio to mix solvents. Some of the numerical mathematical models are used to find out the model constants for extraction yield and efficiency [26].

Compared to above mentioned classical techniques, more, and efficient contemporary extraction techniques such as supercritical based fluids, ultrasound, microwave, accelerated solvent-based and RSLE are used in Naviglio extractor. This would be improving the characteristics of extract and extraction efficiency. RSLE is a very much important liquid-solid extraction process, it will work based on the osmosis and diffusion [10]. The major attributes for differentiation of the liquid-solid based extraction methods are presented in Table 1.1.

TABLE 1.1 Attributes for Differentiation of Liquid-Solid based Extraction Types

Extraction Methods	Solvent	Extraction Time	Extraction Yield	Quality of Extract	Stability of Extract	References
Squeezing	Polar	More	Definitive	Fair	Fair	[5]
Maceration	Basic	Maximum	Definitive	Excellent	Excellent	[7]
Decoction	Basic	More	Definitive	Excellent	Excellent	[10]
Percolation	Basic	Medium	Restricted	Superior	Superior	[11]
Soxhlet	Basic	Prolong	Definitive	Deficient	Deficient	[17]
RSLE	Nonpolar	Minimum	Exhaustive	Great	Great	[26]

1.3 FUTURE PROSPECTIVES

In this chapter, it was explored the major concept of conventional and advanced extraction methods used in the food industries. It is very much important to understand the molecular scale opportunities of these extraction techniques in food industries. Solvent-free extraction method has become the main issue in the question of industrial application. Present and past literature data conform to the solvent-free original extraction methods. In analytical chemistry, the degree of maturity and molecular scale process are relevant to the solvent-free extraction methods. This chapter will be ambitious about the challenges and widen potential commercial applications. Hence, novel solvent-free extraction techniques can be used in future food and nutraceutical applications.

1.4 SUMMARY

Extraction is the most important unit operation in the processing industries. These industries adopted the most convenient extraction technique for extraction of the food-industrial colorants, essence, important bioactive compounds, essential oils, and oleoresins. Soxhlet method of extraction is widely used in commercial industries for extraction of oil from biological material. Usually, food materials are subjected to mechanical shear using expellers to release the volatiles in a virgin state. Polar solvents were involved in the new solvent extraction methods. Major disadvantages in the new solvent extraction methods are hazardous and flammable solvents, emission of toxic substances during extraction; it is not selective and consumes more time. It is very much important to increase the extraction yield of volatile substances, physiochemical, and functional properties by selecting the alternative correct solvent. Microwaves, SFs, and ultrasounds were used in the solid-liquid extraction process for extraction of essential oils. Application of RSLE is very much useful for herbal, cosmetic, pharmaceutical, food, and beverage industries.

KEYWORDS

- **accelerated solvent extraction**
- **bioactive compounds**
- **extraction technology**
- **food industry**
- **functional properties**
- **microwave**
- **solvent extraction**

REFERENCES

1. Al-Jitan, S., Alkhoori, S. A., & Yousef, L. F., (2018). Phenolic acids from plants: Extraction and application to human health; Chapter 13. In: Atta-Ur-Rahman, (ed.), *Studies in Natural Products Chemistry* (Vol. 58, pp. 389–417). Amsterdam, Netherlands: Elsevier.
2. Armenta, S., Garrigues, S., & De La Guardia, M., (2015). The role of green extraction techniques in green analytical chemistry. *Trends in Analytical Chemistry, 71*(1), 2–8.
3. Baldwin, E. A., Bai, J., Plotto, A., Cameron, R., Luzio, G., Narciso, J., Manthey, J., et al., (2012). Effect of extraction method on quality of orange juice: Hand-squeezed, commercial-fresh squeezed and processed. *Journal of the Science of Food and Agriculture, 92*(10), 2029–2042.
4. Barba, F. J., Zhu, Z., Koubaa, M., Sant'Ana, A. S., & Orlien, V., (2016). Green alternative methods for the extraction of antioxidant bioactive compounds from winery wastes and by-products: A review. *Trends in Food Science & Technology, 49*, 96–109.
5. Battcharya, S., (2014). *Conventional and Advanced Food Processing Technologies* (1st edn., p. 744). Hoboken, New Jersey: Wiley Press Inc.
6. Bicking, M. K. L., (2000). Extractions: Analytical extractions. In: Wilson, I. D., Adlard, E. R., Cooke, M., & Poole, C. F., (eds.), *Encyclopedia of Separation Science* (pp. 1371–1382). Michigan, USA: Academic Press Inc.
7. Cai, Z., Qu, Z., Lan, Y., Zhao, S., Ma, X., Wan, Q., Jing, P., & Li, P., (2016). Conventional, ultrasound-assisted, and accelerated-solvent extractions of anthocyanins from purple sweet potatoes. *Food Chemistry, 197*(1), 266–272.
8. Chan, C. H., Yusoff, R., Ngoh, G. C., & Kung, F. W. L., (2011). Microwave-assisted extractions of active ingredients from plants. *Journal of Chromatography A, 1218*(37), 6213–6225.
9. Chanda, S. V., & Kaneria, M. J., (2012). Optimization of Conditions for the extraction of antioxidants from leaves of *Syzygium cumini L.* Using different solvents. *Food Analytical Methods, 5*, 332–338.
10. Chemat, F., Rombaut, N., Meullemiestre, A., Turk, M., Perino, S., Fabiano-Tixier, A. S., & Abert-Vian, M., (2017). Review of green food processing techniques: Preservation, transformation, and extraction. *Innovative Food Science and Emerging Technology, 41*, 357–377.
11. Chemat, F., Rombaut, N., Sicaire, A. G., Meullemiestre, A., Fabiano-Tixier, A. S., & Abert-Vian, M., (2017). Ultrasound assisted extraction of food and natural products. Mechanisms, techniques, combinations, protocols and applications: A review. *Ultrasonics Sonochemistry, 34*, 540–560.
12. Chizoba, E. F. G., Sun, D. W., & Cheng, J. H., (2017). Acceleration of microwave-assisted extraction processes of food components by integrating technologies and applying emerging solvents: A review of latest developments. *Trends in Food Science & Technology, 67*, 160–172.
13. Ćujić, N., Šavikin, K., Janković, T., Pljevljakušić, D., Zdunić, G., & Ibrić, S., (2016). Optimization of polyphenols extraction from dried chokeberry using maceration as traditional technique. *Food Chemistry, 194*, 135–142.
14. Da Silva, R. P. F. F., Rocha-Santos, T. A. P., & Duarte, A. C., (2016). Supercritical fluid extraction of bioactive compounds. *Trends in Analytical Chemistry, 76*, 40–51.
15. Dinesha, B. L., Nidoni, U., Ramachandra, C. T., & Naik, N., (2016). Qualitative and quantitative analysis of bioactive compounds from supercritical fluid and Soxhlet

extracted moringa (*Moringa oleifera Lam.)* seed kernel oil. *Research Journal of Agricultural Sciences, 7*(2), 339–343.
16. Dinesha, B. L., Nidoni, U., Ramachandra, C. T., Naik, N., & Hugar, A., (2015). Optimization of supercritical fluid extraction process for moringa (PKM-1) seed kernel oil. *International Journal of Agricultural Science and Research, 5*(5), 95–102.
17. Dinesha, B. L., Nidoni, U., Ramachandra, C. T., Naik, N., & Sankalpa, K. B., (2018). Effect of extraction methods on physicochemical, nutritional, antinutritional, antioxidant and antimicrobial activity of moringa (*Moringa oleifera Lam.*) seed kernel oil. *Journal of Applied and Natural Science, 10*(1), 1–9.
18. Dinesha, B. L., Nidoni, U., Ramachandra, C. T., Naik, N., & Sankalpa, K. B., (2017). Effect of ecofriendly supercritical carbon dioxide temperature and pressure on anti-nutritional compositions of moringa seed kernel oil. *Research in Environment and Life Sciences, 10*(3), 228–231.
19. Dinesha, B. L., Nidoni, U., Ramachandra, C. T., Naik, N., & Sankalpa, K. B., (2016). Effect of supercritical carbon dioxide conditions on extraction of food phytochemical constituents from *Moringa oleifera.* Lam seed kernels. *International Journal of Food Fermentation Technology, 6*(2), 373–380.
20. Dinesha, B. L., Nidoni, U., Sankalpa, K. B., Ramachandra, C. T., & Naik, N., (2016). Determination of bioactivity for ecofriendly supercritical CO_2 extracted moringa (*Moringa Oleifera Lam.*) seed kernel oil. *The Ecoscan, 36*(3), 567–572.
21. Fotakis, C., Tsigrimani, D., Tsiaka, T., Lantzouraki, D. Z., Strati, I. F., Makris, C., Tagkouli, D., et al., (2016). Metabolic and antioxidant profiles of herbal infusions and decoctions. *Food Chemistry, 211*, 963–971.
22. Galanakis, C. M., (2012). Recovery of high added-value components from food wastes: Conventional, emerging technologies and commercialized applications. *Trends in Food Science & Technology, 26*(2), 68–87.
23. Jensen, W. B., (2007). The origin of the Soxhlet extractor. *Journal of Chemical Education, 84*(12), 1913–1914.
24. Kaderides, K., Papaoikonomou, L., Serafim, M., & Goula, A. M., (2019). Microwave-assisted extraction of phenolics from pomegranate peels: Optimization, kinetics and comparison with ultrasounds extraction. *Chemical Engineering and Processing-Process Intensification, 137*, 1–11.
25. Khaw, K. Y., Parat, M. O., Shaw, P. N., & Falconer, J. R., (2017). Solvent supercritical fluid technologies to extract bioactive compounds from natural sources: A review. *Molecules, 22*(7), 1186.
26. Kumar, S. P. J., Garlapti, V. K., Dash, A., Scholz, P., & Banerjee, R., (2017). Sustainable green solvents and techniques for lipid extraction from microalgae: A review. *Algal Research, 21*, 138–147.
27. Luque De, C. M. D., & Priego-Capote, F., (2010). Soxhlet extraction: Past and present panacea. *Journal of Chromatography A, 1217*(16), 2383–2389.
28. Manousi, N., Sarakatsianos, I., & Samanidou, V., (2019). Extraction techniques of phenolic compounds and other bioactive compounds from medicinal and aromatic plants; Chapter 10. In: Victoria, F., & Samanidous, L., (eds.), *Engineering Tools in the Beverage Industry* (Vol. 1, pp. 283–314). Sawston – UK: Woodhead Publishing Inc.
29. Naviglio, D., (2000). *Rapid and Dynamic Solid–Liquid Extractor Working at High Pressures and Low Temperatures for Obtaining in Short Times Solutions Containing*

Substances That Initially Were in Solid Matrixes Insoluble in Extracting Liquid. Italian Patent no. 1,303,417.

30. Naviglio, D., (2003). Navigilo's principle and presentation of an innovative solid-liquid extraction technology: Extractor naviglio®. *Analytical Letters, 36*(8), 1647–1659.
31. Nayak, B., Dahmoune, F., Moussi, K., Remini, H., Dairi, S., Aoun, O., & Khodir, M., (2015). Comparison of microwave, ultrasound and accelerated-assisted solvent extraction for recovery of polyphenols from *Citrus sinensis* peels. *Food Chemistry, 187*, 507–516.
32. Posadino, A., Biosa, G., Zayed, H., Abou-Saleh, H., Cossu, A., Nasrallah, G., Giordo, R., et al., (2018). Protective effect of cyclically pressurized solid-liquid extraction polyphenols from cagnulari grape pomace on oxidative endothelial cell death. *Molecules, 23*(9), 2105.
33. Putnik, P., Lorenzo, J. M., Barba, F. J., Roohinejad, S., Jambrak, A. R., Granato, D., Montesano, D., & Kovacevic, D. B., (2018). Novel food processing and extraction technologies of high-added value compounds from plant materials. *Foods, 7*(7), 106–112.
34. Sharif, K. M., Rahman, M. M., Azmir, J., Mohamed, A., Jahurul, M. H. A., Sahena, F., & Zaidul, I. S. M., (2014). Experimental design of supercritical fluid extraction: A review. *Journal of Food Engineering, 124*, 105–116.
35. Souza-Silva, E. A., Gionfriddo, E., & Pawliszyn, J., (2015). A Critical review of the state of the art of solid-phase microextraction of complex matrices II. food analysis. *Trends in Analytical Chemistry, 71*, 236–248.
36. Tatke, P., & Rajan, M., (2014). Comparison of conventional and novel extraction techniques for the extraction of scopoletin from *Convolvulus pluricaulis. Indian Journal of Pharmaceutical Education and Research, 48*(1), 27–31.
37. Tiwari, B. K., (2015). Ultrasound: A clean, green extraction technology. *Trends Analytical Chemistry, 71*, 100–109.
38. Vongsak, B., Sithisarn, P., Mangmool, S., Thongpraditchote, S., Wongkrajang, Y., & Gritsanapan, W., (2013). Maximizing total phenolics, total flavonoids contents and antioxidant activity of *Moringa oleifera* leaf extract by the appropriate extraction method. *Industrial Crops and Products, 44*, 566–571.
39. Wang, L., & Weller, C. L., (2006). Recent advances in extraction of nutraceuticals from plants. *Trends in Food Science & Technology, 17*(6), 300–312.
40. Wei, Y., Du, J., & Lu, Y., (2012). Preparative separation of bioactive compounds from essential oil of *Flaveria bidentis* (L.) Kuntze using steam distillation extraction and one step high-speed counter-current chromatography. *Journal of Separation Science, 35*(19), 2608–2614.

CHAPTER 2

NOVEL EXTRACTION METHODS: PROFILING OF NATURAL PHYTOCHEMICALS

BHUPINDER KAUR, SALAM MAHESHKUMAR SINGH, and
PREM P. SRIVASTAV

ABSTRACT

In the diversified floral assemblage in tropical and subtropical regions, various plants are harbor reservoirs of valuable substances, viz., antioxidants, polyphenols, starch, pectin, pigment, flavonoids, fat, crude fiber, protein, and minerals, along with secondary metabolites. The recovery of these compounds is dependent on the type of grinding and extraction technique. Cryogenic grinding is a very promising pathway to retain the thermo-labile components, which are normally degraded during ambient grinding. Also, the old extraction techniques have some disadvantages, such as a long extraction time and high solvent consumption. In complying with the greener perspectives, modern extraction techniques (single/combined) are suitable alternatives to conventional ones.

2.1 INTRODUCTION

Nowadays, the natural extracts from plant sources, such as fruits, vegetables, flowers, herbs, shrubs, spices, and their byproducts (seeds, peels, leaves, bark, roots, and stem) have gained imperative attention of researchers owing to their bioactive composition [60]. Plant extracts contain vitamins, minerals,

Novel Processing Methods for Plant-Based Health Foods: Extraction, Encapsulation, and Health Benefits of Bioactive Compounds. Megh R. Goyal, N. Veena & Ritesh B. Watharkar (Eds.)

fiber, therapeutic phytochemicals, amino acids, sugars, and other micronutrients [2], which are necessary for human health, growth, and development (Figure 2.1). Some of these compounds have unique functionalities like antioxidant, antimicrobial, antiproliferative, anti-inflammatory, and metal chelation activities and can be implemented for the treatment and prevention of cancer, cardiovascular, and other chronic diseases [57]. The natural constituents of some fruits make them a probable source of detoxification also.

Antioxidants are basically employed to retard the oxidation process by quenching the free radicals like reactive oxygen species (ROS), superoxide dismutase, and reactive nitrogen species. Synthetic antioxidants like butylated hydroxytoluene, butylated hydroxyanisole, dodecyl gallate, propyl gallate, octyl gallate, ethylene diamine tetra-acetic acid, and tertiary-butylhydroquinone are under strict vigilance for their use in food applications, owing to their potential health hazards [52]. The significance of natural antioxidants has been highlighted by many researchers as important nutraceuticals on account of many health benefits [37, 69]. Therefore, the trend of using natural antioxidants from plant sources has been growing continuously with merits of their easy availability and low cost.

The preliminary and very crucial step for the compositional analysis of medicinal plants is extraction. It is one of the key steps involved in the identification, purification, and recovery of high-value ingredients [45]. Every bioactive compound requires a specific extraction technique, solvent, time, temperature, and pressure with respect to its polarity, chemical affinity, and molecular structure based on the presence of hydroxyl groups and aromatic and aliphatic rings [6].

Recently, a number of advanced extraction techniques have been explored for the efficient isolation of value-added compounds (Figure 2.1). New extractive strategies like the use of co-solvents and a combination of different techniques are being used in various industries in order to improve the recovery of the target molecules and pharmacological profile of extracts along with reducing the extraction time and wastage of solvents. Moreover, this will favor environmental safety; as it will reduce the disposal of solvents. The efficiently extracted phenolic antioxidants can be successfully used as an ingredient in the development of nutraceuticals, functional foods, novel foods, and drugs.

This chapter addresses the importance and applications of cryogenic grinding and novel extraction techniques (such as High-pressure extraction (HPE), Supercritical fluid extraction (SFE), Pressurized liquid extraction

(PLE), Microwave-assisted extraction (MAE), Microwave-ultrasonic assisted aqueous enzymatic extraction (MUAAEE), Ultrasound-assisted extraction (UAE), and Ultrasonic-assisted extraction integrated temperature-induced cloud point extraction (UAE-TICPE) for obtaining the valuable bioactive compounds. It also emphasizes on the bioactivities of phyto-chemical compounds.

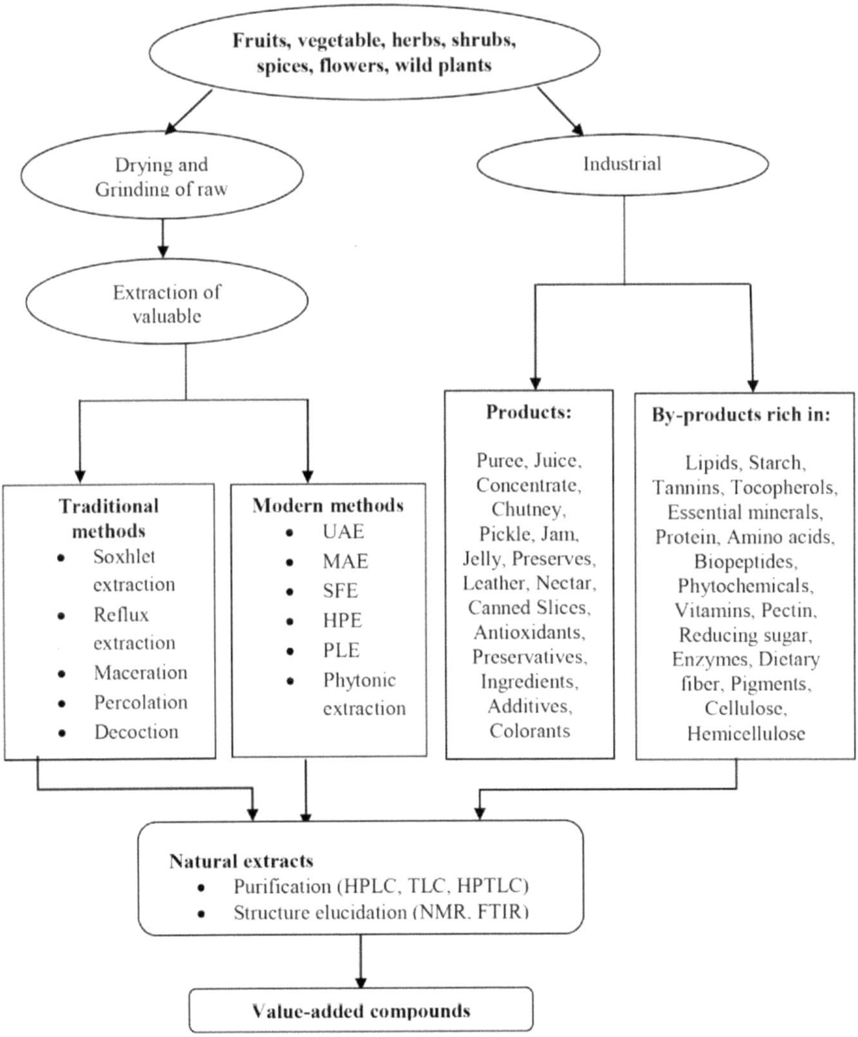

FIGURE 2.1 Processing of various plant extracts.

2.2 CRYOGENIC GRINDING

Fruits and vegetables are rich sources of health-promoting components, which can be degraded during their processing. The different unit operations have an adverse effect on product quality, thermo-labile biomolecules (antioxidants and polyphenols), and volatile oils [16, 18]. Size reduction is a conventional way performed by impact and attrition, which is carried out to reduce the particle size. In ambient or conventional grinding, the temperature rise is quite high, which results in loss of heat-labile bioactive components, leading to partial loss of organoleptic and nutritional properties of food products.

Additionally, there is more probability of heat generation during the production of finer particles due to prolonged grinding time. The traditional grinding at ambient temperature caused an increase in temperature (43–95°C) which resulted into loss of volatile oils (mace: 14%; oregano and cinnamon: 17%; cumin: 18–19%; nutmeg: 37%; coriander: 40%) [62]. The fat/oil content in samples is of high concern, because heat makes the fat to melt, which leaks out of vials and degrades the powder quality with lump formation. This leaked fat can be oxidized and there might be the generation of dark compounds that is ultimately undesirable and unacceptable.

Therefore, cooling can be used to overcome the increased temperature issue; which will have a considerably positive impact on the quality retention of the product. The low temperature during grinding could be achieved with airflow cooling systems like cold air, coolant, or water circulation around the jacket of the grinder. However, this technique is not sufficient to significantly reduce the temperature rise of the product to a level, which is safe enough, so as not to affect its quality characteristics [18].

In contrast, the cryogenic grinding could be used as a solution to limit the effects of heating during the grinding process. Cryogenic grinding can be defined as the production of powdered material accompanied by cryogenic liquids like liquid nitrogen, liquid oxygen, and liquid helium [30]. Liquid nitrogen has been used for many studies in the agricultural field that provides refrigeration effect at – 195.6°C by absorbing heat generated in the grinding operation. The cryogenic grinding not only retains the valuable compounds but enhances the recovery also. The earlier work on the use of cryogenic grinding in an agricultural area is majorly highlighted on spices and mango byproducts by several researchers, such as:

- Barnwal et al. (cinnamon and turmeric) [5];
- Ghodki and Gowsami (black pepper) [16];
- Kaur and Srivastav (mango peel) [31];

- Kaur et al. (mango seed) [30];
- Meghwal and Gowsami (fenugreek) [43];
- Saxena et al. (coriander) [58];
- Sharma et al. (cumin) [62];
- Singh and Goswami (cloves) [63].

The aforementioned studies provide the information on fundamental aspects of powder properties and the process optimization parameters. The cryo-ground products have improved shelf life and enhanced quality due to more retention of color, flavor, aroma, volatile, and bioactive components which is directly proportional to high consumer acceptance [4, 46]. Thus, cryogenic grinding is attractive technology, which avoids the oxidative and evaporation reactions [59] in natural products. The efficiency of cryogenic grinding is very much dependent on certain parameters of grinding equipment like, peripheral speed, amplitude, number of rotor ribs, feed rate, moisture content, sieve mesh number, and moisture content. Hence optimization of these processing parameters is very important. Generally, ball mill is used for cryo-grinding operations and fractures the particles by impact force. There is no chance of choking in ball mill, so it is easy to operate comparatively to other small-scale mills viz. pin, rotor, and hammer mill and less amount of energy is needed for the grinding operation [31].

2.3 TRADITIONAL EXTRACTION TECHNIQUES

Maceration, percolation, infusion, decoction, and Soxhlet extraction (SE) are the well-known traditional methods for extraction of antioxidants from plant sources. The nature of the solvent used for extraction along with the extraction time and temperature plays a dominant role in deciding the antioxidant yield and capacity. To achieve optimum recovery of phenolic antioxidants basically polar solvents like ethanol and methanol are preferred over acetone and ethyl acetate. A combination of solvents have been used to extract bioactives from rosemary, sage, sumac, rice bran, wheat grain bran, mango seed kernel, citrus peel, and many other fruit peels [1, 50].

2.3.1 MACERATION

Maceration is a valuable and simple extraction method which involves soaking of raw material (whole, coarse or fine) in a particular solvent for not less than 3 days with repeated agitation at ambient temperature, until the expected compounds are dissolved [3]. Generally, the biochemical constituents are released by rupturing the cell structure, and as a result, the organized tissues are transformed into a suspension of intact cells. This technique is commonly used in wine processing and has been widely used in medicinal plants research. As compared to other extraction methods, it has the demerit of low efficiency and long extraction time.

2.3.2 PERCOLATION

Percolation is an exhaustive extraction procedure that extracts the soluble constituents of a plant material. The samples are moistened with menstruum prior to percolation followed by addition of more menstruum for proper soaking and extraction (24 h).

2.3.3 INFUSION AND DECOCTION

An infusion and decoction are dilute solution of easily soluble constituents of plant materials. The harder plant materials like seeds, bark, and roots are crushed and immersed in boiling/cold water, allowed to stand 15–30 min and then filtered [21]. They are very suitable for extracting water-soluble and heat-stable constituents of aromatic plants as it prevents the loss of volatile oils. This process is typically used in the preparation of Ayurvedic extracts (quath/kawath). Concentrated infusion and decoction are prepared by modified percolation or maceration and can be used either as whole or after suitable dilution. The main drawback of infusion is that they are quickly susceptible to microbial attack, so it should be disposed within 12 hours of preparation.

2.3.4 SOXHLET EXTRACTION (SE)

The traditional Soxhlet method is a simple, easy-to-use, effective, and well-established standard procedure. A large spectrum of environmental (sediment and soil) and biological (plant and animal tissues) samples can be processed by this method. In this technique, the solvent is heated in a chamber and vaporizes into the sample thimble to extract the desirable compounds, condenses, and drip back. The SE helps to displace the transfer equilibrium, because the sample is infused with the fresh volume of solvent repeatedly throughout the process [35]. The higher yield of oil due to an increase in extraction time and temperature, has been reported by several researchers while using the Soxhlet technique [10, 27, 33, 48]. Lumley et al. [38] explained that it might be because of the solvent's enhanced ability to overcome forces that strongly bind lipids within the sample matrix. PLE gave better results and lower yield compared with SE and UAE [55]. The unfavorable factor is that it is not environmentally friendly, as it involves the use of hazardous and flammable solvents and may contribute to the pollution problem.

2.4 ADVANCED EXTRACTION TECHNIQUES

Extraction methods are widely used for separation of components from chemical and biological mixtures and has a wide range of applications. Advanced methods like SFE, HPE, UAE, MAE, and PLE have been recognized as important extraction tools with several advantages like reducing the solvent volume and extraction time, which in turn improves the quality and wholesomeness of the final product [23]. They can be more advantageous, when used in combination with other extraction techniques (Table 2.1). Additionally, these techniques are eco-friendly in terms of solvent and energy consumption.

2.4.1 MICROWAVE-ASSISTED EXTRACTION (MAE)

MAE is a simple and effective extraction technique that combines microwave energy and traditional solvent extraction. It is based on the absorption of microwave energy by charged particles and polar compounds in the sample. The cell structure of plant materials is disordered by electromagnetic waves via ionic conductance and dipole rotation [64]. The electromagnetic field of microwaves induces the rotation of dipoles to break the hydrogen bonding. This causes the quick movement of dissolved ions and enables solvent diffusion into the matrix, which results in high extraction rates [19].

Its effectiveness is related to the dielectric properties of polarizable materials, higher dielectric constant is directly proportional to the energy absorption by molecules [26]. This interaction between microwave radiation and sample helps to attain the boiling point of the solvent and transfer the heat by conduction. By using appropriate parameters to avoid the loss of thermo-labile components, this technique improved the extraction quality of bioactive compounds and reproducibility. The yield of phenolic and

TABLE 2.1 Conventional and Modern Extraction Techniques for Natural Extracts

Method	Extraction Medium	Yield	Sample	Compounds	Extraction Conditions	References
Conventional Extraction Techniques						
Soxhlet extraction	n-hexane	23.5–27.4 mg/g	Mango peel	Oil	105°C, 60 min	[31]
Maceration	Refined olive oil	1244.5 mg/g	Tomato peels	Lycopene	80°C, 45 min Biomass to oil ratio: 2.5% (w/v) Magnetic stirring: 400 rpm	[32]
Combination of maceration and UAE	water	828 mg/g	Ornithogalum	Total polysaccharides	47.1°C, 5,000 bar, 42.12 min Solvent to solid ratio (37.2 ml/g)	[42]
Modern Extraction Techniques						
HPE	Water	324.89 mg/g	Korean barberry	Total phenols	25°C, 5,000 bar, 5 min Solvent to raw material ratio (1:10)	[56]
SFE	CO_2	330 mg/g	Tomato	*trans*-Lycopene	62°C; 446 bars; 20 min	[28]
PLE	Neutral glass	34.55 mg/g	Green tea leaves	Caffeine	150°C; 40 bars; 10 min	[11]
MAE	Methanol: water (80:20, v/v)	126.3 mg/g	Olive leaf	Luteolin glucoside and isomers	80°C; 6 min	[64]
MUAAEE	Water	852.3 mg/g	Tiger nut	Oil	2% hemicellulase: pectinase: cellulose (1/1/1, w/w/w); Particle size: <600 μm; Microwave and ultrasound power: 300, 460 W; Radiation time and temp.: 30 min, 40°C; Enzymolysis pH and temp.: 4.9, 45°C; liquid-to-solid ratio: 10 ml/g; Extraction time: 180 min	[26]

TABLE 2.1 *(Continued)*

Method	Extraction Medium	Yield	Sample	Compounds	Extraction Conditions	References
UAE	Ethanol	16.1 mg/g	Alfalfa	Total saponins	76.8°C; 2.84 h; Liquid-to-solid ratio: 11.4 ml/g; Power: 112 W; Ethanol concentration: 78.2%	[20]
UAE-enzymatic extraction	Water	5.98 mg/g	Mulberry wine residues	Total anthocyanins	52°C, 315 W; 0.22% enzyme, 94 min	[71]
UAE coupled TICPE	PEG-based aqueous solution	2.61–3.15 mg/g	*Euonymus alatus*	Total flavonoids	90°C, 15 min; PEG-400: 16% (w/w); Particle size: 80 mesh; Liquid-to-solid ratio: 60:1	[39]

flavanones decreases sometimes while using an additional cycle of MAE, and this may be due to the oxidation of compounds [21].

MAE is not suitable for the extraction of bioactive compounds that are non-polar, because energy is transferred by dielectric absorption only. Factors affecting Microwave extraction are solvent, extraction time, microwave power, matrix characteristics, extraction temperature and pressure.

2.4.2 ULTRASONIC-ASSISTED EXTRACTION (UAE)

Ultrasonic-assisted extraction (UAE) does not involve the application of heat. It is one of the easily adapted extraction techniques because it uses organic solvent and water bath with ultrasound assistance at low operating temperature [71]. Basically, it consists of a specific quantity of solvent and high-frequency sounds to obtain maximum extraction of the compounds embedded in the solid mass. The higher interface between solvent and sample area favors the increase in the solubility of compounds and mass transfer rate [39].

In the food industry, high-intensity low-frequency (HILF) having frequency in the range of 20–100 kHz and low-intensity high-frequency (LIHF) with more than 100 kHz frequency are used. HILF causes an increase in permeability of cells via high-intensity shock waves and bubble cavitation. Contrarily, LIHF does not alter the physicochemical properties of the material. During UAE, the choice of solvents like ethanol, methanol, and hexane is an important parameter for better cavitation phenomena that are influenced by its physical properties like polarity, viscosity, vapor pressure, and surface tension. It is less expensive and can extract a large quantity of bioactive compounds within a short period and without using a large amount of solvent as compared to traditional extraction techniques. The cons of UAE related to experimental extraction are its lack of reproducibility and repeatability [20].

2.4.3 SUPERCRITICAL FLUID EXTRACTION (SFE)

Supercritical fluid extraction (SFE) is one of the most interesting extraction techniques, since the extraction of bioactive compounds is carried out near ambient temperature, and prevents the bioactive substance from thermal decomposition. It uses supercritical fluid (SF) as a solvent, which is in its supercritical state, i.e., above its own critical pressure and temperature. The behavior of SF is similar to solubility to liquid as well as diffusivity to gas and also dissolves a wide variety of natural products. However, near their critical points, their solvating properties drastically change due to the small pressure and temperature. Various SF can be exercised based on their pressure and temperature along with variable extraction times, flow rates, and modifiers [6] and literature is available on the use of NH_3, CO_2, N_2O and NH_3 as extraction solvents [23].

CO_2 becomes SF at 73.8 bar and above 31.1°C and has poor solubility for polar compounds. Therefore, to extract polar compounds, a small amount of ethanol or methanol (5–10%) is added and produces analytes at concentrate form as CO_2 vaporizes at ambient temperature [28]. The high initial cost is the major drawback of this extraction technique.

2.4.4 PRESSURIZED LIQUID EXTRACTION (PLE)

PLE is another technique that nowadays is regarded as green extraction technique and commonly referred as accelerated solvent extraction (ASE). However, when water is used as extractant, some terms such as subcritical water extraction, superheated water extraction or hot liquid water extraction are frequently used. It is a method of extracting a sample under high temperature (50–200°C) and pressure (35–207 bar) using conventional solvents [8]. It is carried out in dynamic mode by enabling the continuous flow of solvent through the sample and static mode by applying constant heat and pressure to sample and solvent for a specific time interval [7].

Liquid solvents at elevated temperatures reduce the viscosity of the solvent and can infiltrate without breaking down the thermally labile compounds. The extraction temperature positively influences the diffusion rate, mass transfer, extractability, and decreases the surface tension and viscosity of the solvents. Moreover, at elevated pressure, liquid solvent maintains the liquid phase to disrupt the plant cell by applying pressure on the matrix. Thus, this technique is used to enhance the extraction performance as compared to those techniques carried out near room temperature and atmospheric pressure [11]. Different parameters that influence the extraction process and performances are amount of size and its composition; nature, volume, and flow of the solvent; the number of cycles; and extraction time.

2.4.5 PHYTONIC EXTRACTION

It is an unconventional extraction technique in which a new solvent-based on non-chlorinated fluoro-hydrocarbons, such as hydrofluorocarbon-134a

(HFC-134a) or 1,1,1,2 tetra-fluoroethane is used to extract the high quality natural essential oils, flavors, and bioactive compounds from the medicinal plants. These solvents are non-toxic and eco-friendly refrigerant, commonly used in car air conditioners and nebulizers. The extracted product can be used directly without any clinical trial. HFC-134a is a nonflammable solvent that doesn't mix easily with triglycerides and mineral oils [17]. Additionally, it doesn't dissolve plant wastes, therefore replaced with chlorofluorocarbon, as HFC-134a doesn't deplete the ozone layer. This process is gentle, cool, and requires low energy. Hence, the products are not exposed to higher temperatures.

2.5 POLYPHENOLS

Phenolic compounds are widely distributed phytochemicals in nature and abundant source of antioxidants. The most significant secondary metabolites are synthesized naturally in plants through shikimic acid and phenylpropanoid pathways. They are a group of small molecules having phenol as a basic unit with versatile functionalities. Additionally, Hintz et al. [25] compiled the information regarding the biological activities of bioactive polyphenols viz., phenolic acids, lignans, stilbenes, and flavonoids (anthocyanins, flavanols, and catechins).

2.5.1 PHENOLIC ACIDS

Phenolic acids are a major class of polyphenols available in free, conjugated, and bound form. The main pillars of phenolic acids are hydroxybenzoic acid (vanillic, syringic, gentisic acid, etc.), and hydroxycinnamic acid (chlorogenic, caffeic, *sinapic*, ferulic acid, etc.). Generally, phenolic acids can be digested and absorbed in the upper part of the gastrointestinal (GI) tract in aglycone form [66]. They are capable of modulating metabolic processes and exhibiting *in vitro* antioxidant activity resulting in the promotion of better health. The mechanism of action of these compounds is especially related to reduce risk of chronic diseases.

2.5.2 FLAVONOIDS

Flavonoids are hydroxylated phenolic structures with a C3–C6 aromatic ring linkage. They are classified as flavones, flavanols, flavanones, isoflavones, anthocyanins, and proanthocyanidins. Out of which, 80% are found in ferns and higher plants. Every class of flavonoids is metabolized differently in the body and its physicochemical properties affect the digestion, absorption, and biotransformation, which help the organs to function more efficiently while protecting them against everyday toxins and stressors [39].

Research on flavonoids received an added impulse because they act as estrogen agonists after menopause due to low-estrogen environment and reduce the need for hormone replacement therapy dosage [70]. The mechanism related to the metabolic process of flavonoids is still not elaborated clearly. Nevertheless, it has widely been known for centuries because of its broad spectrum of biological activities. In herbal medicine, anthocyanin rich substances have long been used to treat a number of health conditions involving blood vessel health, high blood pressure, diabetes, and urinary tract infections (UTI).

2.5.3 TANNINS

Tannins are a group of water-soluble, relatively high molecular weight compounds that occur in complexes with alkaloids, polysaccharides, and proteins. They may have long chains of gallic acid coming from central glucose core and are grouped into hydrolysable and condensed tannins [68]. The natural tannins help in giving the structure, texture, and flavor to fermented wine made from fruit skin. It has been cited those tannins caused the decline in feed intake, feed efficiency, protein digestibility, growth rate, and net metabolizable energy in experimental animals. Therefore, tannin-containing foods are categorized as low-nutrition foods.

2.5.4 STILBENES

Stilbenes are a small family derived from the phenylpropanoid pathway and produced in a number of unrelated plant species [9]. They are chromophores that can undergo photoisomerization as do azobenzenes, but they also show photodimerization. An extensive literature suggests that stilbenes are found in inducible and constitutive defense mechanisms produced by biotic elicitation and enzymes released during the elimination of toxic compounds. They exist as cis and trans isomers, namely (*Z*)-stilbene (cis-stilbene), i.e., unstable, and sterically hindered and 1,2-diphenylethylene: (*E*)-stilbene, i.e., stable, and not sterically hindered, respectively. However, the detailed functions of these compounds have not been properly explored.

2.6 BIOACTIVITIES OF PHYTOCHEMICALS

There is enough literature on the biological activities of phytochemicals present in berries, apple, mango, pomegranate, potato, tomato, spinach, giloy, moringa, mint, thyme, curry leaf, ginger, turmeric, and fenugreek extracts [25, 31, 41, 49, 67]. The major studied compounds responsible for these activities are resveratrol, malic acid, β-carotene, lycopene, curcumin, menthol, thymol, berberine, gingerol, capsaicin, eugenol, gallic acid, ellagic acid, and epigallocatechin.

2.6.1 ANTIOXIDANT ACTIVITY

The cellular redox results into a very unstable and uncontrolled production of free radicals. They have the tendency to react readily with organic substrates (DNA, protein, and lipid) and lead to DNA strand breakage, protein denaturation and lipid oxidation [61]. The natural phytochemicals are considered to play an important role against diseases related to oxidative stress like allergies, asthma, arthritis, neurodegenerative diseases, sepsis, coronary heart disease, atherosclerosis, autoimmune disorders, hemodialysis, diabetes, hypertension, and cancer [67]. The free radicals are scavenged by bioactive compounds (majorly from fruits and vegetables) and act as antioxidant agents. More recently, herbs and spices have been recognized as significant sources of phytochemicals too [53].

2.6.2 ANTIMICROBIAL ACTIVITY

The side effects of some synthetic antibiotics namely, benzoyl peroxide, erythromycin, clindamycin, and triclosan have been reported. Hence, the development of novel therapeutic agents with high antibacterial activity but less possible side effects is the need of time [65]. Gallic acid, ellagic acid, gallotannins, methyl gallate, tannic acid and polygalloyl glucose type phenols were mainly investigated in plant-based extracts that effectively suppressed the main spoilage bacteria and fungi [13, 14, 29, 68].

The hydrolysable compounds can interact with proteins to inhibit enzyme activities and iron-complexing properties, which can restore the growth of bacteria. Interestingly some herbs namely thyme, oregano, basil, rosemary, and sage exhibit very strong antimicrobial activity [25].

2.6.3 ANTIPROLIFERATIVE ACTIVITY

Lately, the antiproliferative effect of natural extracts has been gaining the global attention. The anticancer activities of fruits and their polyphenolic compounds have been reported by several researchers on cell lines of prostate, lung, breast, leukemia, colon, cervical and colorectal cancer cells [15, 36, 47, 54]. Moreover, triterpenes lupeol and mangiferin owe their chemopreventive action to the induction of permeability in mitochondria [51]. Interestingly, the cell lines showed specific response with respect to a few factors such as cultivar, tissue, and extract source and growth environment [47]. The antiproliferative activity of extracts also depends on their phenolic and flavonoid contents [34].

2.6.4 METAL CHELATING ACTIVITY

The transition-metal ions (Fe^{2+}, Mg^{2+}, Cu^{2+}) play a very important role in the functionality of free radicals during oxidative phosphorylation. These metal ions are captured by various chelating agents to inhibit their action. Some compounds like gallates (methyl gallate), tannins (ellagitannins and gallotannins) and pentagalloyl glucose, are natural metal chelating agents [40, 44]. Engles et al. [14] reported that gallotannins has almost 10 times higher iron-binding capacity than EDTA, which depends on a number of galloyl groups.

2.7 SUMMARY

For the past few decades, the replacement of artificial antioxidants with natural ones has been becoming a field of mounting interest for researchers. Natural extracts contain numerous health-promoting polyphenolic antioxidants, which should be fractionated very carefully and efficiently. These beneficial components might act as a source of fragrances, flavoring compounds, dyes, bio-sorbents, agrochemicals, and new drugs. Suitable advanced grinding and extraction technologies should be used for the industrial exploitation of such beneficial bioresources, which in the long run will contribute widely to the socio-economic growth of the human race. This chapter highlights the advantages of cryogenic grinding and advanced extraction methods (HPE, SCFE, PLE, MAE, MUAAEE, UAE, and UAE-TICPE).

KEYWORDS

- **cryogenic grinding**
- **microwave-assisted extraction**
- **novel extraction techniques**
- **plant extract**
- **polyphenols**
- **supercritical fluid extraction**

REFERENCES

1. Altemimi, A., Lakhssassi, N., Baharlouei, A., Watson, D. G., & Lightfoot, D. A., (2017). Phytochemicals: Extraction, isolation, and identification of bioactive compounds from plant extracts. *Plants (Basel), 6*(4), 42.
2. Aziz, N. A. A., Wong, L. M., Bhat, R., & Cheng, L. H., (2012). Evaluation of processed green and ripe mango peel and pulp flours (*Mangifera indica* var. Chokanan) in terms of chemical composition, antioxidant compounds and functional properties. *Journal of the Science of Food and Agriculture, 92*(3), 557–563.
3. Azwanida, N. N., (2015). A review on the extraction methods used in medicinal plants, principle, strength and limitation. *Medicinal & Aromatic Plants, 4*(3), 196.
4. Balasubramanian, S., Gupta, M. K., & Singh, K. K., (2012). Cryogenics and its application with reference to spice grinding: A review. *Critical Reviews in Food Science and Nutrition, 52*(9), 781–794.
5. Barnwal, P., Mohite, A., Singh, K. K., Kumar, P., Zachariah, T. J., & Saxena, S. N., (2014). Effect of cryogenic and ambient grinding on grinding characteristics of cinnamon and turmeric. *International Journal of Seed Spices, 4*(2), 26–31.
6. Bhattacharya, M., Srivastav, P. P., & Mishra, H. N., (2014). Optimization of process variables for supercritical fluid extraction of ergothioneine and polyphenols from *Pleurotus ostreatus* and correlation to free-radical scavenging activity. *The Journal of Supercritical Fluids, 95*, 51–59.

7. Carabias-Martinez, R., Rodriguez-Gonzalo, E., Revilla-Ruiz, P., & Hernandez-Mendez, J., (2005). Pressurized liquid extraction in the analysis of food and biological samples. *Journal of Chromatography A, 1089*(1, 2), 1–17.
8. Carr, A. G., Mammucari, R., & Foster, N. R., (2011). A review of subcritical water as a solvent and its utilization for the processing of hydrophobic organic compounds. *Chemical Engineering Journal, 172*(1), 1–17.
9. Chong, J., Poutaraud, A., & Hugueney, P., (2009). Metabolism and roles of stilbenes in plants. *Plant Science, 177*(3), 143–155.
10. Conte, R., Gullich, L. M. D., Bilibio, D., Zanella, O., Bender, J. P., Carniel, N., & Priamo, W. L., (2016). Pressurized liquid extraction and chemical characterization of safflower oil: A comparison between methods. *Food Chemistry, 213*, 425–430.
11. Dawidowicz, A. L., & Wianowska, D., (2005). PLE in the analysis of plant compounds part I. The application of PLE for HPLC analysis of caffeine in green tea leaves. *Journal of Pharmaceutical and Biomedical Analysis, 37*(5), 1155–1159.
12. Dhanani, T., Tapadia, S., Gajbhiye, N. A., & Kumar, S., (2017). Effect of extraction methods on yield, phytochemical constituents and antioxidant activity of *Withania somnifera*. *Arabian Journal of Chemistry, 10*(1), S1193–S1199.
13. Engels, C., Ganzle, M. G., & Schieber, A., (2009). Fractionation of gallotannins from mango (*Mangifera indica* L.) kernels by high-speed counter-current chromatography and determination of their antibacterial activity. *Journal of Agricultural and Food Chemistry, 58*(2), 775–780.
14. Engels, C., Knodler, M., Zhao, Y. Y., Carle, R., Ganzle, M. G., & Schieber, A., (2009). Antimicrobial activity of gallotannins isolated from mango (*Mangifera indica* L.) kernels. *Journal of Agricultural and Food Chemistry, 57*(17), 7712–7718.
15. Garcia-Rivera, D., Delgado, R., Bougarne, N., Haegeman, G., & Berghe, W. V., (2011). Gallic acid indanone and mangiferin xanthone are strong determinants of immunosuppressive anti-tumor effects of *Mangifera indica* L. bark in MDA-MB231 breast cancer cells. *Cancer Letters, 305*(1), 21–31.
16. Ghodki, B. M., & Goswami, T. K., (2016). Effect of grinding temperatures on particle and physicochemical characteristics of black pepper powder. *Powder Technology, 299*, 168–177.
17. Ghosh, U., Haq, M. A. B., & Chakraborty, S., (2011). Application of systematic technologies for the extraction of novel phytoconstituents from pharmacologically important plants. *International Journal of Chemical and Analytical Science, 2*(9), 1153–1158.
18. Goswami, T. K., & Singh, M., (2003). Role of feed rate and temperature in attrition grinding of cumin. *Journal of Food Engineering, 59*(2), 285–290.
19. Guardia, M., & Armenta, S., (2011). Greening sample treatments; Chapter 5. In: Guardia, M., & Armenta, S., (eds.), *Comprehensive Analytical Chemistry* (Vol. 57, pp. 87–120). Amsterdam, Netherlands: Elsevier.
20. Hadidi, M., Ibarz, A., & Pagan, J., (2020). Optimization and kinetic study of the ultrasonic-assisted extraction of total saponins from alfalfa (*Medicago sativa*) and its bioaccessibility using the response surface methodology. *Food Chemistry, 309*, 125786.
21. Handa, S. S., Khanuja, S. P. S., Longo, G., & Rakesh, D. D., (2008). *Extraction Technologies for Medicinal and Aromatic Plants* (p. 260). Trieste–Italy: International center for science and high technology.

22. Herrero, M., Cifuentes, Al., & Ibanez, E., (2006). Sub-and supercritical fluid extraction of functional ingredients from different natural sources: Plants, food-by-products, algae and microalgae: A review. *Food Chemistry, 98*(1), 136–148.
23. Herrero, M., Mendiola, J. A., Cifuentes, Al., & Ibanez, E., (2010). Supercritical fluid extraction: Recent advances and applications. *Journal of Chromatography A, 1217*(16), 2495–2511.
24. Herrero, M., Plaza, M., Cifuentes, A., & Ibáñez, E., (2012). Extraction techniques for the determination of phenolic compounds in food. *Comprehensive Sampling and Sample Preparation, 4*, 159–180.
25. Hintz, T., Matthews, K. K., & Di, R., (2015). The use of plant antimicrobial compounds for food preservation. *BioMed Research International, 2015*, 246264.
26. Hu, B., Li, Y., Song, J., Li, H., Zhou, Q., Li, C., Zhang, Z., et al., (2020). Oil extraction from tiger nut (*Cyperus esculentus* L.) using the combination of microwave-ultrasonic assisted aqueous enzymatic method-design, optimization and quality evaluation. *Journal of Chromatography A, 1627*, 461380.
27. Jahurul, M. H. A., Zaidul, I. S. M., Norulaini, N. N. A., Sahena, F., Jaffri, J. M., & Omar, A. K. M., (2014). Supercritical carbon dioxide extraction and studies of mango seed kernel for cocoa butter analogy fats. *CyTA-Journal of Food, 12*(1), 97–103.
28. Kassama, L. S., Shi, J., & Mittal, G. S., (2008). Optimization of supercritical fluid extraction of lycopene from tomato skin with central composite rotatable design model. *Separation and Purification Technology, 60*(3), 278–284.
29. Kassi, A. B. B., Soro, Y., Fante, B., Golly, K. J., Sorho, S., Toure, A. S., & Coustard, J. M., (2014). Isolation and identification of bioactive compounds from kernel seed cake of the mango (*Mangifera indica* Lam). *International Journal of Biological and Chemical Sciences, 8*(4), 1885–1895.
30. Kaur, B., Pyne, S. K., Srivastav, P. P., & Nag, A., (2019). Characterization and comparison of mango (*Mangifera indica*) processing waste from five Indian cultivars. *International Journal of Chemical Studies, 7*(4), 2844–2849.
31. Kaur, B., & Srivastav, P. P., (2018). Effect of cryogenic grinding on chemical and morphological characteristics of mango (*Mangifera indica* L.) peel powder. *Journal of Food Processing and Preservation, 42*(4), e13583.
32. Kehili, M., Sayadi, S., Frikha, F., Zammel, A., & Allouche, N., (2019). Optimization of lycopene extraction from tomato peels industrial byproduct using maceration in refined olive oil. *Food and Bioproducts Processing, 117*, 321–328.
33. Khoei, M., & Chekin, F., (2016). The ultrasound-assisted aqueous extraction of rice bran oil. *Food Chemistry, 194*, 503–507.
34. Kim, H., Moon, J. Y., Kim, H., Lee, D., Cho, M., Choi, H., Kim, Y. S., Mosaddik, A., & Cho, S. K., (2010). Antioxidant and antiproliferative activities of mango (*Mangifera indica* L.) flesh and peel. *Food Chemistry, 121*(2), 429–436.
35. Kittiphoom, S., & Sutasinee, S., (2013). Mango seed kernel oil and its physicochemical properties. *International Food Research Journal, 20*(3), 1145–1149.
36. Li, F., Li, S., Li, H., Deng, G., Ling, W., Wu, S., Zu, X., & Chen, F., (2013). Antiproliferative activity of peels, pulps and seeds of 61 fruits. *Journal of Functional Foods, 5*(3), 1298–1309.
37. Lobo, V., Patil, A., Phatak, A., & Chandra, N., (2010). Free radicals, antioxidants and functional foods: Impact on human health. *Pharmacognosy Reviews, 4*(8), 118–126.

38. Lumley, I. D., & Colwell, R. K., (1991). Extraction of fats from fatty foods and determination of fat content; Chapter 6. In: Rossell, J. B., & Pritchard, J. L. R., (eds.), *Analysis of Oilseeds, Jars and Fatty Foods* (pp. 238–247). New York, USA: Elsevier Applied Science.
39. Mai, X., Liu, Y., Tang, X., Wang, L., Lin, Y., Zeng, H., Luo, L., Fan, H., & Li, P., (2020). Sequential extraction and enrichment of flavonoids from *Euonymus alatus* by ultrasonic-assisted polyethylene glycol-based extraction coupled to temperature-induced cloud point extraction. *Ultrasonics Sonochemistry, 66*, 105073.
40. Maisuthisakul, P., & Gordon, M. H., (2014). Characterization and storage stability of the extract of Thai mango (*Mangifera indica* Linn. Cultivar Chok-Anan) seed kernels. *Journal of Food Science and Technology, 51*(8), 1453–1462.
41. Mashhadi, N. S., Ghiasvand, R., Askari, G., Hariri, M., Darvishi, L., & Mofid, M. R., (2013). Anti-oxidative and anti-inflammatory effects of ginger in health and physical activity: Review of current evidence. *International Journal of Preventive Medicine, 4*(Suppl. 1), S36–S42.
42. Medlej, M. K., Cherry, B., Naseer, G., Zaviska, F., Hijazi, A., Li, S., & Pochat-Bohatier, C., (2020). Optimization of polysaccharides extraction from a wild species of *Ornithogalum* combining ultrasound and maceration and their antioxidant properties. *International Journal of Biological Macromolecules, 161*, 958–968.
43. Meghwal, M., & Goswami, T. K., (2012). A review on the functional properties, nutritional content, medicinal utilization and potential application of fenugreek. *Journal of Food Processing and Technology, 3*, 1–10.
44. Mila, I., Scalbert, A., & Expert, D., (1996). Iron withholding by plant polyphenols and resistance to pathogens and rots. *Phytochemistry, 42*(6), 1551–1555.
45. Mosic, M., Dramićanin, A. M., Ristivojevc, P., & Milojkovic-Opsenica, D., (2019). Extraction as a critical step in phytochemical analysis. *Journal of AOAC International, 103*(2), 365–372.
46. Murthy, C. T., & Bhattacharya, S., (2008). Cryogenic grinding of black pepper. *Journal of Food Engineering, 85*(1), 18–28.
47. Noratto, G., Bertoldi, M. C., Krenek, K., Talcott, S. T., Stringheta, P. C., & Mertens-Talcott, S. U., (2010). Anticarcinogenic effects of polyphenolics from mango (*Mangifera indica*) varieties. *Journal of Agricultural and Food Chemistry, 58*(7), 4104–4112.
48. Nzikou, J. M., Kimbonguila, A., Matos, L., Loumouamou, B., Pambou-Tobi, N. P. G., Ndangui, C. B., Abena, A. A., et al., (2010). Extraction and characteristics of seed kernel oil from mango (*Mangifera indica*). *Research Journal of Environmental and Earth Sciences, 2*(1), 31–35.
49. Onkar, P., Bangar, J., & Karodi, R., (2012). Evaluation of antioxidant activity of traditional formulation giloy satva and hydroalcoholic extract of the *Curculigo orchioides* gaertn. *Journal of Applied Pharmaceutical Science, 2*(6), 209–213.
50. Ozcan, M., (2003). Antioxidant activities of rosemary, sage, and sumac extracts and their combinations on stability of natural peanut oil. *Journal of Medicinal Food, 6*(3), 267–270.
51. Pardo-Andreu, G. L., Dorta, D. J., Delgado, R., Cavalheiro, R. A., Santos, A. C., Vercesi, A. E., & Curti, C., (2006). Vimang (*Mangifera indica* L. extract) induces permeability transition in isolated mitochondria, closely reproducing the effect of mangiferin, vimang's main component. *Chemico-Biological Interactions, 159*(2), 141–148.

52. Park, P. J., Jung, W. K., Nam, K. S., Shahidi, F., & Kim, S. K., (2001). Purification and characterization of antioxidative peptides from protein hydrolysate of lecithin-free egg yolk. *Journal of American Oil Chemists Society, 78*(6), 651–656.
53. Paur, I., Carlsen, M. H., Halvorsen, B. L., & Blomhoff, R., (2011). Antioxidants in herbs and spices: Roles in oxidative stress and redox signaling; Chapter 2. In: Benzie, I. F. F., & Wachtel-Galor, S., (eds.), *Herbal Medicine: Biomolecular and Clinical Aspects* (2nd edn., pp. 11–36) Boca Raton–FL: CRC Press.
54. Percival, S. S., Bukowski, J. F., & Milner, J., (2008). Bioactive food components that enhance γδ T-cell function may play a role in cancer prevention. *The Journal of Nutrition, 138*(1), 1–4.
55. Peres, V. F., Saffi, J., Melecchi, M. I., Abad, F. C., Jacques, R. A., Martinez, M. M., Oliveira, E. C., & Caramao, E. B., (2006). Comparison of Soxhlet, ultrasound-assisted and pressurized liquid extraction of terpenes, fatty acids and vitamin E FROM *Piper gaudichaudianum* kunth. *Journal of Chromatography A, 1105*(1, 2), 115–118.
56. Qadir, S. A., Kwon, M. C., Han, J. G., Ha, J. H., Chung, H. S., Ahn, J., & Lee, H. Y., (2009). Effect of different extraction protocols on anticancer and antioxidant activities of *Berberis koreana* bark extracts. *Journal of Bioscience and Bioengineering, 107*(3), 331–338.
57. Rice-Evans, C., Miller, N., & Paganga, G., (1997). Antioxidant properties of phenolic compounds. *Trends in Plant Science, 2*(4), 152–159.
58. Saxena, S. N., Sharma, Y. K., Rathore, S. S., Singh, K. K., Barnwal, P., Saxena, R., Upadhyaya, P., & Anwer, M. M., (2015). Effect of cryogenic grinding on volatile oil, oleoresin content and antioxidant properties of coriander (*Coriandrum sativum* L.) genotypes. *Journal of Food Science and Technology, 52*(1), 568–573.
59. Schweiggert, U., Carle, R., & Schieber, A., (2007). Conventional and alternative processes for spice production: A review. *Trends in Food Science & Technology, 18*(5), 260–268.
60. Serafini, M., & Peluso, I., (2016). Functional foods for health: The interrelated antioxidant and anti-inflammatory role of fruits, vegetables, herbs, spices and cocoa in humans. *Current Pharmaceutical Design, 22*(44), 6701–6715.
61. Serna-Cock, L., García-Gonzales, E., & Torres-Leon, C., (2015). Agro-industrial potential of the mango peel based on its nutritional and functional properties. *Food Reviews International, 32*(4), 364–376.
62. Sharma, L. K., Agarwal, D., Sharma, Y., Rathore, S. S., & Saxena, S. N., (2014). Cryogenic grinding technology enhances volatile oil, oleoresin and antioxidant activity of cumin (*Cuminum cyminum* L.). *International Journal of Seed Spices, 4*(2), 68–72.
63. Singh, K. K., & Goswami, T. K., (2000). Cryogenic grinding of cloves. *Journal of Food Processing and Preservation, 24*(1), 57–71.
64. Taamalli, A., Arráez-Román, D., Ibanez, E., Zarrouk, M., Segura-Carretero, A., & Fernández-Gutiérrez, A., (2012). Optimization of microwave-assisted extraction for the characterization of olive leaf phenolic compounds by using HPLC-ESI-TOF-MS/IT-MS(2). *Journal of Agricultural and Food Chemistry, 60*(3), 791–798.
65. Tan, H. H., (2004). Topical antibacterial treatments for acne vulgaris: Comparative review and guide to selection. *American Journal of Clinical Dermatology, 5*(2), 79–84.
66. Vicente, A. R., Manganaris, G. A., Sozzi, G. O., & Crisosto, C. H., (2009). Nutritional quality of fruits and vegetables; Chapter 5. In: Florkowski, W. J., Shewfelt, R. L., Brueckner, B., & Prussia, S. E., (eds.), *Postharvest Handling: A Systems Approach* (pp. 57–106). Cambridge, MA–USA: Elsevier Academic Press Inc.

67. Waisundara, V., & Hoon, L. Y. A., (2013). Comparative study on the antioxidant activity of commonly used south Asian herbs. *Journal of Traditional and Complementary Medicine, 3*(4), 263–267.
68. Widsten, P., Cruz, C. D., Fletcher, G. C., Pajak, M. A., & McGhie, T. K., (2014). Tannins and extracts of fruit byproducts: Antibacterial activity against foodborne bacteria and antioxidant capacity. *Journal of Agricultural and Food Chemistry, 62*(46), 11146–11156.
69. Wilson, D. W., Nash, P., Buttar, H. S., Griffiths, K., Singh, R., De Meester, F., Horiuchi, R., & Takahashi, T., (2017). The role of food antioxidants, benefits of functional foods, and influence of feeding habits on the health of the older person: An overview. *Antioxidants (Basel), 6*(4), 81.
70. Wilson, V. S., Bobseine, K., & Gray, L. E. Jr., (2004). Development and characterization of a cell line that stably expresses an estrogen-responsive luciferase reporter for the detection of estrogen receptor agonist and antagonists. *Toxicological Sciences, 81*(1), 69–77.
71. Zhang, L., Fan, G., Khan, M. A., Yan, Z., & Beta, T., (2020). Ultrasonic-assisted enzymatic extraction and identification of anthocyanin components from mulberry wine residues. *Food Chemistry, 323,* 126714.

CHAPTER 3

EXTRACTION OF VALUE-ADDED AND HIGH-VALUE FOOD PRODUCTS

PRADIP B. DHAMOLE, SHEETAL KOTHAWALE, and S. S. LELE

ABSTRACT

Extraction of the desired product from an aqueous solution or solid biomass requires an efficient extraction method. The selection of a suitable extraction method is important as it not only affects the extraction efficiency but also determines the downstream processing steps. An ideal extraction method would give maximum extraction efficiency and will selectively extract the desired component/product from the source. Principles of various extraction methods (physical, chemical, biological, and hybrid), their performance, advantages, and limitations, along with examples, are presented in this chapter.

3.1 INTRODUCTION

The selection of an appropriate extraction method is of utmost importance as it decides the further downstream processing. There is hardly any production line without an extraction process in food, biotechnology, pharmaceutical, and bioenergy. The challenge in the extraction of a solute/product from a solid matrix/biomass/cell or solution is that the desired molecule is usually embedded within the matrix or dissolved in a complex solution. An important parameter in the selection of the extraction process is the yield which should be maximum without affecting the properties of the product. In addition, it should also extract no/minimal amounts of undesirable components.

Novel Processing Methods for Plant-Based Health Foods: Extraction, Encapsulation, and Health Benefits of Bioactive Compounds. Megh R. Goyal, N. Veena & Ritesh B. Watharkar (Eds.)

Therefore, an ideal extraction method should: (i) give high yield; (ii) preserve the biological activity or other important properties; (ii) require less time; (iii) require less energy; (iv) extract only the target compound (selectivity should be high); (v) have low environmental impact; (vi) be safe; (vii) require less unit operations. Conventional extraction methods (CEMs) lack some of these characteristics, and hence, there is a need of advanced extraction methods, which would fulfill most of the criteria of an ideal extraction method.

This chapter focuses on advanced extraction methods for extraction of high-value/value-added products from various food or food processing waste/agriculture biomass/waste and fermentation broths.

3.2 ADVANCED EXTRACTION METHODS

A number of advanced extraction methods are being explored nowadays. The various methods of extraction can be broadly classified into physical, chemical, biological, and hybrid methods (Figure 3.1). Physical methods include solvent extraction, ionic liquid extraction, supercritical and subcritical fluid extraction, pressurized liquid extraction (PLE), salting out, sugaring out, polyol-induced extraction (PIE), cloud point extraction (CPE), and polymer-salt and polymer-polymer based aqueous two-phase system

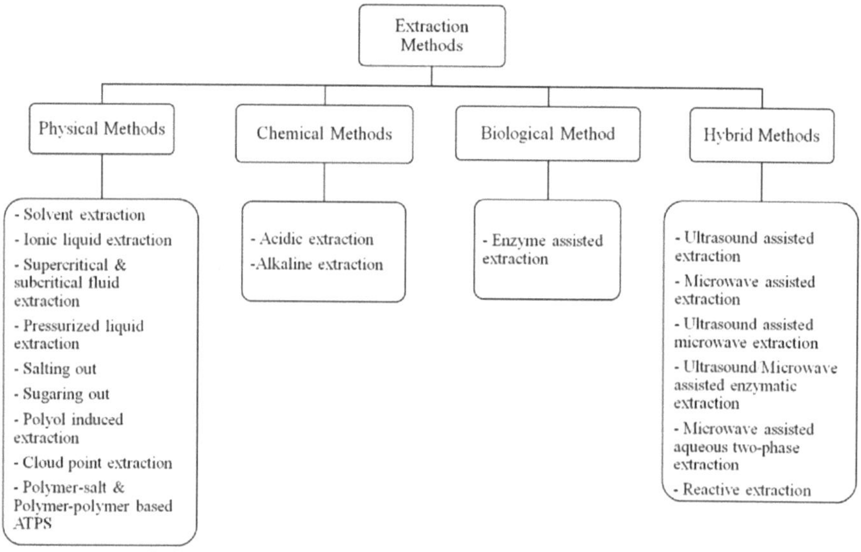

FIGURE 3.1 Types of extraction methods.

(ATPS) (Table 3.1). Chemical methods that are covered in this chapter are acidic and alkaline extraction methods (Table 3.2). Enzyme-assisted extraction (EAE) is the biological method. Hybrid methods combine any two or even three different methods.

TABLE 3.1 Applications of Various Physical Extraction Methods

Solute/Product	Source	Method of Extraction	Efficiency	References
Astaxanthin	Shrimp waste	$SC\text{-}CO_2$	39%	[67]
Bioactives	Food and agriculture byproducts	SFE	70–80%	[30]
Bioactives	Food (fruits, ginger, tea, etc.)	Subcritical	40%	[30]
Bioactives (antioxidants)	Plant and food	Subcritical	~ 80%	[27]
Carotenoids	Fruits, vegetables, industrial waste	Ionic liquid extraction	–	[67]
Carotenoids	Fruits, vegetables, industrial waste	SFE	50%	[67]
Coffee oil	Spent coffee grounds	Solvent extraction	14.7 wt.%	[69]
Essential oil	Plants	Supercritical fluid extraction	27.5%	[82]
Lipids (tocopherol)	Spent coffee grounds	Solvent extraction	93%	[49]
Lycopene	Tomato processing waste	Solvent extraction	75.75%	[63]
Lycopene	Tomato peels	$SC\text{-}CO_2$	60.85%	[65]
Lycopene	Tomato peel residues	Surfactant assisted extraction	25%	[62]
Natural products (flavonoids, terpenoids)	Herbal medicinal plants	Ionic liquid extraction	83.5%	[78]
Phenolics	Spent coffee ground	Subcritical	47%	[43]

Hybrid methods include ultrasound-assisted extraction (UAE), microwave-assisted extraction (MAE), ultrasound-assisted microwave extraction (UAME), ultrasound/microwave-assisted enzymatic extraction, and microwave-assisted aqueous two-phase extraction (MAATPE), and reactive extraction.

Efficient extraction methods are also needed in the analysis of the component in a product or a raw material. For analytical purposes, the important criteria for the selection of extraction method are the efficiency. Further, the method should not lead to the release of undesired component which would hinder the quantification process. Economics is generally not important for analytical methods since the quantity of raw material processed is small, and hence, the energy and other requirements are also low. Thus, a method selected for quantification may not be suitable for production of the compound on a large scale.

TABLE 3.2 Applications of Various Chemical and Biological Extraction Methods

Solute/Product	Source	Method of Extraction	Efficiency	References
Edible oil	Soybean	Enzymatic	90%	[29]
Ginger oleoresin	Ginger	Enzymatic	80%	[59]
Lignans	Flax hulls and seeds	Enzymatic	70%	[64]
Lignin	Softwood kraft pulp	Acidic	50–70%	[44]
Lycopene	Whole tomato, peel, fruit pulper waste and industrial waste	Enzymatic	18-fold higher purification	[89]
Niacin	Cereals	Alkaline	21%	[38]
Pectin	Pomelo peel	Acidic	12.1–20.5%	[83]
Pectin	Pomelo peel	Alkaline	13.9–24.2%	[83]
Phenolic acid	Rice bran	Enzymatic	72.7%	[42]

3.2.1 PHYSICAL METHODS

3.2.1.1 SOLVENT EXTRACTION

Solvent extraction (also known as liquid-liquid extraction (LLE)) is one of the most widely used extraction method for extraction of target compounds from solutions. Compounds are separated based on its solubility in two different immiscible liquids. Hydrophilic compounds are extracted with polar solvents such as ethanol, methanol whereas non-polar solvents, e.g., hexane, are employed for extraction of lipophilic compounds. Solvent should be immiscible with the solution, have high selectivity for the desired solute/compound, should be easy to recover, inert, safe, and stable.

Efficiency of solvent extraction depends on choice of solvent, mass transfer, selectivity of solvent towards the desired solute and solubility of solvent with the solution. Limitations of the use of organic solvents for the recovery of natural products are toxicity of the solvent, poor product quality, energy intensive solvent recovery process [41] and safety of the process. It is applied for the extraction of essential/edible oil. Extraction of seed oil is usually performed using a non-polar solvent (petroleum ether or *n*-hexane). This method is used instead of the mechanical system (i.e., seed pressing) to recover a larger quantity of oil [41]. In the production of coffee oil from the coffee powder [69], carotenoids from tomato processing waste, extraction of spices from plants and lipid component extraction from food [49].

3.2.1.2 IONIC LIQUID EXTRACTION

Ionic liquid extraction uses ionic liquids (ILs) as green solvent in conjugation with other extraction processes like LLE, UAE, MAE, and so on [64]. ILs can be used as an alternative to conventional solvents due to their unique physicochemical properties. ILs can be designed in a large number of cation and anion combinations which helps them to adjust their properties in terms of hydrophobicity and solution behavior [35, 78]. Mechanism of ILs is based on ion exchange which is different than traditional solvents. Low toxic, high selectivity, high ionic conductivity, and good thermal stability are some of the advantages of this process [35]. These qualities of ionic liquid extraction led to its vast applications. Using IL as a solvent, bioactive compounds (flavonoids, alkaloids, etc.), and essential oils can be extracted from plants. Also, carotenoids can be isolated from food byproducts [64, 78].

ILs are also being applied as surface-bonded stationary phases and mobile phase additives in chromatography separations. For selective transport of organic compound IL supported membrane are used [35]. On the other hand, as ILs are mostly obtained from fossil sources, their nature as green solvent is also questionable. Some ILs are toxic (Imidazolium-based ILs) and can cause potential environmental impact [34]. Designing ILs is a high-cost process and there is also difficulty in recovering extracted ions which makes them impractical for industrial application [35, 64, 78].

3.2.1.3 SUPERCRITICAL AND SUBCRITICAL FLUID EXTRACTION

Supercritical and subcritical fluid extractions are the new age effective green methods as it avoids the use of harmful solvents [30]. Supercritical fluid (SF) extraction has been widely applied for the abstraction of biologically active components and essential oils from food byproducts, plants, and algae. The method has high solubility, selectivity, and high extraction yield in a short time (three times higher than distillation) [82]. Supercritical fluid extraction (SFE) is established on the unique property of SF which shows a diffusivity and density values between liquid and gas [30, 82]. With changing pressure and temperature, the density of SF can be changed enormously. Due to such qualities, it is viable to separate out desired constituents from a multicomponent concoction using SF (CO_2, water) as a solvent [64].

The feed containing solute is put into an extraction vessel. Fluid has been constantly added to the vessel so as to mix solute with SF. The ambient pressure and temperature (above the critical point of solute) is supplied to the mixture. Finally, with the help of the separator, SF comprising recovered components are separated out by lowering the pressure and collected in a collection vessel [6, 43, 82]. There is no production of toxic waste and solvent can also be easily recovered.

Currently CO_2 has been applied to carry out almost 90% of SFE because of its low critical temperature (32°C) and critical pressure (71.8 bar) that enables extraction of thermally labile food compounds [64]. Moreover, it has been used in the extraction of essential oils [82], extraction of non-polar bioactive compounds from food and agricultural waste [64, 67], recovery of natural flavors and essence from lime peels, coffee, basil [6] and target compound extraction from plant and food byproduct [30]. Although there are some limitations such as SFs fail in extraction of polar analytes and it requires high pressure and high capital investment.

On the other hand, polar components can be removed efficiently by subcritical water extraction as it has the advantage of using water as the only solvent. Subcritical water extraction is used in isolation of antioxidants from byproducts and wastes from the food industry [6, 30]. It is a safe, environmentally friendly, and suitable alternative to other conventional methods. It is also called pressurized hot water extraction or superheated water extraction [27]. This extraction technique uses water at its subcritical state as a solvent. Subcritical water can be maintained in the liquid form between boiling point temperature (100°C) and critical point temperature (374°C) and a pressure greater than 50 bars to extract polar and semi-polar solutes [6, 30, 43].

Subcritical water extraction is performed in batch or continuous system. Continuous system is most applied because the extraction bed is fixed and direction of solvent flow is up to down which easily cleans up solutes. The water is pumped into the extraction vessel in which food byproduct is present. Oven is used to heat the vessel and pressure restrictor maintain the appropriate pressure. Solute from byproduct is transferred to water by mass transfer principle [27, 30, 43]. Nitrogen is mostly used as an inert gas to purge the system to prevent oxidation during extraction. The method is highly selective, time conserving and cause no loss of bioactivity of extracts. As a result of so many advantages, subcritical water extraction is applied in the separation of phenolics from food byproducts and anthocyanin extraction from coffee pulp [6, 27, 43]. Separation of polysaccharides, proteins, and antioxidants from biomass as well as flavoring compounds from spices can also be carried out using subcritical liquid extraction [27, 30]. Sometimes ethanol has been used as a cosolvent to increase water polarity which can lead to loss of bioactivity and high temperature can cause thermal degradation of component [27].

3.2.1.4 PRESSURIZED LIQUID EXTRACTION (PLE)

PLE is also called accelerated solvent extraction (ASE), pressurized solvent extraction and pressurized hot water extraction (when water is the solvent) [57]. It is considered as highly effective method for extraction of solute from the solid mass as compared to conventional ones. PLE is a technique which provides green extraction method that uses solvents at a high pressure and temperature as compared to conventional extraction (CE). The pressure and temperature must be high enough to maintain solvent in a liquid state so that it penetrates into the solid sample which is being extracted [6, 36].

High temperature and pressure will enhance the mass transfer and solubility. The equipment used in this method needs to support both, high temperature and pressure. The working of the whole process of pressurized liquid extraction is automatic to attain the exhaustive extraction [64]. Pressurized extraction gives high extraction yield with less extraction time. However, further purification will be required if other molecules get extracted with the desired molecule [36, 64].

There are various sizes of extraction cells in which the solid feed is filled with solvent in the vessel and covered with two filtration nozzles. After loading the feed, the system begins to pressurize and heat the feed automatically. The pressure can be retained up to 3,000 psi and temperature up to 200°C. When the system is in a equilibrate state, solvent including desired components and byproducts, is collected into a collector, automatically [6, 90]. The method is employed for carotenoid extraction from food matrices, extraction of bioactive compounds from plant extracts, separation of organic contaminants from food samples and extraction of alkaloids from food waste [6, 36, 64, 75, 90]. PLE uses a low amount of solvent, however, high pressure requires expensive pressurized vessels. Also, one needs to be careful while separating thermally sensitive molecule using PLE [57].

3.2.1.5 SALTING OUT

Salting out process is a separation process, basically applied to proteins but not limited to the extraction of proteins. High salt concentration helps the proteins to precipitate; therefore, it is also known as salt-induced precipitation or salt fractionation [61]. The process of salting-out uses the decreased solubility of components present in suspension containing salt at elevated ionic strength leading to precipitation of molecules. Selective precipitation of desired component is based on the type of salt and its concentration.

The ammonium sulfate is added to a solution of macromolecule to a concentration just below the precipitation point of the component of interest. The mixture is centrifuged to separate precipitated compounds. Unwanted components which are precipitated at the bottom of centrifuge tubes are discarded, and to the aqueous mixture (supernatant), more salt is added to a concentration sufficient to salt out the desired components. The mixture is again centrifuged and protein precipitate is recovered while the supernatant is discarded [14, 61]. Salting out minimizes the unfolding stress to protein encapsulates, the used salt has no harmful effect on protein, no heating process is required and there is no use of hazardous solvents. Sometimes high salt concentration may induce unwanted chemical reactions and can damage biological components.

ATPS consisting of an aqueous solution, solvent, and salt is also known as salting out. Addition of salt into solution containing solvent leads to two-phase formation; upper phase is a solvent rich phase and the bottom phase is aqueous phase containing salts. The target molecule moves from aqueous solution to the solvent rich phase. This phenomenon is widely used for extraction of erythromycin from fermentation broth [45], allicin from garlic [47], succinic acid [71], carboxylic acids and acetoin from broth [11]. Different solvents such as acetonitrile (ACN), ethanol, [45, 47] ethyl acetate and some ILs are also explored. Salts such as NaCl, $MgSO_4$, KH_2PO_4 and K_2HPO_4 are used. Major drawback of salting-out process is the recovery of salt from the lower aqueous phase. Also, some of the salts are corrosive, reactive, change the environment conditions and denature the product.

3.2.1.6 SUGARING OUT

Sugaring out is a newly discovered physical method of extraction reported in 2008, which uses ACN, water, and sugars [76]. The sugaring out method uses monosaccharide (C_5/C_6 sugars) or disaccharides as a phase separating agent [24]. Sugaring out is based on the theory that when sugar is added into a water and ACN solution, hydrogen bonds are formed between sugar and water molecules replacing water-solvent bond leading to new phase formation. It results in formation of two-phase, ACN forming the upper phase while lower phase abounding water and sugar [20, 24]. The phase separation is rapid and it requires no external energy.

Various parameters that affect sugaring out are temperature, type of sugar (C_5 and C_6) used, concentration of sugar, volume ratio of solvent to aqueous solution. Low temperature and high sugar concentration favors sugaring out. Initial reports on sugaring out mainly used ACN as a solvent. However, recent studies have explored various other solvents such as butyl acetate, ethyl acetate, pentanol, butanol, propanol, and acetone. There are some reports on use of ILs [72]:

- 1-Butyl-3-methylimidazolium tetrafluoroborate ([Bmim]BF4);
- 1-Butyl-3-methylimidazolium triflate ([Bmim]OTF);
- 1-Butyl-3-methylimidazoliumbromide ([Bmim]Br);
- 1-Ethyl-3-methylimidazolium tetrafluoroborate ([Emim]BF4);
- 1-Hexyl-3-methylimidazoliumbromide ([Hmim]Br).

The method is used in various food and pharmaceutical applications which include extraction of phenolic compounds (vanillin, syringaldehyde, ferulic acid, p-coumaric acid) from lignocellulosic hydrolysates, fermentation products (acetoin, succinic acid, lactic acid (LA), antibiotics, and 2,3-butanediol) [12, 13, 54, 69, 79]. This method is more suitable for fermentation products since the bottom sugar rich phase can be re-used as a carbon source. It is also used for separation of ACN from reverse phase-HPLC eluent generated during protein separation [24]. Solute molecule extracted into solvent rich phase could be back-extracted by evaporating the solvent by vacuum distillation [12, 13, 79].

3.2.1.7 POLYOL INDUCED EXTRACTION (PIE)

Polyol induced extraction (PIE) is another recently developed method based on ATPS between water/ACN mixture. Polyol is a molecule with more than two hydroxyl groups such as glycerol, ethylene glycol and sorbitol. Similar to salting out and sugaring out, polyol (sugar alcohol) based ATPS uses polyol as a mass separating agent. PIE was reported for extraction of water from organic solvents [18]. PIE is relatively less explored method despite many advantages such as non-toxicity of most of the polyols, biodegradability, and reusability of polyols. This method can be used for extraction of various compounds from aqueous solutions. Polyols such as glycerol, erythritol, xylitol, sorbitol, and maltitol were used as mass separating agent for extraction of vanillin from aqueous solutions [7]. Better phase separation was obtained with polyols having higher number of hydroxyl groups. Phase separating performance of various polyols was ranked as [7]:

maltitol > sorbitol > xylitol > erythritol > glycerol

Essential oils were extracted using glycerol as a mass separating agent without losing the main components of essential oils [17]. Different leaves (dried) were used for extraction of essential oils and the extraction was carried out between – 20°C and 20°C. The highest partitioning and recovery were obtained at – 10°C. Thus, PIE due to its low heat, less time and low cost could be a promising extraction method for extraction of heat sensitive components.

3.2.1.8 CLOUD POINT EXTRACTION (CPE)

Cloud point is the property of non-ionic surfactants. Non-ionic surfactants form micelle in aqueous solution, and on increasing the temperature above cloud point temperature it separates into two phases. During the micelle formation, the hydrophobic tail of nonionic surfactants forms an inner hydrophobic core. A hydrophobic compound gets trapped into the micellar core. The micelles settle with time or heating above a cloud point temperature leads to two phase formation. The lower phase is called coacervate or surfactant rich phase. The upper phase is called lean, aqueous phase or dilute phase. Hence, CPE is also a type of ATPS. The desired component/product is extracted into the surfactant rich phase.

CPE has many advantages over other extraction methods such as concentration of the product in a coacervate phase, reduction in further downstream processing volumes, low amount of surfactant requirement and easy to tune physiochemical properties. CPE is widely used for analytical purposes; however, it is also explored for other non-analytical applications. CPE finds a good number of applications in fermentation processes where substrate or product inhibition occurs, i.e., large concentrations of substrate or product inhibits the fermentation process. Some of the examples are butanol fermentation [23]; lipase [82]; and red monascus pigments [88]. It is also used for the extraction of some phenolic compounds produced during the hydrolysis of lignocelluloses [22] and chlorophyll from spinach [46]. It can be used for extraction of high value low volume products which are hydrophobic in nature.

One of the major limitations of CPE is the back-extraction of solute from the coacervate phase. Some of the methods investigated for back-extraction include changing the pH of the solution and microemulsion formation with other solvents [31]. Further, the aqueous phase contains surfactant levels (close to its critical micellization concentration) that might act as a pollutant if not recovered.

3.2.1.9 POLYMER-SALT AND POLYMER-POLYMER BASED ATPS

Aqueous two-phase extraction method based on salt (sulfate, phosphate)-polymer (dextran, Poly-ethyl glycol (PEG)) or polymer-polymer works as two immiscible aqueous solutions that act as an extractants [37]. In polymer-salt aqueous two-phase extraction, salt takes up most of the water and polymer forms aggregate and starts to separate out. Phase separation is accomplished either by settling under gravity or by centrifugation leading to the formation of two immiscible phases. The upper phase is hydrophobic polymer phase while the lower phase has inorganic salt. Mostly, low molecular weight polymer and high salt concentration favor the partition of proteins [37]. This single-step approach increases fold purification and recovery yield while a polymer-polymer system is formed by mixing two different polymers (PEG, dextran). The system leads to formation of aggregates and thus polymers begin to split amongst two separate phases due of steric exclusion. The exclusion depends on the pH, ionic strength and temperature of the solution and type of polymers. The upper phase is formed by more hydrophobic (PEG) polymer, while the lower phase is formed by hydrophilic polymer (dextran) [37, 81].

In both the methods, the target product, such as biomolecule is concentrated in one of the phases and the contaminants in the other. Both methods are non-toxic and can be operated at large scale. Salt-polymer has an advantage over polymer-polymer by being low operational cost and isolation process while polymer-polymer separation operates at mild conditions than salt-polymer [37, 81]. Separation of drug residues from milk, honey can be carried out by salt-polymer system whereas extractive fermentation of products like LA can be taken out by polymer-polymer separation [37, 81]. Extraction of low molecular weight compounds as high value products such as enzymes and proteins for the food industry can be obtained by both separation methods [81].

3.2.2 CHEMICAL METHODS

3.2.2.1 ACIDIC EXTRACTION

Acid extraction or acid hydrolysis is widely used in the treatment of lignocellulosic materials for biochemical and biofuel production. Dilute acid hydrolysis breaks the lignin structure and the glycosidic bonding between the cellulose molecules by HCl with H_2SO_4 as a catalyst to form monosaccharides [1, 4]. During extraction, the matrix structure of fiber is disrupted by acid into polysaccharide components, and additionally facilitates the degradation of these components into monosaccharide. Such type of extraction is carried out in two ways, low temperature high acid concentration or at high temperature with low acid concentration [1].

Due to faster acting reactions in shorter time, this method leads to higher sugar yield and good reproducibility. Acid extraction has been employed in food and chemical industries for cellulose extraction from biomass to use it for the production of bioethanol and to change cellulose and hemicellulose into valuable products [1]. Despite so many application acid extractions have plenty of drawbacks. The use of acid at high concentration might trigger severe environmental alarms. Acid extraction needs high quality material of construction since acids are corrosive [4]. Also, extreme conditions (high temperature and pressure) under acidic conditions leads to the generation of inhibitors (such as acetic acid, phenolic acids, etc.), which inhibit the fermentation process.

3.2.2.2 ALKALINE EXTRACTION

Alkaline extraction is a low cost and widely used method. It applies alkali (NaOH and KOH) for extraction of proteins, polysaccharides from food [84], soluble dietary fiber from fruit peels and phenolic acids from lignocelluloses [22]. Alkali treatment hydrolyzes the cell wall of biomass (mostly made of hemicellulose) that leads to release of desired product [19].

Alkaline solution is added to dried powdered biomass and mixed for a sufficient time so as to release the target molecule in the solution. Temperature plays an important role in alkaline extraction. Hence, concentration of alkali, type of alkali, time, speed of agitation, temperature, and solid/liquid ratio needs to be optimized for improved extraction of target molecule. The residue (solid biomass) is filtered out from the solution and it is followed by solvent extraction [21]. The process requires a small amount of solvent and lesser time compared to enzymatic extraction. Scale-up to industrial production is also easy due to ease in availability of solvent [19, 39]. Alkaline extraction is also used in the isolation of proteins and starch from canola and cereals like rice [19], for vitamin (niacin) extraction from cereals [39] and can be coupled with spectroscopy for determination of iodine in food sample [28]. Recovery of alkali or neutralization is a major problem with alkaline extraction. Also, high concentration of alkali may damage the product.

3.2.3 BIOLOGICAL METHOD

3.2.3.1 ENZYME ASSISTED EXTRACTION (EAE)

Enzyme assisted extraction (EAE) has been considered widely as one of the most effective, sustainable, and environmentally friendly method into separation technology [29]. It is carried out by enzymes in trivial environment by means of specificity and selectivity while retaining the biological capabilities of bioactive compounds [89]. The basis of the enzymatic extraction is the use of enzyme as a catalyst for the plant cell wall disruption through hydrolyzing it in mild surrounding, so as to discharge the intracellular components [59, 89]. This method has benefits over another method as it lowers impact on the environment, the enzymes used are highly efficient and specific, and also preserves original properties of natural product. EAE reduces solvent consumption, gives higher yield, and requires lower energy as compared to physicochemical methods.

EAE aids in recovery of lycopene from tomato processing waste as well as extraction of biomolecules such as phenols from natural sources for application in food processing [59, 89]. EAE has been widely used for extraction of bioactives from plant material. Some of the applications of EAE include the use of laccase in the pulp and paper industry, vanillin from vanilla green pods, extraction of carotenoids from marigold flowers, oil extraction from different seeds such as tomato seeds, grape seed. Polysaccharides and edible oil can also be extracted efficiently by EAE from biomass and oilseeds, respectively (Table 3.2) [29, 59]. Additionally, enzymes can improve extraction efficiency up to 97% by combining with microwave, ultrasonic, and SF methods [32, 59]. Cost of the enzymes is a major constraint in the application of EAE. So, they are less suitable for industrial application, plus they are not always feasible because their behavior is limited by static environmental conditions [32, 59].

3.2.4 HYBRID METHODS

Hybrid methods are a combination of two or more methods. For example, a biological method can be combined with a physical method or a chemical method (Table 3.3).

TABLE 3.3 Applications of Various Hybrid Extraction Methods

Solute/Product	Source	Method of Extraction	Efficiency	References
Carboxylic acid	Fermentation broth	Reactive extraction	65%	[3]
Lactic acid	Fermentation broth	Reactive extraction	97%	[74]
Lycopene	Dry tomato pomace	UAE	70%	[50]
Pectin	Orange peels	MAE	19.24%	[52]
Pectin	Sisal waste	Enzymatic ultrasound-assisted extraction	31.1%	[80]
Penicillin G	Mycel-containing broth	Reactive extraction	96%	[5]
Soluble dietary fiber	Papaya peels	UAE	36.99%	[84]

3.2.4.1 ULTRASOUND-ASSISTED EXTRACTION (UAE)

UAE is a simple, fast, inexpensive, and has low environmental impact. UAE is capable of efficiently extracting valuable components from biomass by operating with many samples at one time [5, 10]. Hybrid method UAE is centered on the ultrasound pressure wave proliferation through solvent which results in a cavitation phenomenon [53]. Ultrasound waves produce frequency exceeding threshold of nearly 20 kHz. During the process, ultrasound frequency leads to the formation of bubble and develop extreme negative pressure which break down the bubbles formed by cavitation. As the bubbles break down close to cell walls, the cell wall disrupts mechanically, and consequently, mass transfer improves and solvent infiltrates the cells and discharge the desired components [53, 54, 90]. This reduces the consumption of solvent as well as processing time and increase the extraction rate.

Extraction of heat sensitive bioactive components from fruit and vegetable such as carotenoids, phenols, anthocyanins, aroma, natural color, and phenol are some of the applications of UAE [8]. Extraction of antioxidants, volatile compounds, capsaicinoids, and aroma compounds can also be extracted from herbs and spices using UAE. Further, UAE is also applied for extraction of oil from various oleaginous seeds such as soybean, sunflower, almond, flaxseed, etc. [8]. Nonetheless, high frequency ultrasound energy can cause occasional but deleterious effect on compounds.

3.2.4.2 MICROWAVE-ASSISTED EXTRACTION (MAE)

Microwave-assisted extraction (MAE) is another hybrid method which combines microwave and conventional solvent technique. MAE uses non-ionizing electromagnetic waves of frequency up to 300 GHz [53]. Heating occurs in selective manners in a closed system with practically no environmental loss. Energy of an electromagnetic radiation while traveling through a medium may be absorbed and converted into thermal energy. Heating of the medium takes place through two ways, i.e., ionic conduction and dipole rotation [26, 40].

Microwave produces electric field that leads to migration of charge carriers. This phenomenon is referred as ionic conduction [58]. Heating is caused due to the friction of charge carriers and medium [26]. Dipolar molecules try to follow the electric field in the same alignment. Collisions take place between the dipoles and surrounding molecules, resulting in heating. Dipole interaction and ionic conduction takes place together resulting in the conversion of microwave energy into thermal energy. High temperature and pressure are generated inside the oven. This high temperature evaporates the moisture present in cell which reduces mechanical strength of the cell wall and cell ruptures releasing the contents [5].

MAE reduces the solvent consumption, improves product quality, and accelerate extraction rate due to high temperature (37% high extraction efficiency than traditional solvent extraction) [6]. Factors affecting the efficiency of microwave extraction are size of the material, moisture content, selection of solvent and its concentration, microwave frequency, solid/liquid ratio, temperature, and pressure [77].

Choice of solvent is one of the important factors since its capacity to solubilize the desired solute and the amount of energy the solvent can absorb are vital. Also, the concentration of solvent (i.e., solid/liquid ratio) is another parameter having significant effect on extraction efficiency. High solvent volume would solubilize the solute and enhance extraction.

MAE has some limitations, e.g., high capital cost and additional filtration step for removal of the solid residue after the extraction. Operating conditions such as elevated temperatures possibly damage heat-sensitive compounds or just in case when solvents are volatile and non-polar, the efficiency of MAE can be very poor [6, 53, 90].

MAE facilitates isolation of polysaccharides, proteins, and phenolic compounds from food waste and biomass as well as extraction of essential oil from plant leaves and seeds [6, 53, 90]. MAE is also explored for extraction of total phenolics, catechin, hesperidin, and isoflavone from different plant species [87]. It has been commonly employed in food safety analysis as a relatively inexpensive separation method nowadays. It can also be applied in the extraction of herbs and spices and nutraceutical products from plants [6, 53].

3.2.4.3 ULTRASOUND-ASSISTED MICROWAVE EXTRACTION (UAME)

It is a newly developed hybrid technique that combines ultrasound and microwave radiation that exhibits the advantages of both methods by intensifying the traditional process [69]. UAME is very fast and highly efficient and hence has a great potential as a new method for process intensification. UAME has been used in the extraction of polysaccharides from fruits, essential oil extraction from plants [9] and glycoprotein extraction from cereals like barley [33]. UAME was done in totally separate extractor known as ultrasound microwave extractor. For the extraction of compounds, microwaves offer easy and appropriate heating, however it limits the mass transfer while on the other hand ultrasonic waves helps in enhancing mass transfer rate but it cannot produce sufficient amount of sensible heat [9, 33].

Cavitation (due to ultrasound) disrupts the cell wall and significantly enhances the extraction of desired component whereas microwave improves the migration of dissolved molecules due to rapid heating of the entire sample. Microwave radiation also helps in solvent penetration into the biomass/solid mass and helps in enhancing the solubility of the compound [8]. Even though the method requires high power input, it has shortened extraction time and increases the extraction yield as compared with the conventional method [70].

3.2.4.4 ULTRASOUND/MICROWAVE-ASSISTED ENZYMATIC EXTRACTION

Ultrasound and microwave methods are coupled separately with enzymatic extraction which is a recently emerged hybrid method as a way to decrease biochemical reaction time in traditional methods [34, 59]. In enzymolysis ultrasonic-assisted extraction (UAE), enzymes catalyze efficient disruption of cell walls by hydrolysis and trigger the discharge of bioactive compounds. Ultrasonic waves generate sufficient energy which creates super agitation and allows the effective mass transfer between immiscible phases [38].

In the case of the microwave-assisted enzymatic method, microwave heating is used to support and facilitate the natural compound removal from plant resources and enzymes leads to hydrolysis of cell wall of biomass. Thus, the combined capability of microwave and enzymatic extraction utilizes the microwave energy and enzymatic treatment leading to increased recovery of extracts from biomass [2]. Rapid extractions, no use of toxic solvents, and no production of undesirable byproducts are some of the advantages of microwave-assisted and ultrasound-assisted enzymatic extraction methods. It also increases extraction yield up to 10–11% [2]. Both the techniques can be applied for polysaccharides extraction from biomass [9, 59] and protein, polyphenols extraction from food waste [38, 85]. Microwave-assisted enzymatic method can separately aid the extraction of vegetable oil from oilseeds [2] as well as extraction of toxic metals (arsenic, selenium) from food [34]. Enzymolysis ultrasonic method can carry out the extraction of flavonoids of high antimicrobial and antioxidant activity from plant residue [59]. However, cost of the methods and extra purification steps may limit the application of these methods to laboratory scale [59].

3.2.4.5 MICROWAVE-ASSISTED AQUEOUS TWO-PHASE EXTRACTION (MAATPE)

Microwave-assisted aqueous two-phase extraction (MAATPE) combines the advantages of microwave extraction and ATPS. Microwave helps in breaking the bonds between the matrix and the target molecule and ATPS separates the molecule from the solution without altering its biochemical properties [86]. Thus, while acting as an extracting agent in MAATPE; the two-phase system combines extraction and purification and creates a single-stage process [16]. ATPS can be formed by combining either two distinct hydrophilic polymers (polyethylene glycol, ethanol, dextran, etc.), or a hydrophilic polymer and a salt (phosphate, sulfate, etc.), in addition to water [51, 86]. It is a rapid and effective method. There is no need of additional separation step and has great potential for industrial application.

Simultaneous extraction and purification of alkaloids (90%) was achieved using MAATPE consisting of ethanol and ammonium sulfate [86]. In this case, the powdered dry biomass remained at the interface of the bottom saline phase (aqueous phase) and top solvent rich phase. Salt not only helped in mixing the components but also in absorbing the microwave radiations. Higher conductivity of the saline bottom phase could result in strong heating and molecular agitation leading to breaking of bonds between the target molecule and the matrix. The target molecule would get extracted into the bottom phase first (due to microwave heating) and would be transferred to the top phase subsequently (due to salting out) [86]. MAATPE can be used in separation of phenolic compounds from fruit seed [16], isoflavonoids extraction from leaves and to separate active constituents from fermentation broth [16, 51].

3.2.4.6 REACTIVE EXTRACTION

Reactive extraction is a simple, novel, low cost, and environmentally safe process of producing high purity product. It is applied to separate organic acid, mainly carboxylic acids such as LA from fermentation broth and wastewater [15]. Carboxylic acid and its derivatives are used in the food industry as solvents, food additives, antimicrobials, and flavorings [56]. Reactive extraction is a separation process based on the reactions between extractant and the solution from which the desired solute is to be extracted. The solute forms a complex with extractant and is extracted into a solvent phase.

Reactive extraction reactor operates as a single unit of reaction and extraction. Reaction complex formed at the interface of the extractant containing organic phase and the solute containing the aqueous phase is then transferred in the organic phase by diffusion and solubilization mechanism [15, 56]. Both phases are immiscible. Type of solvent plays an important role in reactive extraction as solvent properties affects the structure of complex formed [25]. In recent years supercritical CO_2 has been used as a green replacement for traditional organic extractants. The method is also applied for biodiesel production from Jatropha seeds and food waste [73], for recovery of antibiotic (penicillin G) from fermentation broth [48]. One of the major limitations of reactive extraction is back-extraction of target molecule from the complex.

3.3 SUMMARY

Choice of extraction method is of utmost importance since it decides the downstream processing of the product molecule. Advanced extraction methods are highly efficient in the extraction of desired molecule. Physical methods are simple and easy to use as compared to chemical methods, which usually lead to secondary products. Biological methods are environment friendly. However, they are expensive and slow. Hybrid methods are an attractive alternative, which utilizes the benefits of one or more methods. Hence, a method should be selected based on the nature/properties of biomass/broth from which the compound is to be extracted and the properties of desired compound.

KEYWORDS

- **advanced extraction methods**
- **biological method**
- **chemical method**
- **high-value food products**
- **hybrid extraction method**
- **physical method**

REFERENCES

1. Ahmad, E. K., & Pant, K. K., (2018). Lignin conversion: A key to the concept of lignocellulosic biomass based integrated biorefinery; Chapter 14. In: Bhaskar, T., Pandey, S., Mohan, S. V., Lee, D. J., & Khanal, S. K., (eds.), *Waste Biorefinery: Potential and Perspectives* (pp. 409–444). Amsterdam–Netherlands: Elsevier.
2. Anwar, F., Yaqoub, A., Shahid, S. A., & Manzoor, M., (2013). Effect of microwave-assisted enzymatic aqueous extraction on the quality attributes of maize (*Zea mays* L.) seed oil. *Asian Journal of Chemistry, 25*(11), 6280–6284.
3. Babu, B. V., & Kumar, S., (2008). Process intensification for separation of carboxylic acids from fermentation broths using reactive extraction. *Journal on Future Engineering and Technology, 3*(3), 21–28.
4. Basu, P., (2018). Production of synthetic fuels and chemicals from biomass; Chapter 11. In: *Biomass Gasification, Pyrolysis and Torrefaction: Practical Design and Theory* (3rd edn., pp. 415–443). Amsterdam–Netherlands: Elsevier.
5. Belwal, T., Devkota, H. P., Ramola, S., Andola, H. C., & Bhatt, I. D., (2020). Optimization of extraction methodologies and purification technologies to recover phytonutrients from food; Chapter 10. In: Nabavi, S. M., Suntar, I., Barreca, D., & Khan, H., (eds.), *Phytonutrients in Food: From Traditional to Rational Usage* (pp. 217–235). Duxford-United Kingdom: Woodhead Publishing.

6. Belwal, T., Ezzat, S. M., Rastrelli, L., Bhatt, I. D., Daglia, M., Baldi, A., Devkota, H. P., et al., (2018). A critical analysis of extraction techniques used for botanicals: Trends, priorities, industrial uses and optimization strategies. *Trends in Analytical Chemistry, 100*, 82–102.
7. Cardoso, G. B., Souza, I. N., Mourão, T., Freire, M. G., Soares, C. M. F., & Lima, A. S., (2014). Novel aqueous two-phase systems composed of acetonitrile and polyols: Phase diagrams and extractive performance. *Separation and Purification Technology, 124*, 54–60.
8. Chemat, F., Rombaut, N., Sicaire, A. G., Meullemiestre, A., Fabiano-Tixier, A. S., & Abert-Vian, M., (2017). Ultrasound assisted extraction of food and natural products. Mechanisms, techniques, combinations, protocols and applications: A review. *Ultrasonics Sonochemistry, 34*, 540–560.
9. Cheng, Z., Yang, Y., Liu, Y., Liu, Z., Zhou, H., & Hu, H., (2014). Two-steps extraction of essential oil, polysaccharides and biphenyl cyclooctene lignans from schisandra chinensis baill fruits. *Journal of Pharmaceutical and Biomedical Analysis, 96*, 162–169.
10. Chmelovaa, D., Skulcovaa, D., Legerskáa, B., Horníkb, M., & Ondrejoviča, M., (2020). Ultrasonic assisted extraction of polyphenols and antioxidants from *Picea abies* bark. *Journal of Biotechnology, 314, 315*, 25–33.
11. Dai, J., Guan, W., Ma, L., & Xiu, Z., (2017). Salting-out extraction of acetoin from fermentation broth using ethyl acetate and K_2HPO_4. *Separation and Purification Technology, 184*, 275–279.
12. Dai, J. Y., Liu, C. J., & Xiu, Z. L., (2015). Sugaring-out extraction of 2, 3-butanediol from fermentation broths. *Process Biochemistry, 50*(11), 1951–1957.
13. Dai, J. Y., Ma, L. H., Wang, Z. F., Guan, W. T., & Xiu, Z. L., (2016). Sugaring-out extraction of acetoin from fermentation broth by coupling with fermentation. *Bioprocess and Biosystems Engineering, 40*(3), 423–429.
14. Dai, J. Y., Sun, Y. Q., & Xiu, Z. L., (2014). Separation of bio-based chemicals from fermentation broths by salting-out extraction. *Engineering in Life Sciences, 14*, 108–117.
15. Datta, D., Kumar, S., & Uslu, H., (2015). Status of the reactive extraction as a method of separation. *Journal of Chemistry*. Article ID 853789.
16. Dang, Y., Zhang, H., & Xiu, Z. L., (2013). Microwave-assisted aqueous two-phase extraction of phenolics from grape (*Vitis vinifera*) seed. *Journal of Chemical Technology and Biotechnology, 89*(10), 1576–1581.
17. DelMastro, T., Snow, N. H., Murphy, W. R., & Sowa, J. R., (2017). Polyol-induced partitioning of essential oils in water/acetonitrile solvent mixtures. *Journal of Liquid Chromatography & Related Technologies, 40*(7), 376–383.
18. Deshpande, M., (2014). *Polyol Induced Extraction (PIE) of Water from Organic Solvents* (p. 108). PhD Thesis; Seton Hall University, NJ, USA.
19. De Souza, D., Francisco, S. A., Ziegler, D. R., Marczak, L. D. F., & Tessaro, I. C., (2016). Characterization of rice starch and protein obtained by a fast-alkaline extraction method. *Food Chemistry, 191*, 36–44.
20. Dhamole, P. B., Mahajan, P., & Feng, H., (2011). Sugaring out for separation of acetonitrile and extraction of proteins and antibiotics. *11th International Conference on Food and Engineering*, 1699–1700.
21. Dhamole, P. B., Chavan, S., Patil, R. G., Feng, H., Bule, M., & Kinninge, P., (2016). Extraction of p-coumaric acid from agriculture residues and separation using sugaring out. *Korean Journal of Chemical Engineering, 33*(6), 1860–1864.

22. Dhamole, P. B., Demanna, D., & Desai, S. A., (2014). Extraction of p-coumaric acid and ferulic acid using surfactant based aqueous two-phase system. *Applied Biochemistry and Biotechnology, 174*(2), 564–573.
23. Dhamole, P. B., Wang, Z., Liu, Y., Wang, B., & Feng, H., (2012). Extractive fermentation with non-ionic surfactants to enhance butanol production. *Biomass and Bioenergy, 40*, 112–119.
24. Dhamole, P. B., Mahajan, P., & Feng, H., (2010). Sugaring out: A new method for removal of acetonitrile from preparative RP-HPLC eluent for protein purification. *Process Biochemistry, 45*(10), 1672–1676.
25. Djas, M., & Henczka, M., (2018). Reactive extraction of carboxylic acids using organic solvents and supercritical fluids: A review. *Separation and Purification Technology, 201*, 106–119.
26. Eskilsson, C. S., & Björklund, E., (2000). Analytical-scale microwave assisted extraction. *Journal of Chromatography A, 902*(1), 227–250.
27. Essien, S. O., Young, B., & Baroutian, S., (2020). Recent advances in subcritical water and supercritical carbon dioxide extraction of bioactive compounds from plant materials. *Trends in Food Science & Technology, 97*, 156–169.
28. Fecher, P. A., Goldmann, I., & Nagengast, A., (1998). Determination of iodine in food samples by inductively coupled plasma mass spectrometry after alkaline extraction. *Journal of Analytical Atomic Spectrometry, 13*(9), 977–982.
29. Fullbrook, P. D., (1983). The use of enzymes in the processing of oilseeds. *Journal of the American Oil Chemists' Society, 60*, 476–478.
30. Gallego, R., Bueno, M., & Herrero, M., (2019). Sub-and supercritical fluid extraction of bioactive compounds from plants, food-by-products, seaweeds and microalgae-an update. *Trends in Analytical Chemistry, 116*, 198–213.
31. Gedam, P. S., Raut, A. N., & Dhamole, P. B., (2019). Microemulsion extraction of biobutanol from surfactant based-extractive fermentation broth. *Chemical Engineering and Processing–Process Intensification, 146*, 107691.
32. Gligora, O., Mocana, A., Moldovana, C., Locatellib, M., Crişana, G., & Ferreira, I. C. F. R., (2019). Enzyme-assisted extractions of polyphenols: A comprehensive review. *Trends in Food Science & Technology, 88*, 302–315.
33. Guan, X., Li, L., Liu, J., & Li, S., (2018). Effects of ultrasonic-microwave-assisted technology on hordein extraction from barley and optimization of process parameters using response surface methodology. *Journal of Food Quality, 2018*, 9280241.
34. Guzman, M. J. L., Reyes, L. H., Mizanur, R. G. M., & Skip, K. H. M., (2009). Simultaneous extraction of arsenic and selenium species from rice products by microwave-assisted enzymatic extraction and analysis by ion chromatography-inductively coupled plasma-mass spectrometry. *Journal of Agricultural and Food Chemistry, 57*(8), 3005–3013.
35. Han, D., & Row, K. H., (2010). Recent applications of ionic liquids in separation technology. *Molecules, 15*(4), 2405–2426.
36. Hossain, M. B., Rawson, A., Aguiló-Aguayo, I., Brunton, N. P., & Rai, D. K., (2015). Recovery of steroidal alkaloids from potato peels using pressurized liquid extraction. *Molecules, 20*(5), 8560–8573.
37. Iqbal, M., Tao, Y., Xie, S., Zhu, Y., Chen, D., Wang, X., Huang, L., et al., (2016). Aqueous two-phase system (ATPS): An overview and advances in its applications. *Biological Procedures Online, 18*, 18.

38. Junjian, R., Mingtao, F., Yahui, L., Guowei, L., Zhengyang, Z., & Jun, L., (2013). Optimization of ultrasonic-assisted extraction of polyphenols from apple peel employing cellulase enzymolysis. *International Journal of Food Science & Technology, 48*(5), 910–917.
39. Juraja, S. M., Trenerry, V. C., Millar, R. G., Scheelings, P., & Buick, D. R., (2013). Asia pacific food analysis network (APFAN) training exercise: The determination of niacin in cereals by alkaline extraction and high-performance liquid chromatography. *Journal of Food Composition and Analysis, 16*(1), 93–106.
40. Kaufmann, B., & Christen, P., (2002). Recent extraction techniques for natural products: Microwave-assisted extraction and pressurized solvent extraction. *Phytochemical Analysis, 13*(2), 105–113.
41. Kemper, T. G., (2013). Solvent extraction. In: Hamm, W., Hamilton, R. J., & Calliauw, G., (eds.), *Edible Oil Processing* (2nd edn., pp. 97–126). Oxford-UK: Wiley-Blackwell.
42. Kim, S. M., & Lim, S. T., (2016). Enhanced antioxidant activity of rice bran extract by carbohydrase treatment. *Journal of Cereal Science, 68*, 116–121.
43. Kovalcik, M., Obruca, S., & Marova, I., (2018). Valorization of spent coffee grounds: Review. *Food and Bioproducts Processing, 110*, 104–119.
44. Lachenal, D., Mortha, G., Sevillano, R. M., & Zaroubine, M., (2004). Isolation of residual lignin from softwood kraft pulp. Advantages of the acetic acid acidolysis method. *C. R. Biologies, 327*(9, 10), 911–916.
45. Le, Q., Shong, L., & Shi, Y., (2001). Extraction of erythromycin from fermentation broth using salt-induced phase separation processes. *Separation and Purification Technology, 24*, 85–91.
46. Leite, A. C., Ferreira, A. M., Morais, E. S., Khan, I., Freire, M. G., & Coutinho, J. A. P., (2018). Cloud point extraction of chlorophylls from spinach leaves using aqueous solutions of nonionic surfactants. *ACS Sustainable Chemistry & Engineering, 6*(1), 590–599.
47. Li, F., Li, Q., Wu, S., & Tan, Z., (2017). Salting-out extraction of allicin from garlic (*Allium sativum* L.) based on ethanol/ammonium sulfate in laboratory and pilot scale. *Food Chemistry, 217*, 91–97.
48. Likidis, Z., & Schügerl, K., (1987). Recovery of penicillin G from fermentation broth with reactive extraction in a mixer-settler. *Biotechnology Letters, 9*, 229–232.
49. Loyao, Jr. A. S., Villasica, S. L. G., Dela, P. P. L. L., & Go, A. W., (2018). Extraction of lipids from spent coffee grounds with non-polar renewable solvents as alternative. *Industrial Crops and Products, 119*, 152–161.
50. Luengo, E., Condón-Abanto, S., Condón, S., Álvarez, I., & Raso, J., (2014). Improving the extraction of carotenoids from tomato waste by application of ultrasound under pressure. *Separation and Purification Technology, 136*, 130–136.
51. Ma, F. Y., Gu, C. B., Li, C. Y., Luo, M., Wang, W., Zu, Y. G., Li, J., & Fu, Y. J., (2013). Microwave-assisted aqueous two-phase extraction of isoflavonoids from *Dalbergia odorifera* T. Chen leaves. *Separation and Purification Technology, 115*, 136–144.
52. Maran, J. P., Sivakumar, V., Thirugnanasambandhama, K., & Sridhar, R., (2013). Optimization of microwave assisted extraction of pectin from orange peel. *Carbohydrate Polymers, 97*(2), 703–709.
53. Medina-Torres, N., Ayora-Talavera, T., Espinosa-Andrews, H., Sánchez-Contreras, A., & Pacheco, N., (2017). Ultrasound assisted extraction for the recovery of phenolic compounds from vegetable sources. *Agronomy, 7*(3), 47.

54. Melgar, B., Dias, M. I., Barros, L., Ferreira, I. C. F. R., Rodriguez-Lopez, A. D., & Garcia-Castello, E. M., (2019). Ultrasound and microwave assisted extraction of opuntia fruit peels bio compounds: Optimization and comparison using RSM-CCD. *Molecules, 24*(19), 3618.
55. Moradi, F., & Shahrouzi, J. R., (2020). Phase equilibrium and partitioning of cephalosporins (cephalexin, cefazolin, cefixime) in aqueous two-phase systems based on carbohydrates (glucose, fructose, sucrose, maltose) / acetonitrile. *Fluid Phase Equilibria, 507,* 112388.
56. Mungma, N., Kienberger, M., & Siebenhofer, M., (2019). Reactive extraction of lactic acid, formic acid, and acetic acid from aqueous solutions with tri-N-octylamine/1-octanol/N-undecane. *Chemical Engineering, 3*(2), 43.
57. Mustafa, A., & Turner, C., (2011). Pressurized liquid extraction as a green approach in food and herbal plants extraction: A review. *Analytica Chimica Acta, 703*(1), 8–18.
58. Nadagouda, M. N., Speth, T. F., & Varma, R. S., (2011). Microwave assisted green synthesis of silver nanostructures. *Accounts of Chemical Research, 44*(7), 469–478.
59. Nadar, S. S., Rao, P., & Rathod, V. K., (2018). Enzyme assisted extraction of biomolecules as an approach to novel extraction technology: A review. *Food Research International, 108*, 309–330.
60. Nagendra, C. K. L., Manasa, D., Srinivas, P., & Sowbhagya, H. B., (2013). Enzyme-assisted extraction of bioactive compounds from ginger (*Zingiber officinale* roscoe). *Food Chemistry, 139*(1–4), 509–514.
61. Novak, P., & Havlicek, V., (2016). Protein extraction and precipitation; Chapter 4. In: Ciborowski, P., & Silberring, J., (eds.), *Proteomic Profiling and Analytical Chemistry: The Crossroads* (2nd edn., pp. 51–62). Amsterdam–Netherlands: Elsevier.
62. Papaioannou, E., & Karabelas, A., (2012). Lycopene recovery from tomato peel under mild conditions assisted by enzymatic pre-treatment and non-ionic surfactants. *Acta Biochimica Polonica, 59*(1), 71–74.
63. Poojary, M. M., & Passamonti, P., (2015). Extraction of lycopene from tomato processing waste: Kinetics and modeling. *Food Chemistry, 173*, 943–950.
64. Renouard, S., Hano, C., Corbin, C., Fliniaux, O., Lopez, et al., (2010). Cellulase-assisted release of secoisolariciresinol from extracts of flax (*Linum usitatissimum*) hulls and whole seeds. *Food Chemistry, 122*(3), 679–687.
65. Saini, R. K., & Keum, Y. S., (2018). Carotenoid extraction methods: A review of recent developments. *Food Chemistry, 240*, 90–103.
66. Saini, R. K., Moon, S. H., & Keum, Y. S., (2018). An updated review on the use of tomato pomace and crustacean processing waste to recover commercially vital carotenoids. *Food Research International, 108*, 516–529.
67. Sánchez-Camargo, A. P., Martinez-Correa, H. A., Paviani, L. C., & Cabral, F. A., (2011). Supercritical CO_2 extraction of lipids and astaxanthin from Brazilian red-spotted shrimp waste (*Farfantepenaeus paulensis*). *The Journal of Supercritical Fluids, 56*(2), 164–173.
68. Shoushtari, B. A., Pazuki, G., Shahrouzi, J. R., Shahriari, S., & Hadidi, N., (2020). Separation of erythromycin using aqueous two-phase system based on acetonitrile and carbohydrates. *Fluid Phase Equilibria, 505,* 112360.
69. Somnuk, K., Eawlex, P., & Prateepchaikul, G., (2015). Optimization of coffee oil extraction from spent coffee grounds using four solvents and prototype-scale extraction using circulation process. *Agriculture and Natural Resources, 51*(3), 181–189.

70. Sun, H., Li, C., Ni, Y., Yao, L., Jiang, H., Ren, X., Fu, Y., & Zhao, C., (2018). Ultrasonic/microwave-assisted extraction of polysaccharides from *Camptotheca acuminata* fruits and its antitumor activity. *Carbohydrate Polymers, 206*, 557–564.
71. Sun, Y., Yan, L., Fu, H., & Xiu, Z., (2018). Salting out extraction and crystallization of succinic acid from fermentation broths. *Process Biochemistry, 49*(3), 506–511.
72. Sun, Y., Zhang, S., Zhang, X., Zheng, Y., & Xiu, Z., (2018). Ionic liquid-based sugaring-out and salting-out extraction of succinic acid. *Separation and Purification Technology, 204*, 133–140.
73. Supardan, M. D., Satriana, Fahrizal, & Moulana, R., (2013). Reactive extraction of jatropha seed for biodiesel production: Effect of moisture content of jatropha seed and co-solvent concentration. *International Journal on Advanced Science, Engineering and Information Technology, 3*(3), 218–221.
74. Udachan, I. S., & Sahoo, A. K., (2014). A study of parameters affecting the solvent extraction of lactic acid from fermentation broth. *Brazilian Journal of Chemical Engineering, 31*(3), 821–827.
75. Vazquez-Roig, P., & Picó, Y., (2018). Pressurized liquid extraction of organic contaminants in environmental and food samples. *Trends in Analytical Chemistry, 118*, 709–721.
76. Wang, B., Feng, H., Ezeji, T., & Blaschek, H., (2008). Sugaring-out separation of acetonitrile from its aqueous solution. *Chemical Engineering Technology, 31*(12), 1869–1874.
77. Wang, L., & Weller, C. L., (2006). Recent advances in extraction of nutraceuticals from plants. *Trends in Food Science & Technology, 17*(6), 300–312.
78. Xiao, J., Chen, G., & Li, N., (2018). Ionic liquid solutions as a green tool for the extraction and isolation of natural products. *Molecules, 23*, 1765.
79. Yan, L., Zun, Y. Q., & Xiu, Z. L., (2016). Sugaring-out extraction coupled with fermentation of lactic acid. *Separation and Purification Technology, 161*, 152–158.
80. Yang, Y., Wang, Z., Hu, D., Xiao, K., & Wu, J. Y., (2018). Efficient extraction of pectin from sisal waste by combined enzymatic and ultrasonic process. *Food Hydrocolloids, 79*, 189–196.
81. Yau, Y. K., Ooi, C. W., Ng, E. P., Lan, J. C. W., Ling, T. C., & Show, P. L., (2015). Current applications of different types of aqueous two-phase systems. *Bioresources and Bioprocessing, 2*, 49.
82. Yousefi, M., Nasrabadi, M. R., Pourmortazavi, S. M., Wysokowski, M., Jesionowski, T., Ehrlich, H., & Mirsadeghi, S., (2019). Supercritical fluid extraction of essential oils. *Trends in Analytical Chemistry, 118*, 182–193.
83. Yure, W., Dudsadee, U., & Petra, M., (2019). Yield and structural composition of pomelo peel pectins extracted under acidic and alkaline conditions. *Food Hydrocolloids, 87*, 237–244.
84. Zhang, G., Hu, M., He, L., Fu, P., Wang, L., & Zhou, J., (2013). Optimization of microwave-assisted enzymatic extraction of polyphenols from waste peanut shells and evaluation of its antioxidant and antibacterial activities *in vitro*. *Food and Bioproducts Processing, 91*(2), 158–168.
85. Zhang, H. F., Yang, X. H., & Wang, Y., (2011). Microwave-assisted extraction of secondary metabolites from plants: Current status and future directions. *Trends in Food Science & Technology, 22*(12), 672–688.
86. Zhang, W., Zeng, G., Pan, Y., Chen, W., Huang, W., Chen, H., & Li, Y., (2017). Properties of soluble dietary fiber-polysaccharide from papaya peel obtained through alkaline or ultrasound-assisted alkaline extraction. *Carbohydrate Polymers, 172*, 102–112.

87. Zhang, W., Zhu, D., Fan, H., Liu, X., Wan, Q., Wu, X., Liu, P., & Tang, J. Z., (2015). Simultaneous extraction and purification of alkaloids from *Sophora flavescens* ait. by microwave-assisted aqueous two-phase extraction with ethanol/ammonia sulfate system. *Separation and Purification Technology, 141*, 113–123.
88. Zhao, L., Lu, F., Zhang, X., & Wang, Z., (2017). Isolation of ionizable red *Monascus* pigments after extractive fermentation in nonionic surfactant micelle aqueous solution. *Process Biochemistry, 61*, 156–162.
89. Zuorro, A., Fidaleo, M., & Lavecchia, R., (2011). Enzyme-assisted extraction of lycopene from tomato processing waste. *Enzyme and Microbial Technology, 49*(6,7), 567–573.
90. Zwingelstein, M., Draye, M., Besombes, J. L., Piot, C., & Chatel, G., (2020). Viticultural wood waste as a source of polyphenols of interest: Opportunities and perspectives through conventional and emerging extraction methods. *Waste Management, 102*, 782–794.

CHAPTER 4

ULTRASONIC-ASSISTED EXTRACTION OF POLYPHENOLS FROM FOOD PROCESSING WASTES

MADHULEKHA RAKSHIT and PREM P. SRIVASTAV

ABSTRACT

The waste from the food processing industry contains a high level of polyphenols. In recent years industries have increasingly aimed at a "zero waste economy," which implies reusing wastes as raw material for extraction of existing components or fabrication of new products. Therefore, valorization of such waste can be an alternative of the extraction of polyphenols. On the other hand, classical extraction techniques have problems with high solvent consumption, low extraction yield, long extraction time, and large energy consumption. Ultrasonic-assisted extraction (UAE) is an economically viable and green extraction technique for polyphenol extraction from plant sources, which can be scaled up for industrial production. It involves a combination of mechanisms: fragmentation, detexturation, capillarity, erosion, and sonoporation. This chapter highlights the extraction of natural polyphenols from food processing waste by UAE.

4.1 INTRODUCTION

The food processing industry generates byproducts and wastes, principally, peels, seeds, etc. The food waste generated due to household activity accounts for 42%, while losses at the food manufacturing industry, food service sectors (such as catering and restaurants), and food distribution are 39%, 14%,

Novel Processing Methods for Plant-Based Health Foods: Extraction, Encapsulation, and Health Benefits of Bioactive Compounds. Megh R. Goyal, N. Veena & Ritesh B. Watharkar (Eds.)

and 5%, respectively [26]. These wastes might pose severe environmental problems and thus need to be managed in a sustainable way [4]. The food byproducts are the source of bioactive such as polysaccharides, polyphenols, essential oil, pigments, dietary fibers (DFs), etc. [32]. There is great potential in converting these wastes into valuable functional ingredients. The major challenge in the extraction of bioactive compounds from food processing waste is that these compounds are embedded within the complex cellular matrix. Valorization of waste can be achieved by extraction of the bioactive from these byproducts using a cost-effective, highly productive, and green extraction method [41].

Fruits and vegetables are high source of polyphenols that have several health benefits along with immunity-boosting ability. With the onset of the pandemic, the consumer's acceptance of polyphenolic rich food has increased worldwide. This demand can be fulfilled by eating an edible source of polyphenols or incorporating these polyphenols extracted from inedible parts of these fruits/vegetables into daily meals. Food processing waste is as whole, or part of food and/or drink which is discarded at any point in food supply chain due to being damaged, reached expiry of shelf life, disposing of inedible or distorted parts [35]. The extracts rich in polyphenol can be used to fortify or enhance the nutritional quality of daily meals. Some of the important polyphenols found in food byproducts are listed in Table 4.1.

Several extraction techniques have been reported for the extraction of polyphenols from food source [2, 6, 18, 36, 67]. However, cost-effective, less solvent consuming, highly productive and green method attracts the focus of food processing industry. Ultrasonic-assisted extraction (UAE) is one of such green extraction method. Compared to other non-thermal extraction techniques such as microwave-assisted extraction (MAE), supercritical fluid extraction (SFE), the UAE setup is cheaper and easier to operate. It also holds the potential for rapid return on investment for individual applications [28]. Apart from the extraction process, ultrasonication is an efficient tool for homogenization, emulsification, activation/inactivation of enzymes, microbial inactivation, dewatering, degassing, low-temperature pasteurization, crystallization, defoaming, particle-size reduction and changing viscosity [42].

This chapter emphasizes the role of UAE of bioactive compounds from food processing wastes. This chapter also provides an overview of UAE mechanisms and provides a holistic approach to process parameters, influencing factors, and ultrasound devices used for extraction of polyphenols exclusively from food waste.

TABLE 4.1 List of Food Byproducts with Polyphenolic Compounds

Food Byproduct	Phenolic Compounds	References
Apple byproducts	Catechins, proanthocyanidins, hydroxycinnamates, flavonols, dihydrochalcones, anthocyanins	[35]
Beetroot byproducts	Betacyanin, betaxanthin	[31]
Carrot byproducts	β-carotene	[47]
Citrus byproducts	Neohesperidin, narirutin, nobiletin, hesperidin, naringin, sinensetin, and tangeretin	[35]
Coffee byproducts	Chlorogenic acid, protocatechuic acid	[1]
Grape and wine byproducts	Catechin, gallic acid, ellagic acid, syringic acid, caffeic acid, epicatechin, myricetin, rutin, quercetin, trans-resveratrol, phenolic acids, kaempferol, stilbenes, flavones, flavan-3-ols, and resveratrol	[34]
Olive byproducts	Chlorogenic acid, vanillin, luteolin, luteolin-7-glucoside, apigenin, luteolin, tyrosol, lignans, kaempferol, caffeic acid, hydroxytyrosol, and chlorophyll	[22, 35, 54]
Onion byproduct	Quercetin, cyanidin-3-O-glucoside	[23]
Pomegranate byproducts	Punicalagin, punicalin, ellagic acid, gallic acid	[24]
Potato peel	Chlorogenic acid, ferulic acid	[39]
Sugar beet molasses	Gallic acid, vanillin, hydroxybenzoic acid, cyanidin-3-O-rutinoside, delphinidin-3-O-glucuronide, delphinidin-3-O-rutinoside, cyanidin-3-O-glucoside, catechin, syringic acid, and ferulic acid	[9]

4.2 MECHANISM OF ULTRASONIC-ASSISTED EXTRACTION (UAE)

Ultrasound cavitation is produced by the propagation of ultrasound waves in the liquid medium. As the ultrasound wave propagates in the medium, there is continuous compression and rarefaction due to the alternating pressure. This compression and rarefaction cycle creates the cavitation bubbles, which vary with the frequency of the sound wave pulse. These bubbles are known as cavitation nuclei [65] that facilitate cavitation phenomena. Cavitation bubbles are of two types: stable cavitation and transient cavitation. During stable cavitation, the bubble exists for many compression and rarefaction cycles, whereas, during transient cavitation, the bubble collapses in less than one cycle.

The cavitation bubbles grow by diffusion and/or coalescence since vapors or gas dissolved in the medium enters the bubble during rarefaction phase and eventually, these bubbles grow in size and implodes (upon reaching a critical value of high temperature of 5,000 K and pressure of 100 MPa) to produce shear force and turbulence in the medium. This phenomenon creates hotspots that can increase the reactivity in the solvent leading to higher mass transfer across the cellular structure. The mixing effect takes place at the microscopic and macroscopic levels due to microstreaming and acoustic streaming, respectively.

Acoustic streaming is generated by high-frequency acoustic waves that are propagating through a fluid medium. It is characterized by a steady fluid motion. On the other hand, micro-streaming is generated by the growth and collapse of the cavitation bubble during compression and rarefaction cycle. It is characterized by dynamic circulatory motion.

The physical mechanism of cavitation phenomena caused by the implosion of cavitation bubbles includes fragmentation, detexturation, sonocapillary effect, local shear stress, erosion, and sonoporation [7]. The cavitation phenomenon creates shock waves and causes inter-particle collision, which leads to fragmentation of particles, thus, increasing the surface area. Detexturization is the destruction of the cellular structure. The sonocapillary effect created by cavitation is the increase of solvent penetration into the pores of the cellular matrix. The alternation and implosion of cavitation bubbles create local shear stress within the solvent and near to the solute matrix. Shear force created within the solvent creates acoustic micro-streaming. Erosion is the effect of destruction on the cellular matrix, which increases the accessibility of the solvent into the fragmented particle to solubilize maximum target compounds. The increase in permeability of the cell membrane due to the pores and perforation is called sonoporation effect [33]. Along with cavitation phenomena, ultrasound also enables hydration and swelling of plant tissues leading to enlargement of pores, which increases the mass transfer [61].

4.3 PROCESS PARAMETERS FOR UAE

Ultrasound has been defined as the frequency exceeding 20 kHz (from 20 kHz to 100 kHz) [65]. Ultrasonic wave propagates through an elastic medium by creating longitudinal displacement. The vibrating body is the output of ultrasound, which makes the surrounding medium vibrate, thus, transferring energy to adjacent particles. The cavitation phenomenon depends on the ultrasound properties (e.g., intensity, power, and frequency), solvent characteristics (e.g., surface tension and viscosity) and system conditions (e.g., temperature).

The parameters responsible for the ultrasound process are intensity, power, and frequency. Previous literature shows that ultrasonic power induces greater shear forces in materials, based on the characteristics of both the solute and solvent; nevertheless, in order to obtain maximum extract using minimum power, optimization needs to be done [54, 66]. The energy level of ultrasound is expressed as acoustic energy density (W/mL or W/cm^3), ultrasound intensity (W/cm^2), or ultrasound power (W) [37].

Al-Dhabi et al. [1] investigated the effect of power on extraction yield from waste spent coffee ground and reported that the yield increased with increment in power from 100 to 250 W. In contrast, beyond ultrasonic power of 250 W, the yield decreased. A study conducted by Rabelo et al. [51] reported that ultrasound power higher than 240 W had no influence on process efficiency for the extraction of phenolic compounds from artichoke waste.

The authors Kazemi et al. [24]; Lantzouraki et al. [28]; and Maran et al. [31] have reported for the extraction of bioactive compounds from red beet stalks, pomegranate arils, and pomegranate peel, respectively. The chemical reactivity of the solvent also determines the sonochemical reaction. If the solvent is water, it dissociates into H• and OH• radicals; usually, the stable cavitation is responsible for increasing the OH• radicals with the increase in frequency [61]. Belwal et al. [5] reported a decrease in the extraction of anthocyanin with an increase in power from 150 to 450 W due to water hydrolysis, isomerization, and thermal effect.

Ultrasonic power is related to increase in energy dissipation: power of 44, 79, 102, 125, 168 and 215 W caused dissipation of 20.65, 29.05, 32.9, 38.15, 50.75 and 62.3 W, respectively [29]. Therefore, appropriate power must be optimized for maximum extraction as the dissipation of heat might cause degradation of the target molecule. Purohit et al. [47] studied UAE of β-carotene from waste carrot residue and observed a 50% increase in yield when power was increased from 20 W to 60 W; however, the increase in power from 60 W to 100 W resulted in only 14.8% increase in the β-carotene yield. This was attributed to the dissipation of energy in the form of heat, which resulted in degradation of thermolabile bioactive compounds. Pan et al. [40] investigated UAE of antioxidants from pomegranate peel. The authors noted the increase in total phenolic from 7.6% to 12.4% when the intensity was increased from 2.4 to 7.1 W/cm^2. This study also depicted that the highest yield of 14.5% was achieved at an ultrasound intensity of 59.2 W/cm^2.

The ultrasonic intensity increases with an increase in frequency. Thus, the cohesive force between solute and solvent is overcome to facilitate higher diffusion [61]. The frequency of extraction affects the size of cavitation bubbles [54]. The study conducted by Purohit et al. [47] for extraction of β-carotene from waste carrot residue showed that with an increase in frequency from 25 to 40 kHz, the extraction yield increased.

Ultrasonic intensity is the energy transmitted per second per square meter of the emitting surface. Intensity can be correlated to the amplitude. At higher amplitude, the cavitation bubbles implode vigorously. However, the higher amplitude can also lead to the degeneration of ultrasonic transducer, resulting in solvent agitation instead of creating cavitation.

Kaderides et al. [21] reported that the extraction yield improved with increased amplitude up to 40%; however, on further increase in amplitude above 40% led to a small decrease in yield. This was attributed to the degradation of the plant material. Simultaneously, the higher amplitude is required for high viscous solvents, as it raises the cavitation threshold. Cavitation threshold is reached by applying a minimum value of ultrasonic intensity.

A study conducted by Foujdar et al. [11] showed that the extraction efficiency of phenolic compounds from pomegranate peel increased from 36.18% to 42.24% with increased amplitude (20–30%). This trend could be due to improved mass transfer [11]. Similarly, the increase in kinetic parameters was observed with an increase in amplitude level up to 40%, as reported by Goula and coworkers [15].

In the case of ultrasonic probe, pulse mode can be used in order to prevent the instrument from heat generation. At the same time, it is energy-efficient way of operation, with higher yield and better operation life of the instrument. A study conducted by Purohit et al. [47] showed that maximum β-carotene yield from waste carrot residue was achieved at 60% duty cycle. Similarly, Pan et al. [40] reported that the combination of 10 s cycle time and 50% duty cycle was the best for extracting maximum antioxidants with minimum energy requirement from pomegranate peel.

4.4 FACTORS AFFECTING UAE

4.4.1 SOLVENTS

The factors affecting maximum extraction of cellular content involve type of the solvent, solute to solvent ratio, and concentration of solvent. The cavitation intensity is affected by the physical properties of solvent, including surface tension, viscosity, and vapor pressure. The intensity of bubble collapse is controlled by vapor pressure; whereas, the transient threshold of cavitation is controlled by surface tension and viscosity of the solvent [61]. A solvent with low vapor pressure improves solvent migration through the matrix, thus increases the extraction efficiency [43]. On the other hand, a solvent with low vapor pressure leads to a more intense implosion of the cavitation bubble in contrast to high vapor pressure solvents. Nevertheless, vapor pressure is influenced by the temperature of the solvent system. An increase in temperature of the liquid medium causes decrease in both viscosity and surface tension, while increasing the vapor pressure [42, 54].

Goula et al. [15] showed that higher carotenoid yield was obtained from pomegranate wastes when soy oil was used as solvent for extraction. This was attributed to the lower viscosity of soy oil. Another study conducted by Paleologou et al. [39] showed that the diffusivity (D_e) for water/ethanol and water/glycerol were 0.46 and 0.33 $\times 10^{-11}$ m^2 s^{-1}, respectively, which proved that the diffusion of polyphenols within water/ethanol is comparatively faster. On the other hand, Philippi et al. [41] showed that the maximum sonochemical advantages could be obtained by the use of solvents with low vapor pressure (water/glycerol). The authors reported that the use of 90% (w/v) glycerol at 50°C and 40% (v/v) ethanol at 80°C were equally effective, even though there is a high difference in viscosity and D_e.

The solubility of the target compound in a particular solvent affects the extraction yield. Also, the extraction yield is affected by plant material's plasticity structure or constituent differences [54]. Solvent polarity plays an important role in the extraction process as a solvent of a particular polarity can dissolve a particular phenolic group or alter its antioxidant capacity. A study conducted by Singh et al. [58] showed that the highest antioxidant activity was detected in the solvent combination of ether, ethanol, and water extract, followed by a combination of ether and ethanol extract for extraction of pomegranate polyphenols. The extraction process is governed by the polarities of solvents and the synergistic interaction between target compounds and solvent. There is no single solvent able to extract all phenolic compounds from vegetable samples [33].

For example, the dielectric constant of ethanol ($\epsilon = 25.2$) is lower than that of glycerol ($\epsilon = 42.5$), hence a lesser proportion of ethanol could decrease water polarity, resulting in the solvent system that can appropriately enhance polyphenol solubilization. Thus, the optimal concentration of extraction between water/glycerol extract and water/ethanol extract will be different due to their polarity difference. Moreover, since both ethanol and glycerol have dielectric constants lower than water [22], their aqueous mixtures would dissolve higher amounts of polyphenols. Nevertheless, the solubility of polyphenols could also be governed by hydrogen bonding and steric effects in solvent systems [11]. Thus, affinity of the solvents towards the target compound from the solid matrix is an important criterion for selection of solvent system for extraction.

Belwal et al. [5] showed that 35 and 65% of ethanol concentration gave higher anthocyanin yield, however, further increment in concentration of ethanol resulted in a reduction of anthocyanin extraction yield from starkimson fruit peel. The maximum total phenolic content (TPC) and antioxidant activity (DPPH and ABTS) from lime peel waste were observed at 50–60% ethanol concentration [53]. Similarly, Galván D'Alessandro et al. [14] reported that the use of 50% ethanol produced about 3-fold higher anthocyanins than 100% water for extraction of anthocyanin from black chokeberry.

However, Safdar et al. [56] reported that maximum polyphenols (32.48 ± 0.36 mg GAE/g extract) were extracted with 80% methanol, whereas 100% ethanol yielded minimum phenolics (24.39 ± 0.28 mg GAE/g extract) from kinnow peel. For extraction of polyphenols from onion waste, 90% (w/v) aqueous glycerol was recommended to obtain the highest yield in a study conducted by Katsampa et al. [23]. Also, Venkataramanamma et al. [62] reported that ethanol concentration beyond 50% significantly decreased in antioxidant activity and polyphenol content from pomegranate peel.

Rababah et al. [50] studied the UAE of polyphenols from pomegranate seeds and peels and concluded that the highest extract of total phenolic was found when methanol was used as an extractant. While, Singh et al. [58] studied the effect of solvents on the extraction of polyphenols from pomegranate aril and concluded that the extraction yield using different solvent(s) is in the order:

$$\text{water} > \text{ethanol} + \text{water} > \text{ethanol} = \text{ethanol} + \text{ether} + \text{water} > \text{ethanol} + \text{ether} > \text{acetone} > \text{ether}$$

The authors observed that the highest (13.22%) yield was obtained in aqueous extract and the lowest yield of 1.18% was found in ether extracts. Scanning electron microscope (SEM) depicts the microscopic view of the plant tissues indicating diffusion of target compounds into the solvent by the formation of micro-fissures or microchannels. SEM images showed the more disintegrated and porous structure of orange processing waste powder after ultrasonic-enzymatic treatment by using ethanol solvent versus using water solvent [57].

Also, Venkataramanamma et al. [62] revealed that the SEM topography of the surface of the dried pomegranate peel powder in water extract had a smooth surface, whereas, ethanol extracted particles had shrunken and dehydrated topography. Thus, the authors concluded that water extracted particles might allow the movement of the solvent by absorption/imbibition, while the ethanol extracted particles hinder the free movement of the solvent.

The amount of solvent used for the extraction process plays an important role in controlling the mass transfer of solute into the solvent. Al-Dhabi et al. [1] reported that with the increase in solute to solvent ratio from 1:05 to 1:25 g/ml, the extraction yield from waste spent coffee ground was found to increase, while upon the further increase in solid-liquid ratio, the yield decreased. This was attributed to the higher concentration difference of polyphenols between solvent and spent coffee ground that caused an increase in the rate of dissolution; however, beyond 1:25 g/ml the amount of dissolved impurities (such as polysaccharides and protein) increased which caused hindrance in the dissolution of polyphenols [1].

The study conducted by Kumcuoglu et al. [27] showed that the lycopene yield increased as the solute: solvent ratio increased from 1:20 to 1:35 and a slight decrease was observed as solute-solvent ratio increased from 1:35 to 1:50. Similarly, Qu et al. [49] recommended peel: water ratio of 1:30 g g^{-1} for proanthocyanidins extraction from pomegranate peel. The authors observed that the proanthocyanidins content increased with peel: water ratios of 1:10, 1:20, 1:30 and the highest proanthocyanidins content of 90.9 mg g^{-1} was obtained for peel: water ratios of 1:30 g g^{-1}. On the further increase of peel: water ratios up to 1:50 g g^{-1}, there was no significant change in proanthocyanidins contents. In the study conducted by Pradal et al. [46], the highest extraction yields was obtained for the ratios in between 1:30 and 1:50; and ethanol concentration of 50–60%.

Moreover, Amyrgialaki et al. [3] reported that the addition of pH regulator (citric acid) along with ethanol and water solvent systems could yield a significant amount of polyphenolic phytochemicals with antioxidant properties from pomegranate husk. Kaderides et al. [21] reported that higher solute to solvent ratio causes higher concentration gradient; therefore, more amounts of polyphenols diffuse from solute into the medium leading to the higher concentration of target compound in the solvent.

Activation energy (E_a) is required for the polyphenol extraction process, which depends on various factors, such as the solvent composition and the process temperature. Generally, E_a for phenolic compound extraction lies from 14.54 to 56.00 kJ/mol [23, 48]. The positive E_a denotes the endothermic process; hence polyphenolic extractions are endothermic in nature. Moreover, the solute should overcome E_a, which may be associated with both solvent and solute resistance [23]. Consequently, Wang et al. [64] reported that the E_a greater than 40 kJ/mol depicts dissolution reaction, while, if the E_a is less than 20 kJ/mol, the extraction process is controlled by a diffusion process, and E_a ranging from 20 to 40 kJ/mol denotes that the process is controlled by both diffusion and solubilization reaction.

4.4.2 TEMPERATURE

The optimum amount of temperature is the key to maximize the extraction process. Cavitation is enhanced by increasing the temperature and thus, facilitating diffusion [68]. However, higher temperatures may lead to degradation of bioactive compounds (depending on their types and properties). Therefore, the appropriate temperature for extraction must be studied to enhance maximum mass transfer. Purohit et al. [47] studied the effect of temperature on extraction of β-carotene from carrot waste. The authors noted three stages: (i) when the temperature was increased from 20 to 30°C resulted in 23.08% increment in the β-carotene yield; (ii) increasing the temperature from 30 to 40°C, the extraction yield improved by 12.1%; (iii) increasing the temperature from 40 to 50°C resulted in 5.84% increase in the β-carotene yield.

Mahindrakar et al. [30] revealed that the highest yield of catechin and gallic acid from the cumin seed kernel was obtained at 35C. The authors also reported that the increase in yield was not significant when the temperature was further increased from 35 to 65C. This effect was due to the fact that an increase in temperature increases the mass transfer across cell due to increased solubility. Also, the low vapor pressure of the solvent leads to a violent implosion of cavitation bubbles. During UAE, the temperature causes additional loosening and softening of cellular tissues and aids in better penetration of the solvent [59]. However, cavitation and surface tension decrease at higher temperatures of extraction because the voids are filled up with the solvent vapors, leading to less violent collapse [21, 37, 60, 61]. Another explanation for lower yield at higher temperature could be the fact that the high temperature lowers the viscosity of the solvent, causing increased vapor pressure, thus, leading to the formation of a greater number of cavitation bubbles. These bubbles collapse because of the lesser pressure difference between its inner and outer side; however, this collapse is lesser in intensity [30]. Therefore, sonochemical effects are favorable at low temperatures. The extraction temperature is also dependent on the type of polyphenols to be extracted.

For example, Zhang et al. [66] observed that the total anthocyanin yield decreased with increasing extraction temperatures beyond 50C. The decrease in yield was attributed to the fact that the anthocyanins are heat-labile [49, 66]. Similarly, a study by Goula et al. [15] reported that at temperatures higher than 40°C, a higher amount of impurities got dissolved from pomegranate waste, and some heat labile constituents decomposed during the extraction of carotenoids. While Al-Dhabi et al. [1] reported oxidation of polyphenols from spent coffee waste on exposing the solvent above 45°C. Table 4.2 gives an insight into the effect of temperature in the extraction of polyphenols from food processing wastes, denoting the range of temperature causing an increase in extraction yield and the temperature range of decreased yield.

TABLE 4.2 Effect of Temperature on Extraction of Different Polyphenols

Polyphenols and its Source	Temperature for Increasing Yield	Temperature for Decreasing Yield	References
Anthocyanin from mulberry wine residue	30 to 50°C	Above 50°C	[66]
Anthocyanin from starkrimson fruit peel (Cyanidin-3-O-galactoside)	15 to 70°C	–	[5]
Carotenoid from pomegranate waste	20 to 40°C	40 to 60°C	[15]
Polyphenols and pigments from onion waste	50 to 80°C	–	[23]
Polyphenols from chicory byproduct	20 to 60°C	–	[46]
Polyphenols from pomegranate peel extract	40 to 50°C	Above 50°C	[10]
Polyphenols from pomegranate peel extract	25 to 35°C	35 to 45°C	[21]
Proanthocyanidins from pomegranate peel	40 to 80°C	80 to 90°C	[49]
Waste spent coffee ground	30 to 45°C	Above 45°C	[1]

4.4.3 PARTICLE SIZE

The solute particle size is another factor that affects the extraction yield, as it is related to the surface area exposed to the solvent and the length of the target compound migration path. The size of the sample and the surface area available for extraction display an inverse relationship.

Ganesapillai et al. [13] showed that extraction yield increased from 30 to 70% when size was reduced from 0.4 to 0.2 mm. Similar trend was observed by Qu et al. [48], who reported that as the particle sizes increased: 0.2, 0.5, 0.7, 1.4, and 3.5 mm, their equilibrium antioxidant yields decreased 11.5%, 11.4%, 11.6%, 11.2%, and 9.4%, respectively. The authors also showed that the larger particles took more time to reach equilibrium, attributing to the fact that they provide less surface area for diffusion of solvents into the solute.

The cavitation phenomena cause disruption of cells; however, the change in the chemical structure of the cells needs to be studied to understand the chemical effect caused by ultrasound. Fourier transform infrared spectroscopy (FTIR) study conducted by Belwal et al. [5] revealed that UAE is an efficient way for extraction of anthocyanins without causing any change in its characteristic functional group. Similar FTIR results were observed by Shahram et al. [57], who reported that no change in the functional group was observed during the extraction of β-carotene from orange processing waste powder by the combined treatment of ultrasound and pectinase enzyme. Consequently, Paini et al. [38] studied principle phenolic content grape marc and olive pomace using HPLC with diode array detector (DAD). The authors noted no degradation of phenolic compounds upon subjecting it to UAE. Thus, milder treatment conditions of UAE prevent it from bioactive damage.

4.4.4 TIME

The extraction yield is time dependent and can be divided into two stages:

- Washing stage: the solvent dissolves the maximum solute from the particle surface. This step is characterized by high mass transfer; hence, it has faster rate.
- Diffusion of soluble polyphenols into the solvent is through the disintegration of the residual cellular matrix. This step is slow and rate determining step. This process involves entry and washing; continues till the concentration reaches equilibrium.

SEM study conducted by Khalili et al. [25] indicates that UAE promotes higher cell disruption at 50 min sonication time in the olive-waste cake samples as compared to 10 min sonication time, thus, promoting higher solvent penetration into the sample. Goula et al. [15] found that extraction yield increases from 10 to 30 min; however, yield slowly increased from 30 to 60 min. On the other hand, few studies have reported the degradation of molecules over long extraction time.

Belwal et al. [5] noted an increase in extraction yield when extraction time was increased from 7 to 21 min with maximum yield obtained as 0.29

mg/g. The authors reported a decrease in extraction yield upon the further increase in ultrasonic time. Therefore, the effect of time must be studied as more time of extraction lead to a higher cost of production. Similarly, Kaderides et al. [21]; Kumcuoglu et al. [27]; Pan et al. [40]; and Purohit et al. [47] reported extraction of β-carotene from carrot residues, lycopene from tomatoes, polyphenols from pomegranate peel and antioxidants from dry pomegranate marc, respectively.

4.5 PROCESSING DEVICES

The application of ultrasound can be made through three types of instruments: ultrasonic probe, ultrasonic bath (Figure 4.1), ultrasonic-assisted reactor and all have transducer (commonly piezoelectric transducer) as ultrasonic power generator.

The cavitation intensity decreases both radially and axially for ultrasonic probe. Therefore, the diameter of the container plays an important role in determining the distance between walls of the container as too small may cause damage of container and too big may not distribute intensity evenly. Most of the probes are made of titanium alloy due to its characteristics of thermal resistance. Titanium has low mass and is highly rigid; therefore, lesser energy is lost, and more energy is transmitted to the solution [61].

The probe system is more effective due to the fact that the tip of the probe delivers ultrasonic intensity through a smaller surface as compared to the ultrasonic bath. It uses a transducer attached to the probe, which is submerged into the reactor containing solute-solvent mixture. The direct circulation of ultrasound in the extraction medium causes minimum energy dissipation. The probe designs vary with diverse tip dimensions, lengths, and tip shapes (such as exponential, uniformly cylinder, linear tapered, tapered, cone-shaped or stepped). The type of probe to be selected depends upon the end-use and the volume of the sample to be sonicated. Some of the parameters could affect the ultrasound probe, such as ultrasonic intensity, the shape of the reaction vessel, temperature, probe diameter, shape of probe, and characteristics of the solute [65].

Ultrasonic bath consists of a few basic parts: stainless steel tank, ultrasonic transducers (operating around 40 kHz frequency), and optionally, temperature control. The advantages of the ultrasonic bath are: easy operation, cheap, readily available, and can simultaneously handle a higher number of samples. However, the lesser reproducibility and less power delivery into the samples are its major drawbacks as compared to probes.

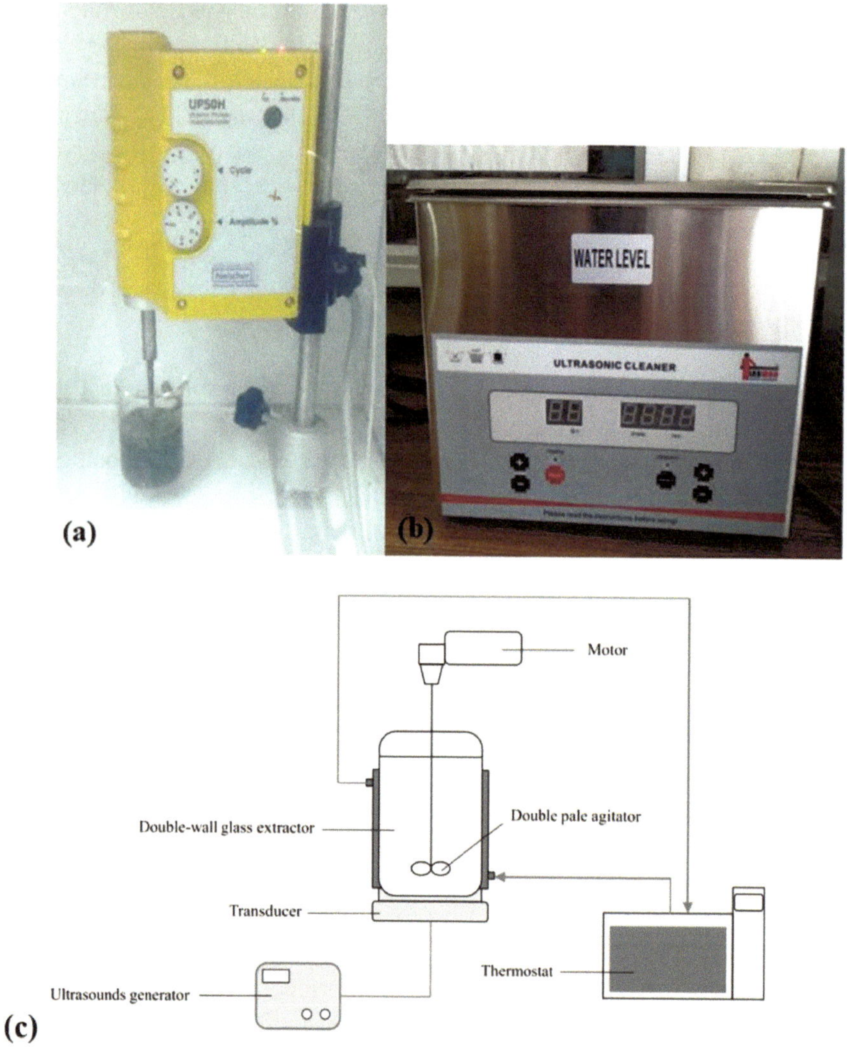

FIGURE 4.1 (a) Ultrasonic probe; (b) ultrasonic bath; (c) ultrasonic reactor.

Source: Reprinted with permission from Pradal et al., 2016b [46]. © Elsevier.

The ultrasonic-assisted reactors consist of laboratory glass extractor fitted with an ultrasound transducer, a temperature regulator, an agitator, and ultrasounds generator. A double wall jacket connected to a thermostatic bath can be provided outside of the reactor for constant temperature maintenance by the circulation of water. This type of set up is easy for temperature control as well as can handle larger volumes of sample. The designing ultrasonic extractor depends on specific parameters as per its application. The two main companies that manufacture large scale ultrasonic extraction devices are Hielscher (Germany) and REUS (France) [61].

Purohit et al. [47] compared ultrasonic-assisted extraction (UAE) of β-carotene from waste carrot residue using ultrasonic horn (or probe) and bath. The authors noted maximum β-carotene extraction yield under optimized operating parameters as 83.32% and 64.66% by using ultrasonic probe and bath, respectively. Thus, it can be concluded that the probe is more effective in the extraction of β-carotene than the bath. Table 4.3 depicts an example of UAE of polyphenols from food processing waste, the type of ultrasound device used and the processing condition studied.

4.6 COMPARISON BETWEEN UAE AND OTHER EXTRACTION TECHNIQUES

UAE has been evidently proved as a high-yielding extraction technique. González-Centeno et al. [14] studied the E_a from grape pomace and noted that E_a for TPC was 12.5 and 4.6 kJ/mol for conventional extraction (CE) and UAE, respectively. Similarly, the authors observed E_a of antioxidant capacity (by ABTS method) as 37.6 and 8.5 kJ/mol and for antioxidant capacity (by FRAP method) as 48.3 and 11.5 kJ/mol for CE and UAE, respectively. From the reported data, it is evident that the acoustic assistance of UAE decreased the E_a of the antioxidant capacity (by both methods) and total phenolics for extraction kinetics. The estimated E_a values suggest that the extraction for CE and UAE is controlled by diffusion regime for total phenolics and antioxidant activity within the temperature range 20 to 50°C [14, 64].

Several other studies have proved that UAE is highly efficient than CE of maceration, solvent-assisted extraction, Soxhlet, etc. For example, a recent study conducted by Mahindrakar et al. [30] showed that lower yield of sequential batch extractions and Soxhlet was obtained for total phenolic, total flavonoid, gallic acid, and catechin from cumin seeds than UAE. Similarly, Belwal et al. [5] reported lesser yield (0.266 ± 0.004 mg/g) of cyanidin-3-galactoside under CE than during UAE (0.343 ± 0.005 mg/g).

Also, UAE of lycopene from tomato wastes yielded 87.25% more in contrast to conventional solvent extraction, as reported by Rahimi et al. [52]. Furthermore, He et al. [18] reported UAE yielded about 2.5-fold higher total anthocyanin and about 3.2-fold total phenolic from blueberry wine pomace than the CE method. Again, González-Centeno et al. [14] reported the TPC (770.9 ± 77.5 mg gallic acid equivalents (GAE)/100 g) of the winery

TABLE 4.3 UAE of Food Processing Waste for Extraction of Polyphenols

Source of Polyphenols	Processing Device	Experimental Conditions	Solvent Composition	References
Apple pomace	Ultrasonic assisted reactor	Frequency 25 kHz Output power 150 W	Malate buffer	[44]
Artichoke solid waste	Ultrasonic probe	Frequency 20 kHz Ultrasonic power 0, 240, 480, and 720 W	Ethanol, water	[51]
Black chokeberry wastes	Ultrasonic assisted reactor	Temperature 20, 45 and 70°C Frequency 30.8 kHz Power 50 and 100 W Output power 100 W Power input density 333 W/L	Ethanol, water	[12]
Blueberry wine pomace	Ultrasonic probe	Ultrasonic power 400 W Temperature 50, 60 and 70°C	Water, ethanol, and hydrochloric acid	[17]
Carrot residue	Ultrasonic bath	Frequency 25 kHz and 40 kHz Power 20 W, 60 W, and 100 W Duty cycle 40% to 80%	Tetrahydrofuran, acetone, hexane, ethanol, and ethyl acetate	[47]
Chicory roots	Ultrasonic assisted reactor	Power 0, 50 and 100 W Duty cycle 1 min ON and 1 min OFF at power of 100 W Temperature 20, 40 and 60°C	Ethanol, water, methanol	[46]
Eggplant peels	Ultrasonic bath	Power 140 W Frequency 37 kHz Acoustic energy density 35 W L^{-1} Time 90 min	Water, glycerol, and ethanol	[41]

TABLE 4.3 *(Continued)*

Source of Polyphenols	Processing Device	Experimental Conditions	Solvent Composition	References
Grape pomace	Ultrasonic probe	Frequency 55 kHz Power density 435 W/L Ultrasonic intensity 22.9 W/cm^2 Duty cycles 0.5 s.	Water	[14]
Grape marc and olive pomace	Ultrasonic probe	Frequency 19.9 kHz Power 100 W	Ethanol, water	[38]
Kinnow peel	Ultrasonic bath	Temperature 35, 45, and 55°C Frequency 35 kHz	Ethanol, methanol, water	[56]
Lime peel waste	Ultrasonic probe	Amplitude 20 and 40%	Water, ethanol	[53]
Mango peel	Ultrasonic probe	Amplitude 100% Power 400 W Frequency 24 kHz Time 15 min	Ethanol, acetone, and hexane	[33]
	Ultrasonic probe	Frequency 20 kHz Temperature 5 and 30°C Ultrasound power intensities 165.87, 331.6 and 497.4 W/cm^2	Ethanol, water	[16]
Olive mill wastewater	Ultrasonic probe	Power 100 W Frequency 30 kHz Amplitude 100%	Ethyl acetate, diethyl ether, methanol,	[20]

TABLE 4.3 *(Continued)*

Source of Polyphenols	Processing Device	Experimental Conditions	Solvent Composition	References
Olive waste	Ultrasonic bath	Frequency 40 kHz Temperature 50°C	Ethanol	[25]
Onion solid wastes	Ultrasonic bath	Frequency of 37 kHz Power of 140 W Acoustic energy density of 35 W L^{-1} Time 60 min	Water, glycerol	[23]
Orange processing waste	Ultrasonic probe	Frequency 20 kHz Power of 500 W	Water, pectinase	[57]
Peach juice waste	Ultrasonic probe	Amplitudes 20, 60, 100% Times 20, 70, 120 s	Ethanol, water	[45]
Pomegranate peel	Ultrasonic probe	Frequency 20 kHz Output power 500 W Ultrasonic intensity 42.34, 105.92, and 211.69 W/cm^2 Amplitude 20, 30 and 40%	Ethanol, water	[10]
Pomegranate wastes	Ultrasonic probe	Output power 130 W Frequency 20 kHz Amplitude 10–100%	Sunflower oil	[15]
Potato peel waste	Ultrasonic probe	Frequency 20 kHz Temperature 15, 25 and 35°C Pulse durations of 5 s on and 5 s off Ultrasound intensities 9.24, 10.16, 13.28, 17.17 and 22.79 W/cm^2	Methanol, water	[19]

TABLE 4.3 *(Continued)*

Source of Polyphenols	Processing Device	Experimental Conditions	Solvent Composition	References
Potato peel	Ultrasonic bath	Power 140 W Frequency 37 kHz Acoustic energy density 35 W L^{-1} Time 90 min	Glycerol, water, and ethanol	[39]
Purple corn bran	Ultrasonic bath	Power 100, 200, 300, 400, and 500 W Times 90 min Temperature 40°C	HCl, ethanol, water	[8]
Red beet stalk waste	Ultrasonic probe	Temperature 40, 50 and 60°C Power 60, 90, and 120 W	Water	[31]
Rice bran	Ultrasonic bath	Frequency 35 kHz Power 140 W Temperature 40, 50 and 60°C	Ethanol, water	[59]
Spent coffee grounds	Ultrasonic probe	Frequency 20 kHz Output power 100–300 W	Ethanol	[1]
Tomato processing wastes	Ultrasonic probe	Ultrasonic intensity 30–70 W/m^2 Time 1.59–18.41 min	Sunflower oil	[52]
Tomato processing wastes	Ultrasonic probe	Ultrasound power 50, 65 and 90 W Frequency 24 kHz	Hexane, acetone, ethanol	[27]
Winery byproducts	Ultrasonic bath	Frequency 40 kHz Power 150 W	Ethanol, water	[63]

byproduct extract was around 3.3 times higher than the extract obtained through CE at 20°C. Similarly, the antioxidant activity was almost 4.0 times greater (722.4 ± 41.0 mg TE/100), and 2.5 times (705.9 ± 41.7 mg TE/100 g) for antioxidant activity by FRAP and ABTS, respectively.

Hossain et al. [19] showed that the recoveries of individual glycoalkoids from potato peel waste using UAE yielded 273, 542.7, 231 and 55.3 μg/g, whereas for solid-liquid extraction yields were 180.3, 337.6, 160.2 and 32.4 μg/g for α-solanine, α-chaconine, solanidine, and demissidine, respectively. Kumcuoglu et al. [27] showed that UAE of lycopene from tomato waste requires less time, lower temperature, and lower solvent than conventional organic solvent extraction. Similarly, Lantzouraki et al. [28] reported that UAE is faster and similarly efficient than classical extraction techniques. Again, Pingret et al. [44] compared CE and UAE and reported that TPC obtained by UAE was 30% higher than the content obtained by CE.

Plazzotta et al. [45] evaluated the efficacy of MAE and UAE on antioxidant activity and phenolic content of dried and frozen peach waste extracts. The authors observed that MAE optimized treatment delivered 2 and 4 times greater energy density (E_v) than by UAE optimal treatment in dried and frozen waste extracts, respectively. Thus, proving that the UAE is a more efficient extraction technique as compared to MAE in terms of the amount of energy distributed into the system.

4.7 SUMMARY

In this chapter, the extraction of polyphenols from food processing waste using UAE technology has been highlighted. Waste management is a global issue and can lead to greater problems if not tackled correctly. The management of post-harvest loss will be financially beneficial for the farmers as well as decrease the burden of waste disposal. Besides, food industry waste disposal is a challenge for waste management. Thus, valorization of food waste by the extraction of polyphenols is the sustainable solution of waste management. Polyphenols are widely known for their health benefits. UAE can be a promising technique for the extraction of polyphenols that can be adopted based on the outcome of the optimization process. Various factors such as amplitude, frequency, ultrasonic power, ultrasonic intensity, duty cycle, solvent type and quantity, extraction temperature, time, and size of solute determine the extraction yield. The efficiency of the UAE is largely dependent on the equipment design and the combination of process parameters. Previous literatures have majorly reported work based on simple equipment such as an ultrasonic probe, bath, or reactor. However, these setups are not suitable for commercial-scale production. Further research may be carried out on designing of UAE reactors, which will be effective for industrial scale production. Lastly, the molecular conformation of each polyphenols plays an important role in determining the optimum parameters of extraction. Therefore, the target compound must be studied in terms of its structure and biological activity in order to maximize its yield.

KEYWORDS

- **diode array detector**
- **food processing waste**
- **Fourier transform infrared spectroscopy**
- **polyphenols**
- **scanning electron microscope**
- **ultrasonic-assisted extraction**
- **valorization**

REFERENCES

1. Al-Dhabi, N. A., Ponmurugan, K., & Jeganathan, P., (2017). Development and validation of ultrasound-assisted solid-liquid extraction of phenolic compounds from waste spent coffee grounds. *Ultrasonics Sonochemistry, 34*, 206–213.
2. Alexandre, E. M. C., Araújo, P., Duarte, M. F., Freitas, V. De., Pintado, M., & Saraiva, J. A., (2017). Experimental design, modeling, and optimization of high-pressure-assisted extraction of bioactive compounds from pomegranate peel. *Food and Bioprocess Technology, 10*, 886–900.
3. Amyrgialaki, E., Makris, D. P., Mauromoustakos, A., & Kefalas, P., (2014). Optimization of the extraction of pomegranate (*Punica granatum*) husk phenolics using water/ ethanol solvent systems and response surface methodology. *Industrial Crops and Products, 59*, 216–222.
4. Arancon, R. A. D., Lin, C. S. K., Chan, K. M., Kwan, T. H., & Luque, R., (2013). Advances on waste valorization: New horizons for a more sustainable society. *Energy Science and Engineering, 1*(2), 53–71.
5. Belwal, T., Huang, H., Li, L., Duan, Z., Zhang, X., Aalim, H., & Luo, Z., (2019). Optimization model for ultrasonic-assisted and scale-up extraction of anthocyanins from *Pyrus communis* 'starkrimson' fruit peel. *Food Chemistry,* 124993.

6. Çam, M., & Hişil, Y., (2010). Pressurized water extraction of polyphenols from pomegranate peels. *Food Chemistry, 123*(3), 878–885.
7. Chemat, F., Rombaut, N., Sicaire, A., Meullemiestre, A., Anne, & Abert-Vian, M., (2017). Ultrasound assisted extraction of food and natural products-mechanisms, techniques, combinations, protocols and applications: Review. *Ultrasonics Sonochemistry, 34*, 540–560.
8. Chen, L., Yang, M., Mou, H., & Kong, Q., (2018). Ultrasound-assisted extraction and characterization of anthocyanins from purple corn bran. *Journal of Food Processing and Preservation, 42*(1), 1–7.
9. Chen, M., Zhao, Y., & Yu, S., (2015). Optimization of ultrasonic-assisted extraction of phenolic compounds, antioxidants, and anthocyanins from sugar beet molasses. *Food Chemistry, 172*, 543–550.
10. Foujdar, R., Bera, M. B., & Chopra, H. K., (2020). Optimization of process variables of probe ultrasonic-assisted extraction of phenolic compounds from the peel of *Punica granatum* var. bhagwa and its chemical and bioactivity characterization. *Journal of Food Processing and Preservation, 44*(1), 1–16.
11. Galanakis, C. M., Goulas, V., Tsakona, S., Manganaris, G. A., & Gekas, V., (2013). A Knowledge base for the recovery of natural phenols with different solvents. *International Journal of Food Properties, 16*(2), 382–396.
12. Galván, D. L., Dimitrov, K., Vauchel, P., & Nikov, I., (2014). Kinetics of ultrasound assisted extraction of anthocyanins from *Aronia melanocarpa* (black chokeberry) wastes. *Chemical Engineering Research and Design, 92*(10), 1818–1826.
13. Ganesapillai, M., Venugopal, A., & Simha, P., (2017). Preliminary isolation, recovery and characterization of polyphenols from waste *Tamarindus indica* L. *Materials Today: Proceedings, 4*(9), 10658–10661.
14. González-Centeno, M. R., Comas-Serra, F., Femenia, A., Rosselló, C., & Simal, S., (2015). Effect of power ultrasound application on aqueous extraction of phenolic compounds and antioxidant capacity from grape pomace (*Vitis vinifera* L.): Experimental kinetics and modeling. *Ultrasonics Sonochemistry, 22*, 506–514.
15. Goula, A. M., Ververi, M., Adamopoulou, A., & Kaderides, K., (2017). Green ultrasound-assisted extraction of carotenoids from pomegranate wastes using vegetable oils. *Ultrasonics Sonochemistry, 34*, 821–830.
16. Guandalini, B. B. V., Rodrigues, N. P., & Marczak, L. D. F., (2019). Sequential extraction of phenolics and pectin from mango peel assisted by ultrasound. *Food Research International, 119*, 455–461.
17. He, B., Zhang, L. L., Yue, X. Y., Liang, J., Jiang, J., Gao, X. L., & Yue, P. X., (2016). Optimization of ultrasound-assisted extraction of phenolic compounds and anthocyanins from blueberry (*Vaccinium ashei*) wine pomace. *Food Chemistry, 204*, 70–76.
18. He, L., Zhang, X., Xu, H., Xu, C., Yuan, F., Knez, Ž., Novak, Z., & Gao, Y., (2012). Subcritical water extraction of phenolic compounds from pomegranate (*Punica granatum* L.) seed residues and investigation into their antioxidant activities with HPLC–ABTS$^+$ assay. *Food and Bioproducts Processing, 90*(2), 215–223.
19. Hossain, M. B., Tiwari, B. K., Gangopadhyay, N., O'Donnell, C. P., Brunton, N. P., & Rai, D. K., (2014). Ultrasonic extraction of steroidal alkaloids from potato peel waste. *Ultrasonics Sonochemistry, 21*(4), 1470–1476.
20. Jerman, K. T., & Mozetič, V. B., (2011). Ultrasonic extraction of phenols from olive mill wastewater: Comparison with conventional methods. *Journal of Agricultural and Food Chemistry, 59*(24), 12725–12731.

21. Kaderides, K., Goula, A. M., & Adamopoulos, K. G., (2015). A process for turning pomegranate peels into a valuable food ingredient using ultrasound-assisted extraction and encapsulation. *Innovative Food Science and Emerging Technologies, 31*, 204–215.
22. Karakashov, B., Grigorakis, S., Loupassaki, S., & Makris, D. P., (2015). Optimization of polyphenol extraction from *Hypericum perforatum* (St. John's Wort) using aqueous glycerol and response surface methodology. *Journal of Applied Research on Medicinal and Aromatic Plants, 2*(1), 1–8.
23. Katsampa, P., Valsamedou, E., Grigorakis, S., & Makris, D. P., (2015). A green ultrasound-assisted extraction process for the recovery of antioxidant polyphenols and pigments from onion solid wastes using box-Behnken experimental design and kinetics. *Industrial Crops and Products, 77*, 535–543.
24. Kazemi, M., Karim, R., Mirhosseini, H., & Abdul, H. A., (2016). Optimization of pulsed ultrasound-assisted technique for extraction of phenolics from pomegranate peel of malas variety: Punicalagin and hydroxybenzoic acids. *Food Chemistry, 206*, 156–166.
25. Khalili, F., & Taghian, D. S., (2018). Extraction of phenolic compounds from olive-waste cake using ultrasonic process. *Journal of Food Measurement and Characterization, 12*(2), 974–981.
26. Kumar, K., Yadav, A. N., Kumar, V., Vyas, P., & Dhaliwal, H. S., (2017). Food waste : A potential bioresource for extraction of nutraceuticals and bioactive compounds. *Bioresources and Bioprocessing, 4*(1), 18.
27. Kumcuoglu, S., Yilmaz, T., & Tavman, S., (2014). Ultrasound assisted extraction of lycopene from tomato processing wastes. *Journal of Food Science and Technology, 51*(12), 4102–4107.
28. Lantzouraki, D. Z., Sinanoglou, V. J., Zoumpoulakis, P., & Proestos, C., (2016). Comparison of the antioxidant and antiradical activity of pomegranate (*Punica granatum* L.) by ultrasound-assisted and classical extraction. *Analytical Letters, 49*(7), 969–978.
29. Leong, T., Ashokkumar, M., & Sandra, K., (2011). The fundamentals of power ultrasound: A review. *Acoustics Australia, 39*(2), 54–63.
30. Mahindrakar, K. V., & Rathod, V. K., (2020). Ultrasonic assisted aqueous extraction of catechin and gallic acid from *Syzygium cumini* seed kernel and evaluation of total phenolic, flavonoid contents and antioxidant activity. *Chemical Engineering and Processing-Process Intensification, 149*, 107841.
31. Maran, J. P., & Priya, B., (2016). Multivariate statistical analysis and optimization of ultrasound-assisted extraction of natural pigments from waste red beet stalks. *Journal of Food Science and Technology, 53*(1), 792–799.
32. Marić, M., Grassino, A. N., Zhu, Z., Barba, F. J., Brnčić, M., & Rimac, B. S., (2018). An overview of the traditional and innovative approaches for pectin extraction from plant food wastes and byproducts: Ultrasound-, microwaves-, and enzyme-assisted extraction. *Trends in Food Science & Technology, 76*, 28–37.
33. Martínez-Ramos, T., Benedito-Fort, J., Watson, N. J., Ruiz-López, I. I., Che-Galicia, G., & Corona-Jiménez, E., (2020). Effect of solvent composition and its interaction with ultrasonic energy on the ultrasound-assisted extraction of phenolic compounds from mango peels (*Mangifera indica* L.). *Food and Bioproducts Processing, 122*, 41–54.
34. Mattos, G. N., Tonon, R. V., Furtado, A. A. L., & Cabral, L. M. C., (2017). Grape byproduct extracts against microbial proliferation and lipid oxidation: A review. *Journal of the Science of Food and Agriculture, 97*(4), 1055–1064.

35. Mourtzinos, I., & Goula, A., (2019). Polyphenols in agricultural byproducts and food waste; Chapter 2. In: Watson, R. R., (ed.), *Polyphenols in Plants: Isolation, Purification and Extract Preparation* (2nd edn., pp. 23–44). London, UK: Academic Press.
36. Mushtaq, M., Sultana, B., Anwar, F., Adnan, A., & Rizvi, S. S. H., (2015). Enzyme-assisted supercritical fluid extraction of phenolic antioxidants from pomegranate peel. *The Journal of Supercritical Fluids, 104*, 122–131.
37. O'Donnell, C. P., Tiwari, B. K., Bourke, P., & Cullen, P. J., (2010). Effect of ultrasonic processing on food enzymes of industrial importance. *Trends in Food Science & Technology, 21*(7), 358–367.
38. Paini, M., Casazza, A. A., Aliakbarian, B., Perego, P., Binello, A., & Cravotto, G., (2016). Influence of ethanol/water ratio in ultrasound and high-pressure/high-temperature phenolic compound extraction from agri-food waste. *International Journal of Food Science and Technology, 51*(2), 349–358.
39. Paleologou, I., Vasiliou, A., Grigorakis, S., & Makris, D. P., (2016). Optimization of a green ultrasound-assisted extraction process for potato peel (*Solanum tuberosum*) polyphenols using bio-solvents and response surface methodology. *Biomass Conversion and Biorefinery, 6*(3), 289–299.
40. Pan, Z., Qu, W., Ma, H., Atungulu, G. G., & Mchugh, T. H., (2011). Continuous and pulsed ultrasound-assisted extractions of antioxidants from pomegranate peel. *Ultrasonics Sonochemistry, 18*(5), 1249–1257.
41. Philippi, K., Tsamandouras, N., Grigorakis, S., & Makris, D. P., (2016). Ultrasound-assisted green extraction of eggplant peel (*Solanum melongena*) polyphenols using aqueous mixtures of glycerol and ethanol: Optimization and kinetics. *Environmental Processes, 3*(2), 369–386.
42. Picó, Y., (2013). Ultrasound-assisted extraction for food and environmental samples. *Trends in Analytical Chemistry, 43*, 84–99.
43. Pierre, F. X., Souchon, I., Athes-Dutour, V., & Marin, M., (2002). Membrane-based solvent extraction of sulfur aroma compounds: Influence of operating conditions on mass transfer coefficients in a hollow fiber contactor. *Desalination, 148*(1–3), 199–204.
44. Pingret, D., Fabiano-Tixier, A. S., Bourvellec, C. L., Renard, C. M. G. C., & Chemat, F., (2012). Lab and pilot-scale ultrasound-assisted water extraction of polyphenols from apple pomace. *Journal of Food Engineering, 111*(1), 73–81.
45. Plazzotta, S., Ibarz, R., Manzocco, L., & Martín-Belloso, O., (2020). Optimizing the antioxidant biocompound recovery from peach waste extraction assisted by ultrasounds or microwaves. *Ultrasonics Sonochemistry, 63*, 104954.
46. Pradal, D., Vauchel, P., Decossin, S., Dhulster, P., & Dimitrov, K., (2016). Kinetics of ultrasound-assisted extraction of antioxidant polyphenols from food byproducts: Extraction and energy consumption optimization. *Ultrasonics Sonochemistry, 32*, 137–146.
47. Purohit, A. J., & Gogate, P. R., (2015). Ultrasound-assisted extraction of β-carotene from waste carrot residue: Effect of operating parameters and type of ultrasonic irradiation. *Separation Science and Technology, 50*(10), 1507–1517.
48. Qu, W., Pan, Z., & Ma, H., (2010). Extraction modeling and activities of antioxidants from pomegranate marc. *Journal of Food Engineering, 99*(1), 16–23.
49. Qu, W., Shi, S., Li, P., Pan, Z., & Venkitasamy, C., (2014). Extraction kinetics and properties of proanthocyanidins from pomegranate peel. *International Journal of Food Engineering, 10*(4), 683–695.

50. Rababah, T. M., Banat, F., Rababah, A., Ereifej, K., & Yang, W., (2010). Optimization of extraction conditions of total phenolics, antioxidant activities, and anthocyanin of oregano, thyme, terebinth, and pomegranate. *Journal of Food Science, 75*(7), C626–C632.
51. Rabelo, R. S., MacHado, M. T. C., Martínez, J., & Hubinger, M. D., (2016). Ultrasound assisted extraction and nanofiltration of phenolic compounds from artichoke solid wastes. *Journal of Food Engineering, 178*, 170–180.
52. Rahimi, S., & Mikani, M., (2019). Lycopene green ultrasound-assisted extraction using edible oil accompany with response surface methodology (RSM) optimization performance: Application in tomato processing wastes. *Microchemical Journal, 146*, 1033–1042.
53. Rodsamran, P., & Sothornvit, R., (2019). Extraction of phenolic compounds from lime peel waste using ultrasonic-assisted and microwave-assisted extractions. *Food Bioscience, 28*, 66–73.
54. Roselló-Soto, E., Barba, F. J., Parniakov, O., Galanakis, C. M., Lebovka, N., Grimi, N., & Vorobiev, E., (2015). High voltage electrical discharges, pulsed electric field, and ultrasound assisted extraction of protein and phenolic compounds from olive kernel. *Food and Bioprocess Technology, 8*(4), 885–894.
55. Roselló-Soto, E., Galanakis, C. M., Brnčić, M., Orlien, V., Trujillo, F. J., Mawson, R., Knoerzer, K., et al., (2015). Clean recovery of antioxidant compounds from plant foods, byproducts and algae assisted by ultrasounds processing: Modeling approaches to optimize processing conditions. *Trends in Food Science & Technology, 42*(2), 134–149.
56. Safdar, M. N., Kausar, T., Jabbar, S., Mumtaz, A., Ahad, K., & Saddozai, A. A., (2017). Extraction and quantification of polyphenols from kinnow (*Citrus reticulate* L.) peel using ultrasound and maceration techniques. *Journal of Food and Drug Analysis, 25*(3), 488–500.
57. Shahram, H., & Dinani, S. T., (2019). Optimization of ultrasonic-assisted enzymatic extraction of β-carotene from orange processing waste. *Journal of Food Process Engineering, 42*(4), 1–16.
58. Singh, M., Jha, A., Kumar, A., Hettiarachchy, N., Rai, A. K., & Sharma, D., (2014). Influence of the solvents on the extraction of major phenolic compounds (punicalagin, ellagic acid and gallic acid) and their antioxidant activities in pomegranate aril. *Journal of Food Science and Technology, 51*(9), 2070–2077.
59. Tabaraki, R., & Nateghi, A., (2011). Optimization of ultrasonic-assisted extraction of natural antioxidants from rice bran using response surface methodology. *Ultrasonics Sonochemistry, 18*(6), 1279–1286.
60. Tiwari, B. K., & Mason, T. J., (2012). Ultrasound processing of fluid foods; Chapter 6. In: Cullen, P. J., Tiwari, B. K., & Valdramidis, V. P., (eds.), *Novel Thermal and Non-Thermal Technologies for Fluid Foods* (pp. 135–165). London, UK: Elsevier Inc.
61. Tiwari, B. K., (2015). Ultrasound: A clean, green extraction technology. *Trends in Analytical Chemistry, 71*, 100–109.
62. Venkataramanamma, D., Aruna, P., & Singh, R. P., (2016). Standardization of the conditions for extraction of polyphenols from pomegranate peel. *Journal of Food Science and Technology, 53*(5), 2497–2503.
63. Wang, J., Zhang, Y., Wang, H., & Huo, S., (2019). Evaluation of extraction technologies and optimization of microwave and ultrasonic assisted consecutive extraction of phenolic antioxidants from winery byproducts. *Journal of Food Process Engineering, 42*(4), 1–13.

64. Wang, Y. G., Yue, S. T., Li, D. Q., Jin, M. J., & Li, C. Z., (2002). Kinetics and mechanism of Y(III) extraction with CA-100 using a constant interfacial cell with laminar flow. *Solvent Extraction and Ion Exchange, 20*(3), 345–358.
65. Wen, C., Zhang, J., Zhang, H., Dzah, C. S., & Zandile, M., (2018). Advances in ultrasound assisted extraction of bioactive compounds from cash crops: A review. *Ultrasonics Sonochemistry, 48*, 538–549.
66. Zhang, L., Fan, G., Khan, M. A., Yan, Z., & Beta, T., (2020). Ultrasonic-assisted enzymatic extraction and identification of anthocyanin components from mulberry wine residues. *Food Chemistry, 323*(50), 126714.
67. Zheng, X., Liu, B., Li, L., & Zhu, X., (2011). Microwave-assisted extraction and antioxidant activity of total phenolic compounds from pomegranate peel. *Journal of Medicinal Plants Research, 5*(6), 1004–1011.
68. Živković, J., Šavikin, K., Janković, T., Ćujić, N., & Menković, N., (2018). Optimization of ultrasound-assisted extraction of polyphenolic compounds from pomegranate peel using response surface methodology. *Separation and Purification Technology, 194*, 40–47.

PART II

ENCAPSULATION METHODS FOR BIOACTIVE COMPOUNDS

CHAPTER 5

ENCAPSULATION TECHNOLOGIES: PRINCIPLES AND APPLICATIONS IN THE FOOD INDUSTRY

NAZIA TABASSUM, SWETA JOSHI, VARISHA ANJUM,
Z. R. A. A. AZAD, and SADAF AHMAD

ABSTRACT

In the present era, the increase in health awareness among consumers caused an enhanced demand for healthy and nutritious food options. Subsequently, this has led to continuous and feasible inventions and research that resulted in the globalization of the food industry. Recently, coupled with consumers' interest, the plant-derived bioactive compounds have emerged as potential and alternate therapeutic candidates that substitute synthetic compounds. However, the major caveats that hinder its effective use are issues related to stability and delivery. Encapsulation provides a sustainable solution for the entrapment of core material (active ingredients) within the compatible, food-grade, biodegradable shell resulting in the successful delivery of the bioactive compound into the food matrices. The protective shell creates the barrier between the labile bioactive compound and the environment, thus causing the increased stability and bioavailability of the active ingredients. Encapsulation may thus also be used for the immobilization of cell or enzyme to produce food products during fermentation. Moreover, this technique provides utility in the flavor industry in stabilizing the volatile compounds, masking the bad smell in addition to the bitter taste and astringency of compounds like polyphenols, soy isoflavones in the final product. The pharmaceutical and food sector share the common platform for the implementation of

Novel Processing Methods for Plant-Based Health Foods: Extraction, Encapsulation, and Health Benefits of Bioactive Compounds. Megh R. Goyal, N. Veena & Ritesh B. Watharkar (Eds.)

encapsulation techniques to obtain distinctive bioactive compounds to be used as drugs or functional foods and nutraceutical products.

5.1 INTRODUCTION

Currently, there is an increase in health awareness among consumers, which has led towards their growing interest in what ingredients make up their food and the benefits that they might provide towards maintaining a healthy body. This has led the food industry to work tirelessly to fulfill the consumer's demand for healthy and nutritious food. Recently, the plant derived bioactive compounds have emerged as a potential therapeutic alternative to synthetic compounds, however, issues related to their stability and delivery have emerged. Encapsulation is an emerging technology that entraps active ingredients with food-grade, biodegradable shell materials, thus creating a barrier between the unstable bioactive compound and the environment resulting in increased stability and bioavailability of the active ingredients.

Encapsulation is a process wherein, small-sized droplets of liquid or particles of solid are enclosed within a shell (inert) that further isolates as well as provides protection to the enclosed material from the outside environment. The enclosed material which comprises the core, also referred to as internal phase or fill is released into the food matrix when the shell or wall deteriorates under specific conditions. This helps in accomplishing a controlled release of the core materials at the desired place and time. Controlled release would not only enhance the effectiveness of the food additives but also ensure optimal dosage and broaden their application range [23]. Encapsulation results in the production of microparticles, microcapsules, and microspheres depending on their morphology and internal structure. The microcapsules are mainly of three types; mononuclear, polynuclear, and matrix encapsulation (Figure 5.1). While the shell surrounds a single core in mononuclear capsules, thus the name, it encloses many cores in polynuclear capsules. The matrix form has the homogeneous distribution of the core particles within the shell.

Encapsulation helps achieve various objectives: (i) protects degradation of core material by decreasing their reactivity to the environment (such as air, moisture, light, heat, etc.); (ii) retards the evaporation rate of the core material to the external environment [50]; (iii) controls the release of the core material until the proper stimulus; (iv) masks the undesired properties (such as taste and flavor) of the core materials; (v) converts liquid components into solid form thus improving material handling; (vi) separates incompatible components and also reduces volatility, flammability, and toxicity; (vii) stabilizes emulsions and dispersions, and finally; (viii) increases the stability and life of the product being encapsulated.

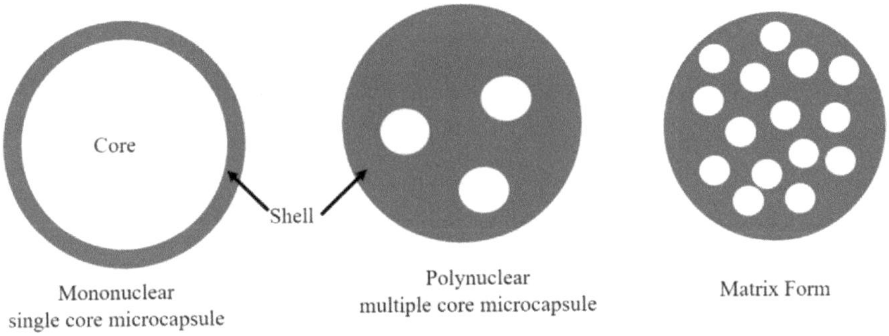

FIGURE 5.1 Morphological representation of microcapsules.

Encapsulation, often referred to as microencapsulation has been used in the past to simply convert liquids into solids or to mask the unpleasant taste of certain ingredients. However, with technological advances, the perception of controlled release of the encapsulated material has become more intriguing.

This chapter addresses the principle and methods of encapsulation, its role and application in the protection, stabilization, and as targeted delivery system for enhanced nutritional health benefits.

5.2 PRINCIPLE OF ENCAPSULATION

Encapsulation involves the incorporation of various bioactive compounds such as cells, enzymes, flavors, etc., into small capsules. These capsules provide protection to the sensitive components or rather provide an extra layer between such labile compounds and the external environment. They help in transforming liquids or gases into easily handled solid ingredients and control their release into product formulations. The basic principle of encapsulation therefore is the enclosing or covering of sensitive compounds into shells that offer protection against the outside environment. These shells, better known as capsules are small asymmetrically or variably shaped particles containing even smaller core particles that are released in the outside environment under some external stimulus such as heat, light, pH, etc.

5.3 ENCAPSULANT MATERIALS

Encapsulation mainly entails knowledge of common properties of microcapsules, including the nature of the encapsulating materials (core and shell), their stability as well as their release characteristics. The final encapsulated product obtained, its physical characteristics and its intended usage in the food industry also needs to be apprehended. They most definitely should be food-grade if have to cater the need of the food industry. Most frequently used encapsulants include carbohydrates, cellulose, gums, lipids, and proteins. The molecular behavior of the materials involving assembly and formation highly influences the adequacy of them being used for encapsulation. Their properties that prove their functionality as encapsulating materials include emulsification, gelling, and viscosity. Another point to keep in mind is the ability of these materials to undergo transition from solid to liquid phase reversibly in response to external stimuli.

Comprehending the encapsulation techniques is also essential to determine the encapsulant materials to be used, however, this topic will be discussed later in the chapter. Food industry employs encapsulation majorly to either mask undesirable flavor/taste or to protect labile components. The capsules formed are usually water-soluble and hence easily release the core materials when dissolved into water.

5.3.1 CARBOHYDRATES

Sugar such as glucose or sucrose and polysaccharides such as starch, and their derivatives, cellulose, and its derivative, dextrins, pectin, alginate, carrageenan, or chitosan are the carbohydrates that find important application as constituents of the encapsulating matrix. These encapsulants have the ability to act as carriers for bioactive compounds and often end up forming glassy solids wherein, the active compounds get trapped after dehydration [48, 51]. Food carbohydrates are known to play a crucial part in the stabilization of emulsion. This feat could be achieved owing to their ability to increase the viscosity of the continuous phase of the emulsion systems. Nevertheless, their performance improves in emulsion-based encapsulation systems when used in combination with other compounds such as proteins that have good emulsifying properties [2].

Starch and its derivatives have the ability to bind, protect, and retain flavor in compounds and release them at high temperature or moisture conditions [61]. However, among the two components, amylose-based gels have better mechanical properties than the amylopectin-rich gels. The latter also is more prone to chemical and enzymatic degradation. Role of maltodextrins was discussed in formulating dehydrated encapsulation systems owing to its low viscosity at high solids [29]. Maltodextrins helps in providing structural integrity to the final encapsulated product and also reduces stickiness during drying due to their higher glass transition temperature.

5.3.2 PROTEINS

Proteins possess the characteristics of self-assembly due to their amphiphilic nature. They also possess good binding properties especially for the flavor compounds [35]. Most commonly used food proteins as encapsulant materials include soy proteins, milk proteins (casein and whey proteins), and gelatin. Whey proteins were successfully utilized as wall material for the encapsulation of anhydrous milk fat using spray drying [49]. The same authors also worked to successfully improve the microencapsulation efficiency by replacing whey proteins by lactose [18]. Oxidation of caraway essential oil was prevented by encapsulation using milk protein as wall ingredient [8].

Gelatin is known to be a good emulsifying and stabilizing agent and also tends towards the formation of a fine dense network on drying. Therefore, they have been proven to be effective entrapping agent in contrast to maltodextrin, glucose, maltose, and pullulan [25]. In combination with anionic polysaccharides, gellan has also been used to encapsulate *Lactococcus lactis* [24].

Chickpea protein along with alginate was used to formulate microcapsules using emulsion technology [58]. They provided excellent protection to *Bifidobacterium adolescentis* in synthetic gastric juice. This study therefore suggested that chickpea protein-alginate microcapsules could effectively serve as probiotic carrier and used in various food applications.

5.3.3 LIPIDS

Lipids though difficult to disperse in food products, can be utilized as solvents to solubilize hydrophobic substances such as volatile aromatic compounds [18]. Lipids/fats have excellent moisture barrier property and hence prove very useful in providing protection against moisture ingress. However, their physical and chemical properties dictate their moisture barrier, rheological, microstructural, or colloidal properties [2]. For example, a reduced hydrocarbon chain length or increased degree of unsaturation of the fatty acid chains results in lowering of the melting point, thereby decreasing the moisture barrier property. Consequently, it becomes essential to match the physical properties of the fat to the trigger temperature to facilitate the release of active components embedded within. Various lipids that can be used for encapsulation purposes include natural fats and oils, mono, and di-glycerides, phospholipids, waxes, etc. Benefits of lipid encapsulation are [39]:

- Retards auto-oxidation;
- Enhances stability;
- Masks taste of lipid-soluble substances;
- Controls the release of lipid-soluble flavor compounds; and
- Protects against enzyme hydrolysis [39].

5.3.4 GUMS

Gums are better wall materials than the other carbohydrates (such as maltodextrin and modified starch) [32]. This is essentially due to the fact that gums have good emulsion stabilization properties and also encourages film formation. Unlike gums, most of the food carbohydrates have to be used in combination with other emulsifying agents to achieve better results while formulating emulsion-based encapsulation systems. Cumin oleoresin was successfully encapsulated using gum Arabic [27].

Gum Arabic is most often preferred for the purpose of lipid encapsulations as they produce emulsions that are highly stable and that too with oils that have a wide range of pH [29]. However, later studies conducted on encapsulation of various monoterpenes (limonene, linalool, citral, β-myrcene, and β-pinene) established the inefficiency of gum Arabic as wall material due to their limited barrier property against oxidation [4]. Similarly, high cost, limited supply, and variations in quality has put a constrain on the application of gum Arabic as a wall material for encapsulation purposes and made researchers look for alternative encapsulants.

5.4 ENCAPSULATION TECHNIQUES

Various techniques (Figure 5.2) are available for the encapsulation of food ingredients into coating materials. The selection of these techniques is overseen through various properties of the encapsulating materials, namely, core, and wall and also their envisioned application [13]. However, the recent technological advances in the encapsulation techniques have led to the development of shell materials with wider functionalities. The technique mainly relies on the tangible or physical nature of material that has been subjected to encapsulation and gets released by any kind of trigger, such as change in pH, time, or temperature. Other forces such as osmotic pressure, mechanical stress and enzymic activity also have been reported to play a role in triggering the release of the encapsulants [13]. The different encapsulation technologies include spray drying and spray chilling, fluidized bed coating, coacervation, liposome entrapment, extrusion, and inclusion complexation.

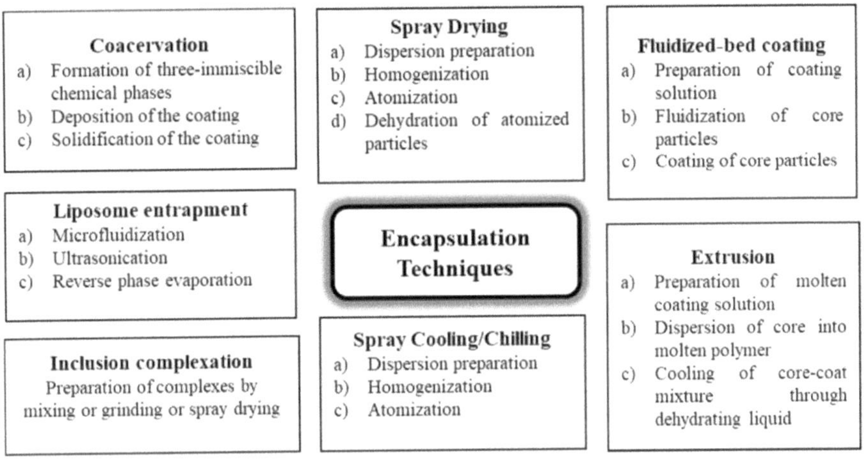

FIGURE 5.2 Different encapsulation techniques and their processing steps.

5.4.1 *SPRAY DRYING*

Spray drying is the most widely employed encapsulation method of food industry. It principally aims to convert liquid material into a solid form of powder. It is economical and also serves to effectively protect materials, particularly flavors, oils, and fragrances from degradation or oxidation [23]. Encapsulation using spray drying aspires to enclose a material that is essentially inert to its surrounding. The purpose is to provide a protection to this active material while ensuring that it does not react with the encapsulating materials. However, this technology faces a major drawback in the form of a limited range of choices between wall and shell materials. Another important factor to keep into consideration is the solubility of the shell material in water as the spray drying process mainly involves aqueous feed formulations. Most common carrier or wall materials used for the purpose of encapsulation include hydrocolloids such as gelatin, maltodextrin, modified starch, or gum acacia. Owing to their low solubility in water, the usage of these hydrocolloids becomes expensive and tedious. However, their addition does show some beneficial effects in regards to the stability of the encapsulating materials [31, 39].

The process encompasses three basic steps: first is the formulation of an emulsion that has to undergo processing, followed by its homogenization. The last step is the atomization of liquid droplets inside the drying chamber. The material to be encapsulated or the core material is homogenized with the wall material to obtain particles or droplets of smaller size which further undergoes atomization while passing through a nozzle inside a spray dryer [19, 41]. The spray velocity, viscosity, or the pressure drop across the nozzle and surface tension of the liquid hugely affects the size of the atomized droplets [41], which further helps in the determination of the drying time. As the atomized droplets come in contact with hot air inside the chamber, evaporation of the solvent takes place followed by the solidification of the wall materials on the core particles leaving behind the dried or encapsulated particles. The obtained particles were of polynuclear or matrix type and in the form of powder, granulates or agglomerates.

5.4.2 SPRAY COOLING/CHILLING

This technique involves usage of both chilled as well as cool air to perform atomization of materials unlike the heated air used in spray drying [44]. Atomization provides a vast surface area for the droplets to experience a quick and intimate mixing with the cooling medium. Various forms and derivatives of vegetable oil are examples of a few typical coating materials used in spray cooling. Henceforth, evaporation of water is no longer required like in the case of spray drying. The outer materials include fat and stearin with melting points of 45–122°C and/or hard mono- and diacylglycerols with melting points of 45–65°C [54]. Contrarily, spray chilling involves fractionated or hydrogenated vegetable oil as coating material with melting point in the range of 32–42°C [5]. They further find applications as textural or functional ingredients to enhance heat stability or perfect the timely release of the encapsulated materials. They can also be used for flavor, vitamins, minerals, or enzyme encapsulation. The low temperature encapsulation provides a platform for conversion of a liquid material into powder form for various frozen or heat-sensitive products that may be insoluble in common solvents. The limitation, however, is the special handling and storage conditions required [54]. The result may not yield a perfect encapsulate, however the end product often exhibits delayed release of encapsulant.

5.4.3 FLUIDIZED BED COATING

This technology was originally adopted by many pharmaceutical companies but later on caught the interest of the food sector and since then it has been used to encapsulate a wide variety of food ingredients. While the technology used by the pharmaceutical companies is quite expensive, the food industry works towards cutting the production costs by applying somewhat of a different approach. A fascinating fact about this advanced technology is that it is one of the few technologies that has the ability to coat particles with almost all kinds of wall materials. This ensures a more versatile use of this technology in the controlled release mechanism. The process follows the spraying of shell materials over the coating materials, followed by rapid evaporation and finally resulting in forming of a protective layer over the materials.

Fluid-bed coating may be performed in various manners, namely top spray, bottom spray, and tangential spray. Coating materials fall downwards in the top spray system while moving upwards in the bottom spray systems, also known as 'Wurster's coater system.' Tangential spray system consists of a rotating disc with a tangential nozzle and is raised during the process. Top spray fluidized bed coaters have been reported to perform better than the other two due to high yields of the encapsulating materials [26]. This technique has been reported to encapsulate a number of food materials such as, ascorbic acid [33] and acidulants for processed meat [59]. Few most common applications of this technology include coating of vegetable oil, waxes, fatty acids, emulsifiers, etc. It has also proved to be appropriate for coating of starch, gums, and maltodextrin [23, 56].

5.4.4 COACERVATION

It is the phase separation of a homogeneous polymer solution into a coacervate and coacervation medium; coacervate being the phase rich in polymer, while the medium is poor in polymer. The coacervate phase then surrounds the core material suspended in the initially taken polymer solution. Finally, the hydrocolloid shell thus formed can be cross-linked using some suitable chemical or enzymatic cross-linking agent. Coacervation process could be referred to as simple and complex coacervation. The former is named so as it considers only one type of polymer, while the latter involves two or more than two types of polymers. The microcapsule formation during coacervation involves three basic steps: (i) formation of three immiscible phases namely, core material, polymer-poor phase (liquid medium) and polymer-rich phase (coating ingredient); (ii) depositing the polymer over the core ingredient; and (iii) rigidization of the formed coating.

The first step may be triggered by various mechanisms such as the addition of salts, non-solvents, or incompatible polymer-to-polymer solution, or by changing the temperature of the solution, or by inducing polymer-polymer interaction [26]. The next step follows the coating of the core ingredient by deposition of the liquid polymer. This process oversees controlled physical mixing of the medium containing core and coating material [13, 23, 26, 62]. The final step involves stabilization of the formed microcapsules by thermal treatment, cross-linking or desolvation techniques [13, 26]. The process finally yields self-sustaining microcapsules, which are collected by filtration or centrifugation, washed using some suitable solvent. Finally, various techniques, viz. spray or fluidized bed drying is used to dry the particles. Some common hydrocolloid systems studied for coacervation process include gelatin/gum acacia system, gliadin, heparin/gelatin system, chitosan, soy-proteins, polyvinyl alcohol, gelatin-carboxymethylcellulose (CMC), starch, β-lactoglobulin/gum acacia and many more [23, 26].

5.4.5 LIPOSOME ENTRAPMENT

Liposomes have more versatile properties and they are less fragile and delicate than fat capsules. Initially developed for medical purposes (drug delivery), they have now found application in the delivery of enzymes and vitamins in the body. There use for encapsulation of food components is being researched widely. The important characteristics of liposomes (for example, stability, permeability, surface activity and affinity) could easily be modified by modifying the size and lipid compositions. They are present either in unilamellar or multilamellar configurations with a single or multiple lipid layer, respectively [19].

Generally, phospholipids are present in the outer layers of liposomes. The two facets of the lipids, i.e., hydrophilic and hydrophobic, are directed towards the aqueous phase and hydrophobic group of the other lipid molecules, respectively. Lipid sheets fold easily into spherical shapes as there is no interaction between lipids and water, thereby forming a stable capsule. Liposomes may enclose a number of compartments containing aqueous or lipophilic compounds. Either hydrophilic (aqueous) or hydrophobic (lipid-soluble) materials can be entrapped in liposomes at a time rather than entrapping both types of compounds [19, 23, 42].

Large unilamellar vesicles (ULVs) have been reported as most appropriate liposomes for the food industry due to various reasons: (i) ease in production, (ii) high stability, (iii) high encapsulation efficiency (EE), (iv) controlled delivery at specific temperatures and specific locations, (v) releasing the contents readily in high moisture environments. However, there are some hitches in their use, such as the requirement of low temperatures for storage or high-cost considerations during freeze-drying of liposome suspension.

5.4.6 EXTRUSION TECHNOLOGY

Encapsulation of volatile and unstable flavors is preferably obtained by extrusion. Coating materials usually involved are glucose, sucrose, maltodextrin, etc., which forms the glassy carbohydrate shell. The process is achieved by taking the coating material in a molten state maintaining a temperature of about 115°C and pressure less than 100 psi. It is then followed by the addition of core material by using a series of dies, nozzles, syringe, atomizing disk, coaxial air-flow, electric field, jet cutter, etc., for shaping them into spheres. Encapsulated microspheres are formed which are released into a dehydrating liquid such as isopropanol [13, 23, 42].

Each one of the above techniques is suitable for lab-scale production of encapsulated microspheres, and each of them comes along with their own advantages and disadvantages. Jet-cutter, coaxial air-flow and electrostatic potential are capable of providing narrow size distribution. Coaxial air-flow generally provides oval-shape particles. Also, the productivity of coaxial air flow as well as electrostatic potential is low. Vibrating nozzle techniques gives the highest productivity in comparison to other techniques. Jet-cutter technology is suitable for both lab-scale and large-scale production systems.

The principal advantage of extrusion encapsulation is the increase in shelf-life for flavor oils prone to oxidation: Normally, the flavor oils have a shelf-life of few months. In some cases of extrusion encapsulation, shelf life has been increased even up to 5 years. This is made possible by hydrophilic glassy matrix which provides an excellent impermeable barrier for atmospheric gases. This is significant when compared with a shelf-life of one year for microcapsules produced by spray-drying [13, 62].

Some other variants of extrusion technology for encapsulation are centrifugal extrusion, centrifugal suspension separation and melt extrusion. Centrifugal extrusion, also known as coextrusion is most often used to encapsulate flavor oils. Centrifugal suspension separation, on the other hand provides protection to sensitive foods and food components such as aspartame, vitamins, methionine, etc., from moisture [13, 23, 41, 42]. Melt extrusion involves use of thermo-mechanical mixers, double screw extruder being one of the examples. However, the final payload for melt extrusion is quite low (typically less than 10%), which may affect their cost-in-use [62].

5.4.7 INCLUSION COMPLEXATION

It can be referred to as an association between a cavity-bearing substrate and a ligand (or an active ingredient). This association occurs at the molecular level, and it may be maintained due to various reasons such as hydrogen bonding, Vander Waals forces and entropy-driven hydrophobic interactions.

The most practically feasible cavity-bearing substrates for inclusion complexation include: 4, 5 and 6 glucopyranose-ringed members known as α, β and γ-cyclodextrins, respectively. They have diameters of about 14, 15 and 17 Å, respectively with an inner pocket of about 5–8 Å, which are suitable for reversibly entrapping an active molecule of right size in an aqueous environment. They can entrap the ligands (active ingredients) that can fit dimensionally inside their central cavities. The dimensions of the cavity depend on the type of cyclodextrin used for encapsulation. Partially hydrolyzed starch (maltodextrin) is used to prepare cyclodextrins with the help of enzymes such as cyclodextrin glycosyltransferase [23, 62].

The cyclodextrins are preferable for inclusion complexation because its internal cavity is hydrophobic while its outer surface is hydrophilic in nature. Thus, they can easily encapsulate lipophilic substances such as flavor compounds, and lipophilic vitamins. They have been used for encapsulating essential oils, garlic, and onion oils which form odorless complexes with cyclodextrins. This encapsulation helps in protecting the essential oils for longer times and preventing the volatilization of volatile flavor components [23]. Few other examples of inclusion complexation are: utilization of amylose to entrap lipids or proteins (ligand-binding) for example, β-lactoglobulin. The hydrophobic part of β-lactoglobulin tends to bind the fatty acids as well as aromatic compounds in a pH and temperature dependent manner [62].

5.5 APPLICATIONS OF ENCAPSULATION

Properties of encapsulation could be exploited by the food industry in various fields such as flavor industry, vitamins, minerals, nutraceuticals, bioactive compounds, beverage industry, dairy industry, or packaging sector. The applications of encapsulation in these sectors have been discussed in this section.

5.5.1 ENCAPSULATION IN FLAVOR INDUSTRY

Encapsulation of flavors dates way back and has been well documented for years. The dawn of flavor science and application-specific delivery of flavors might have led to the introduction of encapsulation in the food and beverage industry. Food products having unique flavor changing capability or varying sensational qualities or long-lasting flavors are the new trend and can be easily achieved by the encapsulation process. The protective covering formed by the encapsulating agents not only thwart degradative changes but also eases the handling of encapsulated flavors that are in the form of stable dry powders [40]. One of the examples involves the preservation of color and flavor of a tomato puree using acetone by spray drying [45], and this technology has become the most preferential among all encapsulation technologies for flavor retention.

Chewing gum has a hydrophobic gum base that interacts with several flavor ingredients, binding them, and ensuring slow and reduced release of the flavor compounds [10]. Later research suggested the use of spray drying to inhibit the binding interactions between flavor and gum base [12]. Encapsulated flavors were also utilized in ice cream preparations. A lipid encapsulated flavor was designed to melt at temperatures typical to a consumer's mouth [17]. Keeping in view the convenience of consumers, beverages in dry mix form for ready-to-drink products were formulated. Encapsulation was later employed to improve the flavor performance and ensure rapid dissolution upon hydration. A patent was filed to enhance the forthright aroma release in a dry mix hot coffee which was achieved by incorporation of an encapsulated flavor delivery system [9].

Encapsulation also aims to mask undesirable taste and aroma that may appear naturally, or as a processing hazard. They tend to form barriers in the surrounding of the cause behind offensive odor thus hiding the initial perception. Fish oil, though highly recommended for heart disease patients have an inherent, unpleasant odor, which can be masked by encapsulation.

Similarly, canned fish products have also utilized cyclodextrins to encapsulate offending aroma compounds [52]. Encapsulation again may play a role in hiding the bitter taste associated with caffeine that sometimes becomes undesirable.

Oregano essential oil and citronella aroma extract were encapsulated with the help of spray drying using milk proteins such as, skimmed milk powder (SMP) and whey protein concentrate (WPC) [3]. Similarly, complex coacervation was employed for the preparation of microparticles holding beta-pinene using milk proteins (sodium caseinate and whey protein isolate) and CMC. Response surface methodology was used to optimize the process parameters while the study revealed the possibility of encapsulation of beta-pinene using milk proteins-CMC complex [34]. Recent researchers have tried to produce flavored fermented and non-fermented milk using microencapsulated canthaxanthin along with alginate and high methoxyl pectin. These formulated functional dairy products exhibited desirable color [1].

Heat treatments often result in loss of flavor especially during the baking process. Coacervation process was used to encapsulate flavors during baking process, thus providing heat stability and controlled release of the flavor compounds [57]. Ultrasonic encapsulation was employed to encapsulate cinnamaldehyde which is susceptible to evaporation during baking operations thus inhibiting yeast growth. Encapsulation acted as a tool in imparting heat stability to cinnamaldehyde and thereby reducing its interactions with yeast during baking [22].

5.5.2 ENCAPSULATION OF VITAMINS AND MINERALS

Vitamins and minerals are micronutrients that are essential for proper growth and development of our body as their deficiency results in diseases. Common examples include calcium, iron, zinc, vitamins A, B, C, D, etc. Encapsulation techniques can be utilized for efficient delivery of vitamins and minerals through food fortification. The major objective lies in improving the absorption rates of these micronutrients in the body while maintaining their stability in foods and ensuring effective delivery.

Iodine is quite unstable and therefore requires fortification in salt. This may be achieved by microencapsulation of iodine using a spray of fluidized bed drying. Among various encapsulating agents that were used, dextrin encapsulating potassium iodide gave the best results with respect to appearance, stability, and taste [15]. Iron (Ferric chloride) was also encapsulated in a double emulsion system consisting of water-in-oil-in-water

(W/O/W) emulsion. This process worked towards enriching food products with iron by using double emulsion encapsulation [7].

Vitamin C is the most common water-soluble vitamin and has several health benefits. Though it has high stability in the form of powder, decrease in stability could be observed on dissolution in water and is also affected by various environmental factors. It is highly oxidative and may react with other ingredients, resulting in undesirable effects. Various encapsulation techniques have been used over the years to encapsulate vitamin C. In one such research, Vitamin C was encapsulated in tripolyphosphate-chitosan complex using spray drying [14]. Vitamin C was encapsulated in a sugar-free rebaudioside-sweetened model beverage using W/O/W double emulsion. The study exhibited improved stability of the vitamin [30]. Another study involving co-encapsulation of vitamin C and β-Carotene in liposomes using ethanol injection method was performed, resulting in improved stability of β-Carotene which is a precursor of vitamin A [37].

Vitamin A has the tendency to readily degrade in the presence of environmental factors such as light, and high temperatures. The Food Engineering Group at the University of Toronto designed a stabilizing system containing approved food-grade antioxidants in ultra-rice that not only improved vitamin A concentration under extreme storage conditions but also the formulation cost [36]. Vitamin E retention in heated beverages also increased with the help of encapsulation performed on orange O/W emulsion beverages [43].

5.5.3 ENCAPSULATION OF BIOACTIVE COMPOUNDS

Bioactive compounds are simple compounds that are active biologically and provide added nutrients on consumption. They can be classified into various groups, such as micronutrients, phytochemicals, prebiotics, probiotics, and dietary fibers (DFs). The intended use of these bioactive compounds depends hugely on their 'in product' and 'in body' behavior that can be controlled and regulated by the use of appropriate encapsulation systems. Encapsulation provides a protective covering to the unstable bioactive compounds during the preparation, processing, and preservation phases. They encourage the controlled release and efficient dispersion of these compounds within the body especially during mastication and gastrointestinal (GI) digestion thus augmenting their bioaccessibility and bioavailability [16].

Phytochemicals are non-essential nutrients which means, they are not required by the body for sustainable development; however, they are reported to have disease preventing properties [28]. Micronutrients are necessary for the body's health but in small quantities. Potassium, calcium, magnesium, and iron comprise some of the micronutrients. DFs include celluloses, hemicelluloses, lignins, β-glucans, etc. Intake of fibers increases satiety, improves bowel activities, and also reduces the risk of some major chronic diseases.

Coacervation technique was used to encapsulate several bioactives, such as: jasmine essential oil [38]; ascorbic acid [11]; sucralose [47]; and lycopene [46] using gelatin and gum Arabic as encapsulating agents. The encapsulated products thus formed showed resistance to high temperatures. Henceforth, it was established that encapsulated forms were better protected than the free form. Riboflavin was also encapsulated with whey protein and alginate beads to evaluate the release characteristics of the bioactive compound in beverages [60]. Similarly, thyme polyphenols [55] and grape-seed polyphenols [20] were also encapsulated within chitosan microbeads to prolong the release of the encapsulated compound.

Electrospraying technique was also used to encapsulate polyphenols with gelatin as the encapsulating agent. EE obtained was very high, and the antioxidant activity of the bioactive compound was also fully preserved upon encapsulation [21]. Increased bioavailability and efficient delivery systems for polyphenols, especially dietary polyphenols in functional foods could be achieved by nanoparticle-based delivery systems [53]. However, additional research needs to be done to optimize the utilization of encapsulated bioactive compounds thus making them more stable as well as bioavailable both *in vitro* and *in vivo*.

5.5.4 MISCELLANEOUS APPLICATIONS

Encapsulation has stretched its roots in various other applications in the food industry as well, such as the beverage sector, dairy sector, packaging industry and many more. For example, lipids have both difficulty in dispersion and susceptibility to auto-oxidation resulting in the development of undesired flavors and toxic compound. Advantages of lipid encapsulation have already been discussed in this chapter under the section of 'lipids' in this chapter.

Encapsulation of oils and oleoresins has also been reported. An example is the encapsulation of cardamom oleoresin [32], or cumin oleoresin [27] that used various wall materials, such as gum Arabic, maltodextrin, or modified starch. The packaging industry has also exploited the encapsulation process, and several researchers over the years have devised specially designed packages, with smart nanocomposites that can play a crucial role towards food preservation and safety [6]. More recently, the 'nano' form of encapsulation is being adopted that has given better results and has also made possible use of more than one encapsulation technology. However, this process has still not garnered enough attention due to the cost considerations.

5.6 SUMMARY

Encapsulation, for several years now, has tried to establish its foot in the food industry with its numerous attributes such as protection against environmental degradation, chemical reactions, controlled, and on-site delivery, masking of undesirable odor, and improving stability of labile compounds such as bioactives during processing and storage. The food industry, however, demands novel technologies that not only provide nutrition and preservation for the food product till it reaches the consumer but also cost-effectiveness to appeal to the masses. Nowadays, this objective can be attained using more than one technology. When encapsulation employs more than one technology, products with superior characteristics, such as nanoencapsulates or double encapsulated products, are obtained. This also solves the dilemma of encapsulating material selection. This initiative has found more application in the pharmaceutical and cosmetic sector than in the food sector, mainly due to low product yield, low EE, and high costs. However, recent technological advances have succeeded to some extent in improving production, organoleptic properties, food fortification, and the development of novel food products.

KEYWORDS

- **bioactive compounds**
- **carboxymethylcellulose**
- **delivery system**
- **encapsulation**
- **functional foods**
- **nutraceuticals**

REFERENCES

1. Arab, M., Razavi, S. H., Hosseini, S. M., Nayebzadeh, K., Meybodi, N. M., Khanniri, E., Mardi, P., & Mortazavian, A. M., (2019). Production and characterization of functional flavored milk and flavored fermented milk using microencapsulated canthaxanthin. *LWT, 114*, 108373.
2. Augustin, M. A., & Hemar, Y., (2009). Nano-and micro-structured assemblies for encapsulation of food ingredients. *Chemical Society Reviews, 38*(4), 902–912.
3. Baranauskienė, R., Venskutonis, P. R., Dewettinck, K., & Verhé, R., (2006). Properties of oregano (*Origanum vulgare* L.), citronella (*Cymbopogon nardus* G.) and marjoram (*Majorana hortensis* L.) flavors encapsulated into milk protein-based matrices. *Food Research International, 39*(4), 413–425.
4. Bertolini, A. C., Siani, A. C., & Grosso, C. R. F., (2001). Stability of monoterpenes encapsulated in gum Arabic by spray-drying. *Journal of Agricultural and Food Chemistry, 49*(2), 780–785.
5. Blenford, D., (1986). Fully protected food flavor ingredients. *Process Packaging, 8*(8), 43–45.
6. Brody, A. L., Bugusu, B., Han, J. H., Sand, C. K., & McHugh, T. H., (2008). Innovative food packaging solutions. *Journal of Food Science, 73*(8), 107–116.
7. Buyukkestelli, H. I., & El, S. N., (2019). Development and characterization of double emulsion to encapsulate iron. *Journal of Food Engineering, 263*, 446–453.
8. Bylaite, E., Nylander, T., Venskutonis, R., & Jönsson, B., (2001). Emulsification of caraway essential oil in water by lecithin and β-lactoglobulin: Emulsion stability and properties of the formed oil–aqueous interface. *Colloids and Surfaces B: Biointerfaces, 20*(4), 327–340.
9. Ceriali, S., Wragg, A., Gaonkar, A. G., & Zeller, B. L., (2008). *Method of Making Aromatization Particles Containing Coffee Aroma Constituents*. US Patent No. 7,470,443.
10. Cherukuri, S. R., Chau, T. L., Raman, K. P., & Orama, A. M., (1991). *Multiple Encapsulated Flavor Delivery System and Method of Preparation*. US Patent No. 5,004,595.
11. Comunian, T. A., Thomazini, M., Alves, A. J. G., De Matos, Jr. F. E., De Carvalho, B. J. C., & Favaro-Trindade, C. S., (2013). Microencapsulation of ascorbic acid by complex coacervation: Protection and controlled release. *Food Research International, 52*(1), 373–379.
12. De Roos, K. B., (2000). Physiochemical models of flavor release from foods; Chapter 11. In: Roberts, D. D., & Taylor, A. J., (eds.), *Flavor Release* (Vol. 763, pp. 126–141). Washington–DC: ACS Publications.
13. Desai, K. G. H., & Jin, P. H., (2005). Recent developments in microencapsulation of food ingredients. *Drying Technology, 23*(7), 1361–1394.
14. Desai, K. G. H., & Park, H. J., (2005). Encapsulation of vitamin C in tripolyphosphate cross-linked chitosan microspheres by spray drying. *Journal of Microencapsulation, 22*(2), 179–192.
15. Diosady, L. L., Alberti, J. O., & Mannar, M. V., (2002). Microencapsulation for iodine stability in salt fortified with ferrous fumarate and potassium iodide. *Food Research International, 35*(7), 635–642.
16. Donsì, F., Sessa, M., & Ferrari, G., (2013). Nanometric-size delivery systems for bioactive compounds for the nutraceutical and food industries; Chapter 37. In: Bagchi, D., Bagchi, M., Moriyama, H., & Shahidi, F., (eds.), *Bio-Nanotechnology: A Revolution in Food, Biomedical and Health Sciences* (pp. 619–666). Oxford–UK: Wiley-Blackwell.

17. Fenner, E., (2011). *The Next Generation of Ice Cream: One Bite, Two Flavors.* https://munewsarchives.missouri.edu/news-releases/2011/0909-the-next-generation-of-ice-cream-one-bite-two-flavors/ (accessed on 30 May 2022).
18. Gharsallaoui, A., Roudaut, G., Chambin, O., Voilley, A., & Saurel, R., (2007). Applications of spray-drying in microencapsulation of food ingredients: An overview. *Food Research International, 40*(9), 1107–1121.
19. Gibbs, B. F., Kermasha, S., Alli, I., & Mulligan, C. N., (1999). Encapsulation in the food industry: A review. *International Journal of Food Sciences and Nutrition, 50*(3), 213–224.
20. Gibis, M., Ruedt, C., & Weiss, J., (2016). *In-vitro* Release of grape-seed polyphenols encapsulated from uncoated and chitosan-coated liposomes. *Food Research International, 88*, 105–113.
21. Gómez-Mascaraque, L. G., Lagarón, J. M., & López-Rubio, A., (2015). Electrosprayed gelatin submicroparticles as edible carriers for the encapsulation of polyphenols of interest in functional foods. *Food Hydrocolloids, 49*, 42–52.
22. Gong, C., Lee, M. C., Godec, M., Zhang, Z., & Abbaspourrad, A., (2020). Ultrasonic encapsulation of cinnamon flavor to impart heat stability for baking applications. *Food Hydrocolloids, 99*, 105316.
23. Gouin, S., (2004). Microencapsulation: industrial appraisal of existing technologies and trends. *Trends in Food Science & Technology, 15*(7, 8), 330–347.
24. Hyndman, C. L., Groboillot, A. F., Poncelet, D., Champagne, C. P., & Neufeld, R. J., (1993). Microencapsulation of *Lactococcus lactis* within cross-linked gelatin membranes. *Journal of Chemical Technology & Biotechnology, 56*(3), 259–263.
25. Imagi, J., Yamanouchi, T., Okada, K., Tanimoto, M., & Matsuno, R., (1992). Properties of agents that effectively entrap liquid lipids. *Bioscience, Biotechnology, and Biochemistry, 56*(3), 477–480.
26. Jyothi, N. V. N., Prasanna, P. M., Sakarkar, S. N., Prabha, K. S., Ramaiah, P. S., & Srawan, G. Y., (2010). Microencapsulation techniques, factors influencing encapsulation efficiency. *Journal of Microencapsulation, 27*(3), 187–197.
27. Kanakdande, D., Bhosale, R., & Singhal, R. S., (2007). Stability of cumin oleoresin microencapsulated in different combination of gum Arabic, maltodextrin and modified starch. *Carbohydrate Polymers, 67*(4), 536–541.
28. Kanazawa, K., (2011). Bioavailability of non-nutrient for preventing lifestyle-related diseases. *Trends in Food Science & Technology, 22*(12), 655–659.
29. Kenyon, M. M., (1995). Modified starch, maltodextrin, and corn syrup solids as wall materials for food encapsulation. In: Risch, S. J., & Reineccius, G. A., (eds.), *Encapsulation and Controlled Release of Food Ingredients* (Vol. 590, pp. 42–50). ACS symposium series Washington, DC: American Chemical Society.
30. Kheynoor, N., Hosseini, S. M. H., Yousefi, G. H., Gahruie, H. H., & Mesbahi, G. R., (2018). Encapsulation of vitamin C in a rebaudioside-sweetened model beverage using water in oil in water double emulsions. *LWT, 96*, 419–425.
31. Kim, Y. D., & Morr, C. V., (1996). Microencapsulation properties of gum Arabic and several food proteins: Spray-dried orange oil emulsion particles. *Journal of Agricultural and Food Chemistry, 44*(5), 1314–1320.
32. Krishnan, S., Bhosale, R., & Singhal, R. S., (2005). Microencapsulation of cardamom oleoresin: Evaluation of blends of gum Arabic, maltodextrin and a modified starch as wall materials. *Carbohydrate Polymers, 61*(1), 95–102.

33. Knezevic, Z., Gosak, D., Hraste, M., & Jalsenjako, I., (1998). Fluid-bed microencapsulation of ascorbic acid. *Journal of Microencapsulation, 15*(2), 237–252.
34. Koupantsis, T., Pavlidou, E., & Paraskevopoulou, A., (2014). Flavor encapsulation in milk proteins–CMC coacervate-type complexes. *Food Hydrocolloids, 37*, 134–142.
35. Landy, P., Druaux, C., & Voilley, A., (1995). Retention of aroma compounds by proteins in aqueous solution. *Food Chemistry, 54*(4), 387–392.
36. Li, Y. O., Lam, J., Diosady, L. L., & Jankowski, S., (2009). Antioxidant system for the preservation of vitamin a in ultra rice. *Food and Nutrition Bulletin, 30*(1), 82–89.
37. Liu, X., Wang, P., Zou, Y. X., Luo, Z. G., & Tamer, T. M., (2020). Co-encapsulation of vitamin c and β-carotene in liposomes: Storage stability, antioxidant activity, and *In vitro* gastrointestinal digestion. *Food Research International, 136*, 109587.
38. Lv, Y., Yang, F., Li, X., Zhang, X., & Abbas, S., (2014). Formation of heat-resistant nanocapsules of jasmine essential oil via gelatin/gum Arabic based complex coacervation. *Food Hydrocolloids, 35*, 305–314.
39. Matsuno, R., & Adachi, S., (1993). Lipid encapsulation technology-techniques and applications to food. *Trends in Food Science & Technology, 4*(8), 256–261.
40. Meyers, M. A., (2014). Flavor release and application in chewing gum and confections; Chapter 34. In: Gaonkar, A. G., Vashisht, N., Khare, A. R., & Sobel, R., (eds.), *Microencapsulation in the Food Industry: A Practical Implementation Guide* (pp. 443–453). San Diego, CA–USA: Academic Press.
41. Mishra, M., (2015). Overview of encapsulation and controlled release; Chapter 1. In: Mishra, M., (ed.), *Handbook of Encapsulation and Controlled Release* (1st edn., pp. 27–44). Boca Raton: CRC Press.
42. Poshadri, A., & Aparna, K., (2010). Microencapsulation technology: A review. *Journal of Research ANGRAU, 38*(1), 86–102.
43. Raikos, V., (2017). Encapsulation of vitamin E in edible orange oil-in-water emulsion beverages: Influence of heating temperature on physicochemical stability during chilled storage. *Food Hydrocolloids, 72*, 155–162.
44. Risch, S. J., (1995). Encapsulation: Overview of uses and techniques; Chapter 1. In: Risch, S. J., & Reineccius, G. A., (eds.), *Encapsulation and Controlled Release of Food Ingredient* (pp. 2–7). Washington, DC: ACS Publications.
45. Risch, S. J., & Reineccius, G. A., (1995). *Encapsulation and Controlled Release of Food Ingredients* (p. 214). Washington, DC: ACS Publications.
46. Rocha-Selmi, G. A., Favaro-Trindade, C. S., & Grosso, C. R. F., (2013). Morphology, stability, and application of lycopene microcapsules produced by complex coacervation. *Journal of Chemistry, 2013*, 982603.
47. Rocha-Selmi, G. A., Theodoro, A. C., Thomazini, M., Bolini, H. M., & Favaro-Trindade, C. S., (2013). Double emulsion stage prior to complex coacervation process for microencapsulation of sweetener sucralose. *Journal of Food Engineering, 119*(1), 28–32.
48. Roos, Y. H., (2008). The glassy state; Chapter 5. In: Aguilera, J. M., & Lillford, P. J., (eds.), *Food Materials Science: Principles and Practice* (pp. 67–81). Verlag–New York: Springer.
49. Rosenberg, M., & Sheu, T. Y., (1996). Microencapsulation of volatiles by spray-drying in whey protein-based wall systems. *International Dairy Journal, 6*(3), 273–284.
50. Shahidi, F., & Han, X. Q., (1993). Encapsulation of food ingredients. *Critical Reviews in Food Science & Nutrition, 33*(6), 501–547.
51. Slade, L., Levine, H., Ievolella, J., & Wang, M., (1993). The glassy state phenomenon in applications for the food industry: Application of the food polymer science approach

to structure-function relationships of sucrose in cookie and cracker systems. *Journal of the Science of Food and Agriculture, 63*(2), 133–176.

52. Sobel, R., Gundlach, M., & Su, C. P., (2014). Novel concepts and challenges of flavor microencapsulation and taste modification; Chapter 33. In: Gaonkar, A. G., Vashisht, N., Khare, A. R., & Sobel, R., (eds.), *Microencapsulation in the Food Industry: A Practical Implementation Guide* (pp. 421–442). San Diego, CA–USA: Academic Press.
53. Squillaro, T., Cimini, A., Peluso, G., Giordano, A., & Melone, M. A. B., (2018). Nano-delivery systems for encapsulation of dietary polyphenols: An experimental approach for neurodegenerative diseases and brain tumors. *Biochemical Pharmacology, 154*, 303–317.
54. Taylor, A. H., (1983). Encapsulation systems and their applications in the flavor industry. *Food Flavor Ingredients Packaging and Processing, 5*(9), 48–51.
55. Trifković, K. T., Milašinović, N. Z., Djordjević, V. B., Krušić, M. T. K., Knežević-Jugović, Z. D., Nedović, V. A., & Bugarski, B. M., (2014). Chitosan microbeads for encapsulation of thyme (*Thymus serpyllum* L.) Polyphenols. *Carbohydrate Polymers, 111*, 901–907.
56. Tsutsumi, A., Nakamoto, S., Mineo, T., & Yoshida, K., (1995). A novel fluidized-bed coating of fine particles by rapid expansion of supercritical fluid solutions. *Powder Technology, 85*(3), 275–278.
57. Wampler, D. J., & Soper, J. C., (2000). *Aqueous Liquid Flavor Oil Capsules, Method of Making and using in Foods*. EP Patent 0,633,732.
58. Wang, J., Korber, D. R., Low, N. H., & Nickerson, M. T., (2014). Entrapment, survival and release of *Bifidobacterium* adolescentis within chickpea protein-based microcapsules. *Food Research International, 55*, 20–27.
59. Weiss, H. D., & Reynolds, R. G., (1989). *Method of Acidulating a Comminuted Meat Product* (p. 5). US Patent No. 4,803,092.
60. Wichchukit, S., Oztop, M. H., McCarthy, M. J., & McCarthy, K. L., (2013). Whey protein/alginate beads as carriers of a bioactive component. *Food Hydrocolloids, 33*(1), 66–973.
61. Wulff, G., Avgenaki, G., & Guzmann, M. S., (2005). Molecular encapsulation of flavors as helical inclusion complexes of amylose. *Journal of Cereal Science, 41*(3), 239–249.
62. Zuidam, N. J., & Nedovic, V., (2010). *Encapsulation Technologies for Active Food Ingredients and Food Processing* (p. 400). Verlag and New York: Springer.

CHAPTER 6

ENCAPSULATION OF NATURAL POLYPHENOLS FOR FOOD APPLICATIONS

MADHULEKHA RAKSHIT, SOUBHAGYA TRIPATHY, and PREM P. SRIVASTAV

ABSTRACT

The health benefits and immune-boosting the ability of polyphenols have attracted attention in the food (as functional and nutraceuticals food), pharmaceutical, and cosmetic industries, especially during pandemic. However, the application of polyphenols in food matrix is limited because of its low solubility, stability, and bioavailability. Further, astringent taste and strong odor has also been reasons for restricting the incorporation of polyphenols in food. To overcome these drawbacks, encapsulation can be a promising approach for designing appropriate delivery systems. The process of encapsulation involves choosing the correct wall material, which should be food grade, biodegradable, and could form a barrier around the core material.

6.1 INTRODUCTION

Encapsulation is a technique of trapping a solid/liquid/gaseous particle (known as core/active material) in a matrix surrounded by a wall material or coating material [106]. The wall material can be made of sugars, gums, proteins, natural and modified polysaccharides, lipids, and synthetic polymers [66]. Depending upon the size of the encapsulated particles, there are microparticles (with size 100 nm–1,000 μm) and nanoparticles (with size

Novel Processing Methods for Plant-Based Health Foods: Extraction, Encapsulation, and Health Benefits of Bioactive Compounds. Megh R. Goyal, N. Veena & Ritesh B. Watharkar (Eds.)

1–100 nm). The objectives of encapsulation are: (i) protection of active material; (ii) modification of the characteristics of active material for easy transport and storage; (iii) masking off-taste of some polyphenols; (iv) controlling the release of active material in appropriate site in gastrointestinal (GI) tract; (v) separation of two active material which could react with each other [63]. Encapsulation involves the following process:

1. **Identification of Core Material:** Includes understanding the physicochemical characteristics of the active agent such as chemical formula, molecular formula, density, refractive index, melting point, boiling point, diffusion coefficient, solubility, dissociation constant (pK_a) value, etc.
2. **Determination of End Product:** Each of the commercial products has its own properties; hence, it is important to understand the final product properties such as its compositional analysis, physical state, optical state, economic factors, and desired shelf life.
3. **Determination of Wall Material and Encapsulation Technique:** Once the properties of both active agent and end products are well defined, the appropriate wall material and encapsulation technique has to be determined. The selection of wall material and encapsulation technique are dependent on each other, in combination, they will determine the physical property of the particle.
4. **Particle Formation:** The active material is coated with appropriate wall material through any encapsulation technique to obtain desired particle.
5. **Evaluation of Particle Stability:** This is an important step in determining the characteristics of the end product formed and its shelf-life stability. Many analytical tools have been developed to assess the same such as dynamic light scattering systems, scanning electron microscope (SEM), atomic force microscope, zeta sizer, spectrophotometer, disc scanning calorimeter, thermo-gravimetric analyzer, small-angle X-ray scattering, micro CT scanner, etc. Apart from physicochemical characterization of the product, it is essential to determine its stability such as pH, temperature, ionic, light, and oxygen stability. The shelf life of any commercial product is of utmost importance for predicting the performance of the delivery system; therefore, short-term accelerated screening of the end product must be carried out.
6. **Evaluation of Release Properties:** The particle must be able to deliver the active agent to the appropriate site in the GI tract. Thus, *in vitro* and *in vivo* release study should be carried out.

The encapsulation methods can be broadly classified into: physical (e.g., spray drying, lyophilization, supercritical fluid (SF) precipitation, solvent evaporation), physicochemical (e.g., coacervation, liposome, ionic gelation, and emulsion) and chemical (e.g., molecular inclusion complexation) [106].

Polyphenols exhibit wide range of health benefits including anti-cancer, anti-diabetics, anti-osteoporosis, etc. [15, 148, 149, 162]. However, the bioavailability of each polyphenol reduces these effects [63]. This is the big challenge faced by the food industry for incorporation of polyphenol into any food formulation as small amount of polyphenol remain available following the oral administration. Also, the off-flavor of certain polyphenol limits its incorporation into food. Consequently, instability during storage (temperature, pH, light, and oxygen), and low permeability of active agent within appropriate site at GI tract may limit polyphenol availability. Moreover, the challenge lies in ensuring the polyphenol stability throughout the food chain from external degradation factors. Therefore, encapsulation provides a viable solution to protect these polyphenols alongside incorporation into food, pharmaceutical, and cosmetic products [106].

This chapter aims to explore the recent advancements in the various encapsulation technologies (such as: emulsion, spray drying, freeze-drying, spray-chilling, coacervation, extrusion, liposomal formation, and molecular inclusion) to entrap various plant polyphenols.

6.2 POLYPHENOL CHARACTERISTICS

Polyphenols are one of the most abundant and ubiquitous groups of plant metabolites that are an integral part of diets, possessing a high range of biological activities [27, 75, 133]. These compounds range from simple phenolic compound to complex high molecular weight compounds. There are various types of polyphenols, such as: phenolic acid, flavonoids, stilbenes, coumarins, lignans, and tannins. Dietary polyphenols have effectively imparted biological activities and several health benefits against oxidative stress-related diseases. Therefore, numerous health benefits for dietary polyphenols have been reported in reducing neurological, cardiovascular diseases (CVDs), and cancer. It also possesses anti-allergic, antibacterial, anti-hypertensive, skin wound healing, anti-inflammatory, and anti-viral effects.

There are numerous techniques for extraction that have been studied, ranging from conventional extraction (CE) process (such as maceration, solvent extraction, etc.), to improved and innovative techniques (such as: ultrasonic-assisted extraction (UAE), microwave-assisted extraction (MAE), supercritical fluid extraction (SFE), etc.). However, all these techniques work under the same principle where initially the plant tissues swell and hydrate followed by mass transfer of solute from plant materials to the solvent by osmosis and diffusion.

Each method has its own advantages and disadvantages; therefore, an appropriate extraction technique can be chosen depending upon the nature of the target compound. The choice of solvent is an essential factor for any extraction process, and depends upon the characteristics of the desired polyphenol to be extracted. The polarity of the target compound decides the appropriate solvent as polarity is the factor responsible for diffusion of polyphenols [43].

There have been various solvents used for extraction purposes: ethanol, methanol, acetone, diethyl ether, isopropanol, ethyl acetate, and their mixtures with water. Recently, the use of ionic liquids (ILs) and eutectic solvents has been suggested over the traditional solvent in order to avoid pollution and health risk.

Due to safety concerns and immunity boosting effects, there have been demands on natural antioxidants over synthetic ones. Also, there has been increased awareness on "clean label" product; these dietary polyphenols have a pivot role. Moreover, the pandemic issue has led to the increased acceptance of these polyphenols as a part of healthy diet. However, their poor bioavailability through oral administration limits their application in food matrix. Recent studies have revealed encapsulation of polyphenols from various sources with different encapsulation techniques (Table 6.1).

6.3 EXTRUSION TECHNIQUE

Extrusion is a technique of encapsulating polyphenols by dropping the liquid mixture into a gelling bath through a small orifice [182]. The droplet formed is immediately solidified into capsules due to the chemical or physical interactions [21, 116]. This technique is usually used for creating hydrogel.

The most common extrusion is carried out using sodium alginate and calcium chloride. Alginate is widely available natural hydrophilic polymer, derived from brown bacteria and seaweed. Chemical structure comprises of alternative blocks of (1–4)-linked β-D-mannuronic acid (MUA) and α-L-guluronic acid (GUA) [92]. The multivalent ions bind with cavities connecting GUA blocks of alginate molecule [161] to form gel. This structure is termed as "egg-box" [76, 115, 159]. Also, this technique is known

TABLE 6.1 Polyphenols Encapsulated with Different Wall Materials and Encapsulation Techniques

Active Ingredient	Source	Wall Material	Encapsulation Technique	References
Anthocyanin	Black rice	Gum Arabic, whey protein isolate	Ionic gelation	[120]
	Blueberry	Maltodextrin (dextrose equivalent (DE) 20) and hi-maize	Spray drying	[48]
	Black carrot	Chitosan/gelatin	Electro-spraying	[16]
	Blackberry	Maltodextrins with 10 and 20 DE	Freeze drying	[185]
	Pomegranate juice	Gum Arabic and maltodextrin	Freeze drying	[83]
	Pomegranate	Gum Arabic and modified starch	Spray drying	[52]
	Saffron	Gum Arabic and maltodextrin (M7 and M20)	Freeze drying	[88]
	Raspberry	Soy protein isolates	Electro-spraying	[178]
	Blueberry juice	Hydroxypropyl-β-cyclodextrin (HP-β-CD) and β-cyclodextrin (β-CD)	Freeze drying and spray drying	[183]
	Black carrot	Maltodextrin with 10 DE and 20 DE	Freeze drying	[113]
	Hibiscus sabdariff	Sodium alginate, pectin	Ionic gelation	[54]
	Brassica oleracea L. var. *capitata*	Gum Arabic and polydextrose	Spray drying	[28]
Betacyanin and polyphenols	Beetroot leaves and stems	Sodium alginate, sucrose, gum Arabic, and guar gums, and low and high methoxyl pectins	Ionic gelation	[5]
Bitter melon aqueous extract	*Momordica charantia* L.	Maltodextrin and gum Arabic	Spray drying	[164]
Curcumin	–	Poly (D, L-lactic-co-glycolic acid)	Emulsification-solvent evaporation	[121]
	–	Sodium alginate and ZnO	Ionic gelation	[177]
	–	Zein, alginate, and gelatin	Anti-solvent	[186]
	–	γ-Zein hydrolysate	Anti-solvent	[126]
	–	Soy soluble polysaccharide and maltodextrin	Spray drying	[41]

TABLE 6.1 *(Continued)*

Active Ingredient	Source	Wall Material	Encapsulation Technique	References
	–	Soy lecithin and β-sitosterol	Liposome	[163]
	–	Aloe vera mucilage	Spray drying	[107]
	–	Polylactic acid	Electro-spraying	[103]
	–	Lecithin	Anti-solvent	[51]
	–	Amaranth protein isolate and pullulan	Electro-spraying	[30]
	–	Gelatin	Electro-spraying	[68]
Curcuminoids	Turmeric	Eudragit® L100 and Pluronic® 127	Supercritical anti-solvent (SAS)	[12]
Gallic acid	–	Polyvinyl alcohol fibers	Electro-spraying	[44]
	–	Calcium alginate	Electro-spraying	[95]
Gallic acid, catechin, chlorogenic acid, tannic acid	–	Gelatin-coated ι-carrageenan	Extrusion (electric field aided)	[69]
Gentisic acid	–	Sunflower phosphatidylcholines and β-sitosterol	Liposome	[84]
Gingerol	*Zingiber officinale Roscoe*	Maltodextrin and gum Arabica	Spray drying	[151]
Polyphenols	*Bougainvillea spectabilis*	Sodium alginate	Extrusion	[122]
	Thymus serpyllum L	Alginate, chitosan, and inulin	Electrostatic extrusion	[132]
	Olive pomace	L-α-Phosphatidylcholine	Supercritical assisted liposome formation	[172]
	Beet greens	Sodium alginate	Extrusion	[72]
	Taraxacum officinale L. leaf	Sodium alginate, whey protein isolates, cocoa powder, and carob	Ionic gelation	[35]

TABLE 6.1 *(Continued)*

Active Ingredient	Source	Wall Material	Encapsulation Technique	References
	Olive leaf	Whey protein	Electro-spraying	[158]
	Olive leaf	Maltodextrin and trehalose, alone	Freeze drying	[71]
	Blackberry pulp	Arrowroot starch/gum Arabic	Spray drying	[119]
	Yacon leaf	Maize starch	Anti-solvent	[46]
	Fermented tea leaf wastewater	Maltodextrin-gum Arabic	Freeze drying	[112]
	Red pepper waste	Whey protein	Freeze drying	[152]
	Elderberry (*Sambucus nigra* L.)	Modified chitosan, sodium alginate and gum Arabic	Spray drying	[143]
	Olive leaf	Sodium alginate	Spray drying	[70]
	Mango leaves	Polyvinylpyrrolidone	Anti-solvent	[74]
	Sour cherry (*Prunus cerasus* L.)	Gelatin-lactalbumin	Electro-spraying	[81]
	Lemon by product	Maltodextrin-soybean protein	Freeze drying and spray drying	[127]
	Blackberry	Polyvinylpyrrolidone	Spray drying, freeze-drying, SAS	[100]
	Spent coffee powder	Maltodextrin	Freeze drying	[22]
	Camellia sinensis L.	Maltodextrin	Spray drying	[168]
	Camellia sinensis	Soy phosphatidylcholine, cholesterol, and lauroyl polyoxylglycerides	Spray drying	[151]
	Olive pomace	Maltodextrin	Anti-solvent	[10]
	Blackberry	Gum Arabic and polydextrose	Spray drying	[144]
	Grape pomace	Poly lactic-co-glycolic acid	Anti-solvent	[110]
	Olive pomace	Maltodextrin	Spray drying	[125]

TABLE 6.1 *(Continued)*

Active Ingredient	Source	Wall Material	Encapsulation Technique	References
Pro-anthocyanidin	*Cinnamomum zeylanicum*	Gelatin with gum Arabic, pectin, cashew gum, carboxymethylcellulose, and κ-carrageenan	Complex coacervation	[55]
Quercetin	–	β-Lactoglobulin	Anti-solvent	[39]
	–	Polycaprolactone, polyethylene oxide, polylactic acid, and polylactic-co-glycolic acid	Electro-spraying	[62]
	–	Amaranth protein isolate and pullulan ultrathin fibers	Electro-spraying	[3]
Resveratrol and α-tocopherol	–	Heat-denatured whey protein isolate, sodium alginate	Emulsion filled gel	[64]

as ionic gelation because the positive multivalent ionic (such as Ba^{2+}, Ca^{2+}, Al^{3+}) interaction with anionic polymer to trap active material [124].

Similarly, another example of ionic gelation is chitosan tripolyphosphate beads where the interaction of cationic polymer takes place with anionic sodium tripolyphosphate to entrap core material. Zheng et al. [193] studied both types of curcumin filled hydrogel (calcium alginate and chitosan tripolyphosphate) for their stability at pH 3 and 7. Alginate beads proved to be least effective at both 3 and 7 pH, whereas chitosan beads were effective at neutral pH.

Extrusion technique can induce internal gelation and external gelation. In external gelation, the solution of active compound and polymer is dropped into the gelling solution where the gelling agent externally diffuses into the polymeric network to form gel. On the other hand, in internal gelation, emulsion is formed with active compound and polymer, however, gel is formed by acidification [99]. In internal gelation, alginate can be gelled by lowering the pH below the pKa value of the uronic acid residues [61], thus effecting the polymer charge density. The capsules formed by internal gelation were reported to have compact and homogeneous structure with delayed-release.

Alginate beads were prepared to encapsulate insulin by internal gelation [155]. Internal gelation can also be used to simultaneously entrap both hydrophilic (polyphenols from dandelion) polyphenols and lipophilic (β-carotene) compounds because this technique provides ways for trapping oil inside the hydrogel [25]. Internal gelation also yields particles with lesser size due to the emulsion formation [36]. Alginate bead formed by external gelation showed improved digestibility of the loaded active compound. Similarly, the encapsulation efficiency (EE) of >88% was obtained for encapsulation of bougainvillea extract in alginate beads by external gelation [122].

Zazzali et al. [188] studied the effect of pH in the formation of calcium alginate beads. The authors reported that at pH 3.8 (pKa for sodium alginate), the chain interactions increase resulting into a larger particle, thus changing the microscopic (density, size, and interconnectivity) and macroscopic attributes (strength and roundness) of the particle. The other parameters that affect the efficiency of the process are flow rate and needle diameter, whereas the loading efficiency depends upon polymer hydrophilicity, porosity, cross-linking and interaction between core material and polymer [24, 160]. The parameter of extrusion such as sodium alginate (molecular weight, ratio of mannuronate residues to guluronate residues (M/G ratio), and concentration), calcium chloride concentration, and active material concentration influence EE and stability in GI tract [128].

Alginate of different molecular characteristics can also have an effect on EE. Alginate beads with higher molecular weight yielded better EE owing to the fact that higher molecular weight reduces porosity [14]. Alginate with lower M/G ratio and the bead with lower water content showed higher antioxidant activity [38]. Moreover, the freeze dried capsules released lesser content as compared to hydrogel [4, 47]. Higher viscosity polymer solution results in higher density and larger bead with higher EE [26, 45]. However, the shape and size of the droplet is affected by the surface tension, viscosity, and composition of polymer solution [106]. In addition, the storage matrix of hydrogels also plays an important role in determining the retention of active molecule in the capsule. The choice of the correct solution of storage helps in delayed diffusion of active compound through the polymeric gel. The driving force of diffusion can be eliminated by providing a same concentration of active compound in the storage matrix [4]. The hardening solution containing stevia extract lowered the concentration gradient for diffusion, thus, lowering down the losses during encapsulation [14].

Bajpai et al. [20] showed that the retention of anthocyanin in microcapsules formed by extrusion was lower as compared to the microcapsules formed by spray drying. This was attributed to the water solubility of anthocyanin and the porous structure of alginate hydrogel. Therefore, there is a need to formulate hydrogel with higher retention by reducing porosity. Such challenge can be solved by introducing filler material as a part of wall material [90]. The recent literature focuses on the use of filler material in the formulation of hydrogel to provide active material barrier by using natural polymers [14]. A study showed that composite of alginate and chitosan exhibited 6.3- and 2.7-times lower diffusion coefficient, respectively, as compared to plain alginate and alginate-inulin composite particle for encapsulation of thyme extract [132].

The nature of filler material plays an important role in enhancing the EE and the release of core material. Alginate combined with whey proteins and calcium caseinate retained the highest tea polyphenols (up to 80%). On the other hand, chitosan, and pectin improved the polyphenol release profile; however, it had no significant effect on EE. Some proteins like bovine serum albumin (BSA) and whey proteins provided softer and spherical surface to the capsules, whereas some others such as hemp and soy proteins provided harder and larger capsules [26]. Also, whey protein isolate as filler material yielded > 93% EE and cocoa powder and carob yielded > 88% EE [35]. Besides, the addition of cationic polymer in the alginate beads can improve storage stability of the hydrogels [49]. Inorganic compounds such as zinc peroxide (ZnO_2) increased the density of alginate beads, thus, prolonging the release time, and the protection of curcumin from ultraviolet (UV) degradation [177].

There are certain polyuronates that can also form an "egg-box" structure with calcium ion, resulting in gel formation [73], which can be used to encapsulate polyphenols. The ability of polyuronates in the formation of gel depends upon the degree of esterification. The lower degree of esterification makes stronger gel [98]. Huynh et al. [79] studied the ionic gelation of rutin using polygalacturonic acid, amidated low methoxyl pectin, and low methoxyl pectin. The viscoelastic properties of these polycarbonates were correlated for the formation of stable gel. Amidated low methoxyl pectin formed smaller and more flexible gel owing to its lowest viscoelastic moduli and lesser affinity to bind calcium. Thus, it proved to be least effective in holding core material during storage. Gorbunova et al. [72] formulated multilayer alginate beads by dipping the hydrogel into a solution containing active material (betacyanins, betaxanthins, and beet green polyphenols), followed by transferring into calcium chloride bath. The multilayer alginate beads retained a higher amount of core material under simulated gastric conditions than the single layered beads.

The advantages of alginate are that they are simpler in preparation, non-toxic in nature and most importantly, it is cost-effective [189]. Alginate hydrogels formed by external gelation usually varies from 1.0 to 1.2 mm, are shows monomodal distribution, indicating uniform bead size [49]. Larger capsules provide better protection to core material but shows poor dispersion in the food matrix, while, small capsules provide less protection to core material [192].

The major drawback of extrusion is that it yields capsules with higher particle diameter, thus, causing problem in incorporating it into any food formulation [25]. The way of reducing particle size is electrostatic extrusion, or by using jet cutter or atomizer. Electrostatic extrusion is a variation of extrusion that uses an electric field to reduce the particle size of the drop falling from the orifice into the gelling bath [166]. Electrostatic extrusion is based on the same principle as electro-spraying.

Gómez-Mascaraque et al. [69] studied electrostatic extrusion of polyphenols mixed with ι-carrageenan into a gelling bath containing gelatin and calcium chloride. The study showed that the polyphenol with higher molecular weight formed cross-links with gelatin and delayed its release. The EE for smaller molecules (gallic acid, catechin, chlorogenic acid, etc.), was found to be less as they could easily diffuse out of the hydrogel. Rijo et al. [145] studied extrusion encapsulation of the extract from medicinal plants, such as: *Plectranthus grandidentatus*, *Plectranthus porcatus*, *Plectranthus ecklonii*, *Plectranthus ornatus*, and *Plectranthus saccatus,* majorly containing caffeic acid and rosmarinic acid. The authors observed 98.64 and 100.0% EE of rosmarinic acid, and caffeic acid, respectively.

6.4 MOLECULAR INCLUSION

Molecular inclusion is carried out by using β-cyclodextrin (β-CD) with internal hydrophobic part and external hydrophilic part [56]. β-CD is naturally occurring cyclic oligosaccharides derived from starch. This is a means of encapsulating less polar molecules into apolar cavity through hydrophobic interactions. Thus, it can help in increasing the water solubility of hydrophobic molecules [53]. The affinity towards CD is dependent on the size and hydrophobicity of the core material. The ratio of the core material bond to β-CD determines the ability to form stable inclusion complex. Spectral investigation provides an insight into the chemical structure [59].

The process of inclusion complex formation was enhanced by coupling ultrasound for simultaneous extracting and encapsulating polyphenols from *Polygonum cuspidatum* [105]. The study also revealed that the encapsulated polyphenols had higher stability in methanolic extract. The β-CD selectively encapsulated resveratrol and other stilbenes because of the structural confirmation. Several studies have reported CDs to retain flavoring compounds, improved their oxidative stability and provided control release [42, 140]. However, the loading capacity of β-CD is low as compared to its cost [176]. β-CD is thermo-protectant [87].

Nanofiber containing quercetin β-CD inclusion complex was prepared by electrospinning [17]. Moreover, β-CD can also be used to increase the solubility of hydrophobic polyphenols. For example, Mangolim et al. [104] encapsulated curcumin in β-CD and observed higher stability and solubility. Addition to this, the inclusion complex of hesperitine, hesperidin [171], olive leaf extract rich in oleuropein [111], myricetin, kaempferol, quercetin [109], and rutin [62] has been studied to improve the water solubility of these polyphenols. Again, this type of encapsulation is advantageous for the polyphenols that are susceptible to acidity.

Gallic acid loaded poly lactic acid (LA) nanofibers were formulated by first forming an inclusion complex of gallic acid with HP-β-CD, then electrospinning into poly LA. The release of the formulated nanofibers was evaluated in: water, 10% ethanol and 95% ethanol. The nanofiber showed least release of the gallic acid in 95% ethanol [18]. The use of sulfobutyl ether-β-CD sodium in encapsulating tea polyphenols along with chitosan has also been investigated [96]. The authors have reported the use of sulfobutylether-β-CD as cross-linking agent to bind chitosan and tea polyphenol.

6.5 LIPOSOME

Liposome is a concentric spherical structure having two or more phospholipid bilayers surrounding the core material. Due to its amphiphilic nature, it can encapsulate both hydrophobic and hydrophilic molecules simultaneously. These are mainly used as a delivery medium of polyphenols because of its amphiphilic nature, where hydrophilic polyphenol can be entrapped inside the liposome and hydrophobic polyphenol can be entrapped in between the lipid bilayer. Liposomes improve the solubility and the release of polyphenols at the target location. However, sometimes the arrangement of phospholipid in liposome can be affected by its preparation method, conditions, and size of the liposome. This leads to fusion or agglomeration of liposome [129]. This aggregation damage the sensitive nature of liposome limits its wide applications.

Liposomes are classified into small to large vesicles based on their structure (Table 6.2). The size of the liposome can be used to know the half-life. However, the EE is mainly affected by both size and number of bilayer(s) present in the liposome. The unilamellar vesicle (ULV) has a single phospholipid bilayer, whereas the multilamellar vesicle (MLV) has a number of bilayers.

TABLE 6.2 Liposome Types

Class	Sub-Class	Number of Bilayer(s)	Particle Size
Unilamellar vesicles (ULV)	Small ULV	1	20–100 nm
	Medium ULV	1	>100 nm
	Large ULV	1	>100 nm
	Very large ULV	1	>1 μm
Oligolamellar vesicles	–	5	0.1–1 μm
Multilamellar vesicles (MLV)	–	5–25	>0.5 μm
Multi vesicular vesicles	–	Contains more than one ULV	>1 μm

The common liposome preparation methods are mechanical dispersion, solvent dispersion, and detergent removal method. The mechanical dispersion method includes sonication, extrusion, freeze-thawing, and homogenization [78, 142]. For small ULV preparation, sonication is most commonly used method. A bath type or probe-type sonicator is used to sonicate MLV. The solvent dispersion methods include solvent vaporization, ethanol injection and reverse-phase evaporation method. The reverse-phase evaporation method is an evolutionary method as it provides high aqueous area to lipid ratio and can entrap more core material. In the detergent removal method, non-encapsulated materials are removed. This method includes dialysis, gel permeation chromatography, and bio-beads:

1. **Dialysis:** In the dialysis method, a detergent having high critical micelle concentration is used. This detergent helps in solubilizing the lipid. The micelles are combined to form large ULVs after the removal of detergent. The most common detergents used in this method are bile salt and octyl glucoside [11, 50, 89].
2. **Gel Permeation Chromatography:** It can be carried out by allowing the solution to pass through Sephadex G-50 or Sephadex G-100 column. The liposomes pass through the inter bead space at a slow rate, thereby, removing the detergent effectively from the liposome [7].
3. **Bio-Bead:** In this method, the detergent micelle mixture is shaken properly in the presence of beads like bio-beads SM-2 or XAD-2. Here the detergent-like Triton X-100, alkyl glycoside and cholate with very low critical micelle concentration can be removed by absorbing in the beads [29].

Liposomes are also used to deliver a wide variety of substance to a specific location and used as an immunity enhancer. Now a day, cholesterol is replaced by plant sterol as it reduces the animal-based diseases, decrease the harmful cholesterol of body and most importantly it can be consumed by both vegetarians and non-vegetarians. Gentisic acid, generally found in cereal grains, was encapsulated in sunflower lecithin and β-sitosterol [84]. A sterol concentration of 50 mol% showed maximum entrapment efficiency and the low pH of the GI fluid enhanced the release of gentisic acid. Similarly, the β-sitosterol concentration affected the thermal, photo, and physical stability of curcumin loaded liposomes and improved the bioavailability of curcumin by increasing its release [163].

6.6 ELECTRO-SPRAYING

Electro-spraying and electro-spinning have been an effective technique for incorporating polyphenols. Electro-spraying is a technology where high-voltage is applied on a liquid while passing through a small nozzle. The droplets break up to nanoparticles due to the interaction of electric charge developed and the surface tension of the droplet. This technology is applied to increase the stability of the polyphenols by reducing its degradation, and increasing shelf life by reducing exposure to atmospheric oxygen [9]. The electro-spraying technology has advantages over other technologies [32, 118]:

- Emulsion is not required to be prepared; however, this is an optional step;
- In this process particles dry instantaneously. Therefore, further drying is not required;
- It can be done at low temperature unlike spray drying (under Section 6.8 in this chapter);
- More sensitive and susceptible polyphenols can also be processed by this method;
- Smaller particles are produced and they do not coagulate or agglomerate due to their own dispersing characteristics;
- This method is versatile and flexible with respect to various configurations.

The principle of electro-spraying and electro-spinning is based on the principle of charged droplets, as described by Rayleigh [139] and examined by Zeleny [190]; and Taylor [167]. An electro-spraying machine consists of four important components: syringe pump, nozzle, high voltage power source and collector [118]. The sample is dissolved in a volatile solvent and loaded onto a glass syringe, equipped with a flow rate controller. High voltage power source is used to develop certain charge or polarity in the solution. The viscous solution forms small spherical droplets at the tip of the needle due to its surface tension. When the supplied voltage is increased, the surface tension of droplets is reduced and forms a cone, called Taylor's cone. At or above this critical voltage, the particles that are ejected from Taylor's cone carry positive charge. These charged particles are collected at the oppositely charged collector and the solvent is evaporated during this process. The morphology and dimensions of the electro-sprayed particles depends on the concentration of solution, applied voltage, flow rate, tip target distance and needle gauge.

Both direct current (DC) and alternating current (AC) can be used for electro-spraying method. In DC electro-spraying, small micron sized particles (or drops) are formed. The major drawback of DC electro-spraying is that it requires more voltage than AC electro-spraying, i.e., the chances of accident are higher due to high voltage. The particles thus formed are not electrically neutral. This is the reason that the drops are more prone to surface absorption. Moreover, due to electroporation or ionization, the encapsulated drugs are susceptible to destabilization [187]. In AC electro-spraying, the particles are larger and electroneutral. AC electro-spraying is quicker than DC electro-spraying and it can encapsulate organic solvent-soluble compounds as well as liposomes. Since the particles are electroneutral, negligible current passes through it, thus, power requirement is low [187].

Electro-spinning has been widely used to develop controlled release of bioactives [30]. Li et al. [94] encapsulated quercetin in ethyl cellulose nanofibers by electro-spinning. Due to the favorable interaction among the core and coating material, both of them exist in an amorphous physical state because of the formation of hydrogen bond between ethyl cellulose and quercetin. The authors also observed that the particles formed by Teflon nozzle had better ability of sustained release of quercetin than the particle produced from metal nozzle.

Another study by Aceituno-Medina et al. [3] showed that the thermal stability of quercetin decreased when encapsulated in a mixture of amaranth protein isolates and pullulan via ultrathin fibers. This was due to the dispersion of the polyphenols. The encapsulation also efficiently preserved the bioactivity of the quercetin twice than that of free form. Similarly, the oral administration of the encapsulated form of quercetin in polycaprolactone (PCL) and with polyethylene oxide, polylactic acid, or polylactic-co-glycolic acid increased the cytotoxicity towards the MCF-7 cancer cell line [62].

Deldar et al. [57] encapsulated chrysin, a flavanone [131], in PCL and polyethylene glycol to preserve the adipose derived stem cell. Fourier-transform infrared spectroscopy and field emission scanning electron microscopy confirmed that it successfully preserves the adipose derived stem cell. The interaction of phenolics increased when the anthocyanins of red raspberry were encapsulated in soy protein isolate electro spun nanofibers [178]. This was due to the interaction of the hydrogen bond of the phenolics of red raspberry and soy protein isolates. Therefore, higher bioactivity of anthocyanin and antibacterial activity were seen in these encapsulated nanofibers. Black carrot anthocyanin was also encapsulated in the electro-sprayed chitosan-gelatin to study the effect of the chitosan on the structure. A perfect structure was formed due to the electrostatic interaction of the chitosan and gelatin. The release of the bioactive compounds was inversely related to the chitosan content [16].

Isik et al. [81] compared the uniaxial and coaxial electro-spraying method to encapsulate the sour cherry concentrate in gelatin or gelatin-lactalbumin. It was confirmed by the authors that the coaxial method is more effective than a uniaxial method for the encapsulation. Therefore, alginate hydrogel was used to encapsulate the polyphenols to protect and to increase the release of water-soluble phenolics like gallic acid in intestinal fluid [95]. Later, Chuysinuan et al. [44] encapsulated gallic acid in electro-spun polyvinyl alcohol fiber-based hydrogels to study the thermal stability of the gallic acid. This nanofiber enhanced the thermal resistance of the gallic acid up to 200°C, preserved the antioxidant activity and controlled the release of gallic acid. Ferulic acid was incorporated into gliadin fiber by electro-spraying process by Sharif et al. [153, 154]. The nanofiber successfully enhanced the photostability and solubility of ferulic acid. Another polyphenol, curcumin was encapsulated by electro-spraying, which found its application in various ways [135, 136].

In order to enhance the thermal stability and maintain the antioxidant potential of curcumin, Blanco-Padilla et al. [30] encapsulated it in an electrospun amaranth protein isolate and pullulan. The authors observed that the bioaccessibility was increased by 11.3 times than the free curcumin after encapsulation. Similarly, the solubility and bioavailability of the curcumin increased by 38.6 and 11 times, respectively, when encapsulated in gelatin nanofibers [68]. The polylactic acid-curcumin microfiber enhanced the release of curcumin, where the initial burst occurred at 12 h and release continued up to 200 h. It also proved to be a good antioxidant agent with biocompatibility and antibacterial agent towards *Escherichia coli* and *Staphylococcus aureus* [103].

The effect of surfactant on the electro-spun gelatin curcumin nanofibers was studied by Deng et al. [58]. The authors found that the addition of surfactants like Tween 80, anionic sodium dodecyl sulfonate decreased the antioxidant activity of curcumin; however, detergents such as cationic cetyltrimethylammonium bromide did not affect the curcumin antioxidant activity as compared to the control. The EE of the olive leaf phenolics increased when the concentration of whey protein increased from 5 to 30% [158]. The particle size depended upon the concentration of olive leaf phenolics. The phenolics of broccoli showed better morphology and thermal stability when encapsulated in zein [134]. This nanocapsules also showed anti-tumor activity against galia tumor cells without affecting the normal cells.

6.7 COACERVATION

The term coacervation means separation of phase due to the changes in the ionic strength, pH, solubility, temperature of the encapsulating media. In coacervation process two phases exists. The one which is colloids rich is called coacervated process and the other one having a smaller number of colloids called equilibrium phase [114]. This process is of two types: simple coacervation and complex coacervation. In simple coacervation, coacervates are formed due to the involvement of one polymer by the mechanism of dehydration where a coacervating agent like salt is added [19, 101].

However, in case of complex coacervation, the phase separation and formation of coacervates occurs due to the addition of two oppositely charged polymers [150, 184]. For food materials, generally two types of oppositely charged particles are used: protein and polysaccharides. In general, this involves three components like the core material, the wall material, and the solvent. The coacervation process involves [123]:

- Aqueous solution is prepared of more than two polymers above the isoelectric point and gelling temperature;
- The core material is added into the polymer solution followed by homogenization to get a stable emulsion;
- The phase separation occurs by changing the temperature and pH;
- The polymer matrix is hardened by increasing temperature or by adding a coacervating agent/cross-linker.

Initiation of coacervation is done by lowering the temperature or by changing the pH or by adding a coacervation agent. In the beginning of the process, very small particles are formed, increasing the turbidity of the solution. This is called micro-coacervation. This process continues to form larger particles by the aggregation of the small particles. Since the density of the larger particles is higher, therefore, they settle down in the solution, forming distinct two-phases. This is referred as macro-coacervation [170].

According Voorn-Overbeek theory (1957), the electrostatic interaction between the polymers is the driving force for coacervation [33]. Veis and Aranyi theory explained the process in a better way and stated that for low charge density polymers at low temperature, the process of coacervation increased significantly [33, 34]. The Veis and Aranyi theory modified by Tainaka reported that aggregate's attractive force is responsible for phase separation and it is not only applicable to low charge density polymers [33, 102]. The Tainaka theory is considered as the most effective theory but failed to explain the complex coacervation process. The complex coacervation process was studied by Burgess and he showed the effect of ionic strength, pH, and polyionic concentration in the process [34].

The major factor that influences the coacervation process is the electrostatic energy reduction due to the interaction of oppositely charged ions [150]. Besides, pH of medium, ionic strength, temperature, concentration, charge density, polymer ratio and polymer molecular weight are the other factors that affect the coacervation process. Pressure plays an important role when SF is used [150, 174]. The size of the coacervates is regulated by the rate of stirring [94]. To induce the interaction, the charge should be high; however, it should not be too high to precipitate the coacervates [173]. The ionic strength of the solution is affected by the amount of coacervating agent like salt. The electrostatic force between the polymers was decreased by adding a small amount of salt and the dissociation of the coacervates occurs at a high concentration of salt [53, 146].

The pH of the medium affects the degree of ionization of the functional groups; thus, appropriate pH is required for the initiation of the complex coacervates [181]. Another factor that affects the coacervation process is mixing ratio of polymers in the solution. The ionic strength and pH are dependent on the mixing ratio of polymers [181]. Temperature is another factor that affects the coacervation process. At low temperature, the interaction of solvent-solvent, solute-solvent, and solute-solute increases, thereby increasing the coacervation process [33, 157]. According to a study conducted by De Kruif et al. [53] the process of complex coacervation is independent of temperature and depends mainly on the entropy gain. Another study reported that the high polymer concentration prevents the free movement of coacervates by reducing the energy gain and affects negatively on coacervation process [33].

de Souza et al. [55] evaluated the effect of coating material (gelatin with gum Arabic, pectin, cashew gum, carboxymethylcellulose (CMC), and κ-carrageenan) on the cinnamon extract rich in proanthocyanin. They found that the coating materials successfully preserved the bioactivity of the cinnamon extract and masked the undesirable sensory characteristic.

6.8 SPRAY DRYING

Among many encapsulation technologies, spray drying is most used and oldest technology. The major advantage of this technique is low operational cost. Besides, it is a simple, fast, continuous flexible operation with better encapsulation and release of the polyphenols [138]. Broadly, the different wall materials used are: lipid (like glycerides and stearic acid), proteins (like soy protein, casein, and gelatin), and carbohydrates (like gum Arabic, corn syrups, maltrodextrin, and starches) [147]. It is used to encapsulate many polyphenols like anthocyanins [82], phenolic acids, etc.

For encapsulation, the homogenized mixture of core material and wall material is fed into the dryer and is atomized at the nozzle. The hot air aids in the evaporation of water when it comes in contact with the atomized mixture. The powders are then collected in the bottom [138]. The inlet and outlet temperatures are two important factors of spray drying and they should be maintained to get the desired product. At low inlet temperature, evaporation of water is not adequate in a short period and this gives low encapsulation yield (EY). Similarly, cracks are formed at the high inlet temperature [175].

At high outlet temperature, the protein denaturation might occur. Drying temperature and moisture content are the factors that are responsible for the morphological changes occurred during the spray drying [8]. The encapsulated powder dried by conventional spray drying, release the core material when added into the water. Therefore, hydrophobic materials like denatured protein and cross-linked biopolymers are added to delay the release of core compound in water [194].

Both water and oil-soluble material can be encapsulated in the spray drying process. It gives protection to the polyphenols from light, oxygen, and temperature. The shelf life of the powder is more in this case because of its low moisture content [138]. The cooling effect is produced due to rapid evaporation of water and thus, the final droplets are produced at a relatively low temperature than outlet drying gas temperature. Therefore, heat sensitive bioactive core material can also be encapsulated in spray drying technology [13].

The process of spray drying includes heating of drying gas, atomization of the feed mixture at the nozzle end, formation of the particles by drying of the atomized mixture and collection of the powdered particle [13]. In addition, some other factors to be considered during spray drying are:

- Ability of the nozzle to produce smaller particle size;
- A laminar and co-current drying gas flow;
- Efficient collection of the encapsulated particles.

Bernstein et al. [28] studied the effect of drying temperature (140 and 160°C) and concentration of coating material on the characteristics of red cabbage using spray drying and found that there was no effect of increase in temperature on the anthocyanin content of the encapsulated extract. Moreover, the authors found that both gum Arabic and polydextrose (10 and 15%) equally retained the bioactive compound and formed a water-soluble powder. The thermal protection of polyphenols of olive pomace extract increased when encapsulated with maltodextrin (100 g/L) using spray drying. The powders were stable up to 70 days at 5°C. Consequently, when storage temperature increased to 25°C, only 21% degradation of core material occurred; however, 66% loss occurred when exposed to UV light for 48 h [125].

Similarly, the process parameters for encapsulation of bitter melon (*Momordica charantia* L.) were optimized and highest yield (71.4 ± 1.4%) and high antioxidant activity (≥ 87.9 ± 2.6%) was obtained at 140°C inlet temperature and 80°C outlet temperature. These powders were stable at –20°C and 10°C when stored for 150 days [164].

Maltodextrin and gum Arabica are the most used coating material for encapsulation of polyphenols. Gum Arabica and modified starch capsule (1:1) enhanced the retention of anthocyanin from pomegranate. The encapsulated anthocyanin could be stored up to 3 months at 25°C [52]. However, *Zingiber officinale* Roscoe encapsulated with maltodextrin and gum Arabica (4:1) decreased the 6-gingerol quantity after encapsulation [156]. Moreover, the phenolics and anthocyanin retention of blackberry extract increased from 878.32 to 1300.83 mg/100 g and 2106.56 to 2429.22 mg gallic acid equivalent/100 g, respectively, when encapsulated using gum Arabica and polydextrose [144]. Maltodextrin and gum Arabica have proved to be potential coating material for encapsulation of fruits like acerola, by preserving their phenolic compounds even at increased shelf-life [141].

Green tea has many health benefits such as higher total antioxidant, cancer prevention, and anti-irritant. Encapsulation with maltodextrin (40%) with a ratio of core to coating material of 1:2, significantly preserved the antioxidant activity, total phenolics content at optimum spray drying condition [168]. Similarly, maltodextrin (15% w/v) also preserve the bioactivity of Asian pear juice, when encapsulated by spray drying at high inlet air temperature (170°C) [91].

Turkish oregano is a rich source of ursolic, rosmarinic acids and carvacrol. Baranauskaite et al. [23] found that maltodextrin and gum Arabica at a ratio of 8.74:1.26, inlet air temperature of 170°C and ratio of core to wall material of 3:1, preserved the Turkish oregano polyphenols significantly. In another study conducted by Seconlin et al. [151] showed encapsulation of green tea polyphenols in lipid-based soy lecithin cholesterol by spray drying. The effect of wall material composition and drying condition were studied to form water-soluble, low density powder containing high amount of green tea polyphenol [151]. The stability and release of the anthocyanin rich extract of berries such as elderberry and blueberry were improved by encapsulating them with different wall materials like maltodextrin, hi-maize, modified starch, gum Arabic and sodium alginate by spray drying. Their release time ranged from 600 s to 1,140 s under simulated GI condition and storage period was from 230–240 days [48, 143].

Whey protein is another promising coating material for the encapsulation of bioactive compounds like curcumin. The 100% retention of curcumin was achieved when sprayed dried with whey protein [97]. The polyphenol retention of espresso spent coffee was improved by encapsulating with whey protein isolate in combination with maltodextrin, gum Arabica and inulin (1:1:1). The whey protein act as a better wall material for the maintenance of antioxidants [2]. Goëlo et al. [67] suggested that maltodextrin and inulin successfully improved the bioavailability and stability of curcumin in human body.

Aloe vera mucilage is also a good wall material for the retention of the bioactivity of curcumin by preserving the total phenolic content (TPC) and delaying the release of polyphenols up to 65% at 24 h at 150°C inlet air temperature and atomization speed of 27,500 rpm [107]. Olive leaf extract (oleuropein) encapsulated with sodium alginate at a ratio 1:1.6 and inlet air temperature 135°C gave the highest bioaccessibility with 90% release of oleuropein in intestinal conditions [70]. Higher stability of antioxidant was seen during storage when curcumin was encapsulated with skim milk, showing skim milk was suitable for the encapsulation of curcumin [117]. The curcumin showed a good cytotoxicity activity against HepG2 cells and increased bioavailability when spray-dried with soy protein isolate by Chen et al. [41].

6.9 FREEZE DRYING

Freeze-drying is a process for preservation of the thermolabile bioactive compounds by the process of sublimation [138]. This process can preserve the original properties of the materials like shape, size, texture, flavor, color, and bioactivity [37]. After encapsulation, freeze-drying is used to get a stable dry product as it preserves the bioactivity of heat sensitive core material and facilitates easy handling and storage of the final product. A freeze-drying process consists of basically three steps: freezing, primary drying and secondary drying [165].

Freezing is the first step of freeze-drying, where the sample comes in contact with very low temperature where ice crystal starts to form due to the aggregation of the samples [165, 179]. Few parameters that affect the freezing process are eutectic temperature (T_{eu}) and glass transition temperature (T_g). At eutectic point a crystalline mixture formed has same physical properties and at T_g an amorphous material transforms into glass like structure. The T_g depends on moisture content. Therefore, T_g increases during drying as moisture is removed during the process [65, 108, 165]. The change in the structure occurs when the temperature of the process rises above the T_{eu} and T_g [108]. Slow freezing leads to large ice crystal formation and small crystals are formed due to fast freezing [80].

The product is initially kept at a subfreezing temperature (below T_{eu} and above T_g) for proper crystallization, this process is called annealing. This process leads to the growth of ice crystals and accelerates the primary drying [165]. In primary drying, at low temperature and pressure, sublimation of the ice crystal occurs from the top surface. The vapor is removed by diffusion phenomenon. Slow freezing reduces the primary freezing time. This is because the mass transfer is high in slow freezing due to the formation of the larger ice crystal [165]. Ice sublimation is also accelerated by the low pressure as pressure is the driving force for the water vapor removal [80]. The residual water is then removed during the secondary drying period, leading to the process of desorption [179]. The adsorption-desorption equilibrium of moisture is the most important phenomenon of the secondary drying process. Therefore, to get a good quality product both moisture content and temperature needs to be controlled [180].

To get the desirable final product, the process parameters should be maintained. The most important factors that affect the freeze-drying process are surfactant nature, solubility of the core material and the type of cryoprotectant used [40]. Besides, the type of container used, and the thickness of the material also affects the freezing rate. Larger container and thinner material is desirable for proper sublimation of the water [60]. Cryoprotectants are the compounds that are used to protect the material from physical and chemical damage, caused due to the stress that occurred during freezing [1]. The cryoprotectant stabilizes the product by increasing the distance between the hydrocarbon chains. The most common cryoprotectants used are mainly sugars like sucrose, glucose, maltose, and mannitol [191].

The anthocyanins present are unstable form in saffron. Khazaei et al. [88] showed that the stability of anthocyanin present in saffron was increased when encapsulated in gum Arabic and maltodextrin (M7 and M20) and there was no degradation of encapsulated anthocyanin up to 10th week of storage. In another study conducted by Jafari et al. [83] revealed that the total anthocyanin content and color remained the same during 10-week storage when core to wall material used was 1:5. They also studied the effect of different coating materials (cress seed gum, maltodextrin, and gum Arabic) on color parameters. The gum Arabic/maltodextrin mixture successfully retained the color parameters than cress-seed gum/maltodextrin mixture and only maltodextrin.

Berries, especially blackberry and blueberry, has high nutritional value due to the presence of phenolic compounds like anthocyanins. Anthocyanins are sensitive to temperature, light, pH, and oxygen [130]. Wilkowska et al. [183] showed that the encapsulation of anthocyanin rich blueberry juice with HP-β-CD and β-CD followed by freeze-drying could retain 1.5 times more anthocyanin content and antioxidant activity as compared to spray dried anthocyanin loaded inclusion complex. Similarly, anthocyanin rich extract of blackberry by-product was encapsulated with maltodextrins of 10 and 20 DE by freeze-drying, and it was found that the retention of anthocyanin, was higher in maltodextrin with 10 DE than 20 DE [185]. However, black carrot extract retained maximum anthocyanin and antioxidant activity when freeze dried with maltodextrin 20 DE [113].

The retention of polyphenol and flavonoids of spent coffee powder were 62% and 73%, respectively, when freeze dried with maltodextrin. In this process, 73–86% of antioxidant activity was also preserved [22]. Encapsulation by freeze-drying preserved the TPC and total flavonoid content of lemon byproduct. The highest retention was obtained when encapsulated with maltodextrin-soybean protein than only maltodextrin and maltodextrin-carrageenan. Freeze drying also resulted in lower moisture content and water activity than spray drying [127]. The wastewater of Miang (fermented tea leaf) was able to successfully retain its phenolic compounds, such as, gallocatechin, epigallocatechin, catechin, epicatechin, gallocatechin gallate, gallic acid, and caffeine when it was freeze-dried with maltodextrin-gum Arabic (50:50% w/w) mixture. With the increase in weight ratio of Miang wastewater to maltodextrin-gum Arabic, the antioxidant activity was increased. This was due to the higher phenolic concentration gradient at increased solvent volume during the process [112].

Similarly, Ravichai et al. [137] confirmed that the highest retention was found at 10:1 ratio of Miang wastewater to maltodextrin-gum Arabic mixture. The freeze dried encapsulates of red pepper waste encapsulated with whey protein showed better physiological characteristics like moisture content, flow property, color, and solubility. Moreover, after the application of encapsulated red pepper waste in yogurt, there was 71.43% carotenoids retention and 123.73% increase in the polyphenol retention, thereby increasing the sensory and acceptability of red pepper waste bioactive encapsulate fortified yogurt [152]. The thermal property (T_g) of olive leaves extract encapsulated with maltodextrin, and trehalose increased with the increase in maltodextrin concentration, while, olive leaves extract had a plasticizing effect, thus, enhancing the functionality of the freeze-dried powder [71].

6.10 ANTI-SOLVENT METHOD

Anti-solvent technique is also called de-solvation or solvent displacement liquid anti-solvent method. This method was first reported by Bleich and co-workers [31]. Non-solvent is added in the anti-solvent method to induce polymer precipitation. Consequently, supersaturation is caused by changing the solubility of the polymers, which act as a driving force for the formation of microparticles. The supersaturation is followed by nucleation and coagulation [85]. A wide range of materials like bioactive compounds, proteins, polymers, and drugs are used in an anti-solvent method to prepare microparticles [169]. Selection of appropriate solvent and anti-solvent is essential as they should be miscible within a range in which they are used, and the polymer should be insoluble in anti-solvent [85, 86]. The imbalance between anti-solvent, solvent, and solute results in the formation of particles. Therefore, the particles formation depends on the selection of anti-solvent, solvent and and their mixing ratio [6].

The supercritical anti-solvent (SAS) process has been used to formulate microparticles and nanoparticles for encapsulation of the active compounds. The most common SAS, i.e., CO_2 must be soluble in solvent and insoluble with active component. However, encapsulation of lipid-soluble or CO_2 soluble material is not suitable in this method. In the SAS process, the active material mixed with organic solvent is sprayed into CO_2 chamber through a nozzle and diffusion of solvent takes place from solution to CO_2 phase. The solvent is evaporated from the CO_2 phase and the supersaturated solute precipitate as microparticles. The CO_2 is finally removed by depressurization process. The bioavailability of polyphenols increased due to the growth of surface area during supercritical process [6]. This method is suitable for the production of encapsulated microparticles and is a better regulated system for controlling pressure, temperature of CO_2 and flow rate, etc. [86].

SAS process was used to encapsulate virgin coconut oil that showed better EE. The change in pressure from 12 MPa to 16 MPa positively affected the EE and negatively affected the surface oil content [77]. A winery byproduct, grape pomace, contains a high amount of antioxidants, which can be used in any value-added product. In SAS, a good precipitation efficiency of grape pomace (94.4 ± 0.6%) obtained showed higher stability in comparison to the crude extract [110]. Olive pomace is another byproduct containing polyphenols.

Aliakbarian et al. [10] studied the effect of the ratio of maltodextrin content to total solid content of the extract and drying temperature on TPC and antioxidant properties of the powdered product. The authors found that SAS encapsulated phenolic compounds efficiently, thus, these encapsulated powders can be used as a functional component for the production of new food. Similarly, the stability of curcumin loaded liposome nanoparticles fabricated by the anti-solvent method was higher when subjected to thermal treatment, UV irradiation and high ionic strength [51].

Machado et al. [100] compared three encapsulation technique, spray drying, freeze-drying and SAS. For SAS, CO_2 was used as anti-solvent and ethanol as solvent to encapsulate blackberry extract in polyvinylpyrrolidone. They found that SAS achieved higher anthocyanin yield and antioxidant capacity than spray drying and freeze-drying method as supercritical CO_2 did not have any affinity towards the anthocyanin. The stability of antioxidants of mango leaves against light and oxygen increased after encapsulation with polyvinylpyrrolidone by SAS. The release of the mangiferin, quercetin 3-D-galactoside and penta-O-galloyl glucose was also appropriately controlled by the coating material [74].

Quercetin loaded β-lactoglobulin nanoparticles showed a controlled release under simulated gastric fluid at pH 2 and in simulated intestinal fluid. This was attributed to the fact that quercetin-loaded β-lactoglobulin nanoparticles acted as a resistance to the pepsin, thereby, increasing the release in gastro-intestinal condition [39]. The SAS preserved the antioxidant activity of the polyphenol extract of Yacon leaf when encapsulated with maize starch and the study showed no degradation of the extract during encapsulation [46].

6.11 SUMMARY

The core and wall material characteristics and the encapsulation technique influence encapsulation properties, such as active compound retention, antioxidant activity, and stability. The process parameters related to the technique applied and physicochemical characterization of the encapsulated material must be optimized to ensure better core retention. The study of *in vitro* digestion of the encapsulated material is an important aspect for formulation of any encapsulated material. Each method provides unique characteristics to the capsules. Hence, the choice of appropriate technique is one of the factors to influence the property of the end product.

KEYWORDS

- **encapsulation**
- **food application**
- **multilamellar vesicle**
- **polyphenols**
- **wall material**
- **α-L-guluronic acid**

REFERENCES

1. Abdelwahed, W., Degobert, G., & Fessi, H., (2006). A pilot study of freeze-drying of poly (epsilon-caprolactone) nanocapsules stabilized by poly (vinyl alcohol): Formulation and process optimization. *International Journal of Pharmaceutics, 309*(1, 2), 178–188.
2. Abrahão, F. R., Rocha, L. C. R., Santos, T. A., Do Carmo, E. L., Pereira, L. A. S., Borges, S. V., & Botrel, D. A., (2019). Microencapsulation of bioactive compounds from espresso spent coffee by spray drying. *LWT, 103*, 116–124.
3. Aceituno-Medina, M., Mendoza, S., Rodríguez, B. A., Lagaron, J. M., & López-Rubio, A., (2015). Improved antioxidant capacity of quercetin and ferulic acid during in-vitro digestion through encapsulation within food-grade electrospun fibers. *Journal of Functional Foods, 12*, 332–341.
4. Aceval, A. N. D., De Medeiros, P. M., Prudencio, E. S., Olivera, M. C. M., De Mello, C. A. R. D., (2016). Encapsulation of Aqueous leaf extract of *Stevia rebaudiana* bertoni with sodium alginate and its impact on phenolic content. *Food Bioscience, 13*, 32–40.
5. Aguirre, C. T. R., Santagapita, P. R., & Perullini, M., (2019). Functional and structural effects of hydrocolloids on Ca(II)-alginate beads containing bioactive compounds extracted from beetroot. *LWT, 111*, 520–526.
6. Ahmadi, P., Jahanban-Esfahlan, A., Ahmadi, A., Tabibiazar, M., & Mohammadifar, M., (2020). Development of ethyl cellulose-based formulations: A perspective on the novel technical methods. *Food Reviews International, 2020,* 1–48.

7. Akbarzadeh, A., Rezaei-Sadabady, R., Davaran, S., Joo, S. W., Zarghami, N., Hanifehpour, Y., & Nejati-Koshki, K., (2013). Liposome: Classification, preparation, and applications. *Nanoscale Research Letters, 8*(1), 102–112.
8. Alamilla-Beltran, L., Chanona-Perez, J. J., Jimenez-Aparicio, A. R., & Gutierrez-Lopez, G. F., (2005). Description of morphological changes of particles along spray drying. *Journal of Food Engineering, 67*(1, 2), 179–184.
9. Alehosseini, A., Ghorani, B., Sarabi-Jamab, M., & Tucker, N., (2018). Principles of electrospraying: A new approach in protection of bioactive compounds in foods. *Critical Reviews in Food Science and Nutrition, 58*(14), 2346–2363.
10. Aliakbarian, B., Paini, M., Adami, R., Perego, P., & Reverchon, E., (2017). Use of supercritical assisted atomization to produce nanoparticles from olive pomace extract. *Innovative Food Science & Emerging Technologies, 40*, 2–9.
11. Alpes, H., Allmann, K., Plattner, H., Reichert, J., Rick, R., & Schulz, S., (1986). Formation of large unilamellar vesicles using alkyl maltoside detergents. *Biochimica et Biophysica Acta (BBA)-Biomembranes, 862*(2), 294–302.
12. Arango-Ruiz, Á., Martin, Á., Cosero, M. J., Jiménez, C., & Londoño, J., (2018). Encapsulation of curcumin using supercritical antisolvent (SAS) technology to improve its stability and solubility in water. *Food Chemistry, 258*, 156–163.
13. Arpagaus, C., (2019). Production of food bioactive-loaded nanoparticles by nano spray drying; Chapter 4. In: Jafari, S. M., (ed.), *Nanoencapsulation of Food Ingredients by Specialized Equipment* (1st edn., Vol. 3, pp. 151–211). London, United Kingdom: Academic Press.
14. Arriola, N. D. A., Chater, P. I., Wilcox, M., Lucini, L., Rocchetti, G., Dalmina, M., Pearson, J. P., & De Mello, C. A. R. D., (2019). Encapsulation of *Stevia rebaudiana* bertoni aqueous crude extracts by ionic gelation–effects of alginate blends and gelling solutions on the polyphenolic profile. *Food Chemistry, 275*, 123–134.
15. Arts, I. C. W., & Hollman, P. C. H., (2005). Polyphenols and disease risk in epidemiologic studies. *American Journal of Clinical Nutrition, 81*, 317S–325S.
16. Atay, E., Fabra, M. J., Martínez-Sanz, M., Gomez-Mascaraque, L. G., Altan, A., & Lopez-Rubio, A., (2018). Development and characterization of chitosan/gelatin electrosprayed microparticles as food grade delivery vehicles for anthocyanin extracts. *Food Hydrocolloids, 77*, 699–710.
17. Aytac, Z., Ipek, S., Durgun, E., & Uyar, T., (2018). Antioxidant electrospun zein nanofibrous web encapsulating quercetin/cyclodextrin inclusion complex. *Journal of Materials Science, 53*(2), 1527–1539.
18. Aytac, Z., Kusku, S. I., Durgun, E., & Uyar, T., (2016). Encapsulation of gallic acid/cyclodextrin inclusion complex in electrospun polylactic acid nanofibers: Release behavior and antioxidant activity of gallic acid. *Material Science Engineering: C, 63*, 231–239.
19. Bachtsi, A. R., & Kiparissides, C., (1996). Synthesis and release studies of oil-containing poly (vinyl alcohol) microcapsules prepared by coacervation. *Journal of Controlled Release, 38*(1), 49–58.
20. Bajpai, S. K., & Tankhiwale, R., (2006). Investigation of dynamic of vitamin b_2 from calcium alginate/chitosan multilayered beads: Part II. *Reactive and Functional Polymers, 66*(12), 1565–1574.
21. Balanc, B., & Trifkovic, K., (2015). Trends in encapsulation technologies for delivery of food bioactive compounds. *Food Engineering Reviews, 7*(4), 452–490.

22. Ballesteros, L. F., Ramirez, M. J., Orrego, C. E., Teixeira, J. A., & Mussatto, S. I., (2017). Encapsulation of antioxidant phenolic compounds extracted from spent coffee grounds by freeze-drying and spray-drying using different coating materials. *Food Chemistry, 237*, 623–631.
23. Baranauskaite, J., Ivanauskas, L., Masteikova, R., Kopustinskiene, D., Baranauskas, A., & Bernatoniene, J., (2017). Formulation and characterization of Turkish oregano microcapsules prepared by spray-drying technology. *Pharmaceutical Development and Technology, 22*(6), 792–803.
24. Belščak-Cvitanović, A., Stojanović, R., Manojlović, V., Komes, D., Juranović, C. I., Nedović, V., & Bugarski, B., (2011). Encapsulation of polyphenolic antioxidants from medicinal plant extracts in alginate-chitosan system enchanced with ascorbic acid by electrostatic extrusion. *Food Research International, 44*(4), 1094–1101.
25. Belščak-Cvitanovic, A., Bušić, A., Barišić, L., Vrsaljko, D., Karlović, S., Špoljarić, I., Vojvodić, A., et al., (2016). Emulsion templated microencapsulation of dandelion (*Taraxacum officinale* L.) polyphenols and β-carotene by ionotropic gelation of alginate and pectin. *Food Hydrocolloids, 57*, 139–152.
26. Belščak-Cvitanović, A., Dordević, V., Karlović, S., Pavlović, V., Komes, D., Ježek, D., Bugarski, B., & Nedović, V., (2015). Protein-reinforced and chitosan-pectin coated alginate microparticles for delivery of flavan-3-ol antioxidants and caffeine from green tea extract. *Food Hydrocolloids, 51*, 361–374.
27. Bennick, A., (2002). Interaction of plant polyphenols with salivary proteins. *Critical Reviews in Oral Biology & Medicine, 13*, 184–196.
28. Bernstein, A., & Noreña, C. P. Z., (2015). Encapsulation of red cabbage (*Brassica oleracea* L. var. Capitata L. f. rubra) anthocyanins by spray drying using different encapsulating agents. *Brazilian Archives of Biology and Technology, 58*(6), 944–952.
29. Betageri, G. V., Jenkins, S. A., & Parsons, D., (1993). *Liposome Drug Delivery Systems* (1st edn., p. 148). Boca Raton: CRC Press.
30. Blanco-Padilla, A., López-Rubio, A., Loarca-Pina, G., Gómez-Mascaraque, L. G., & Mendoza, S., (2015). Characterization, release and antioxidant activity of curcumin-loaded amaranth-pullulan electrospun fibers. *LWT-Food Science and Technology, 63*(2), 1137–1144.
31. Bleich, J., & Müller, B. W., (1996). Production of drug loaded microparticles by the use of supercritical gases with the aerosol solvent extraction System (ASES) process. *Journal of Microencapsulation, 13*(2), 131–139.
32. Bock, N., Dargaville, T. R., & Woodruff, M. A., (2012). Electrospraying of polymers with therapeutic molecules: State of the art. *Progress in Polymer Science, 37*(11), 1510–1551.
33. Burgess, D. J., (1994). Complex coacervation: Microcapsule formation. In: Dubin, P., Bock, J., Davis, R., Schulz, D. N., & Thies, C., (eds.), *Macromolecular Complexes in Chemistry and Biology* (pp. 285–300). Berlin–Heidelberg: Springer.
34. Burgess, D. J., (1990). Practical analysis of complex coacervate systems. *Journal of Colloid and Interface Science, 140*(1), 227–238.
35. Bušić, A., Belščak-Cvitanović, A., Vojvodić, C. A., Karlović, S., Kovač, V., Špoljarić, I., Mršić, G., & Komes, D., (2018). Structuring new alginate network aimed for delivery of dandelion (*Taraxacum officinale* L.) polyphenols using ionic gelation and new filler materials. *Food Research International, 111*, 244–255.
36. Capela, P., Hay, T. K. C., & Shah, N. P., (2007). Effect of homogenization on bead size and survival of encapsulated probiotic bacteria. *Food Research International, 40*, 1261–1269.

37. Ceballos, A. M., Giraldo, G. I., & Orrego, C. E., (2012). Effect of freezing rate on quality parameters of freeze-dried soursop fruit pulp. *Journal of Food Engineering, 111*, 360–365.
38. Chan, E. S., Yim, Z. H., Phan, S. H., Mansa, R. F., & Ravindra, P., (2010). Encapsulation of herbal aqueous extract through absorption with ca-alginate hydrogel beads. *Food Bioproduct Process, 88*, 195–201.
39. Chavoshpour-Natanzi, Z., & Sahihi, M., (2019). Encapsulation of quercetin-loaded β-lactoglobulin for drug delivery using modified anti-solvent method. *Food Hydrocolloids, 96*, 493–502.
40. Chen, C., Han, D., Cai, C., & Tang, X., (2010). An overview of liposome lyophilization and its future potential. *Journal of Controlled Release, 142*(3), 299–311.
41. Chen, F. P., Liu, L. L., & Tang, C. H., (2020). Spray-drying microencapsulation of curcumin nanocomplexes with soy protein isolate: Encapsulation, water dispersion, bioaccessibility and bioactivities of curcumin. *Food Hydrocolloids, 105*, 105821.
42. Choi, M. J., Soottitantawat, A., Nuchuchua, O., Min, S. G., & Ruktanonchai, U., (2009). Physical and light oxidative properties of eugenol encapsulated by molecular inclusion and emulsion-diffusion method. *Food Research International, 42*(1), 148–156.
43. Chua, L. S., (2013). A review on plant-based rutin extraction methods and its pharmacological activities. *Journal of Ethnopharmacology, 150*(3), 805–817.
44. Chuysinuan, P., Thanyacharoen, T., Techasakul, S., & Ummartyotin, S., (2018). Electrospun characteristics of gallic acid-loaded poly vinyl alcohol fibers: Release characteristics and antioxidant properties. *Journal of Science: Advanced Materials and Devices, 3*(2), 175–180.
45. Crittenden, R., Weerakkody, R., Sanguansri, L., & Augustin, M., (2006). Synbiotic microcapsules that enhance microbial viability during nonrefrigerated storage and gastrointestinal transit. *Applied Environmental Microbiology, 72*(3), 2280–2282.
46. Cruz, P. N., Reis, P. M. L., Ferreira, S. R., Masson, M. L., & Corazza, M. L., (2020). Encapsulation of yacon (*Smallanthus sonchifolius*) leaf extract by supercritical fluid extraction of emulsions. *The Journal of Supercritical Fluids, 160*, 104815.
47. Ćujić, N., Trifković, K., Bugarski, B., Ibrić, S., Pljevljakušić, D., & Šavikin, K., (2016). Chokeberry (*Aronia melanocarpa* L.) extract loaded in alginate and alginate/inulin system. *Industrial Crops and Products, 86*, 120–131.
48. Da Rosa, J. R., Nunes, G. L., Motta, M. H., Fortes, J. P., Weis, G. C. C., Hecktheuer, L. H. R., & Da Rosa, C. S., (2019). Microencapsulation of anthocyanin compounds extracted from blueberry (*Vaccinium* spp.) by spray drying: Characterization, stability and simulated gastrointestinal conditions. *Food Hydrocolloids, 89*, 742–748.
49. Da Silva, C. A. G., Da Costa, M. M. T., De Freitas, Q. B., H. D., Cazarin, C. B. B., Maróstica, Jr. M. R., & Hubinger, M. D., (2019). Anthocyanins from Jussara (*Euterpe edulis* martius) extract carried by calcium alginate beads pre-prepared using ionic gelation. *Powder Technology, 345*, 283–291.
50. Daemen, T., Hofstede, G., Ten, K. M. T., Bakker-Woudenberg, I. A., & Scherphof, G. L., (1995). Liposomal doxorubicin-induced toxicity: Depletion and impairment of phagocytic activity of liver macrophages. *International Journal of Cancer, 61*(5), 716–721.
51. Dai, L., Sun, C., Li, R., Mao, L., Liu, F., & Gao, Y., (2017). Structural characterization, formation mechanism and stability of curcumin in zein-lecithin composite nanoparticles fabricated by antisolvent co-precipitation. *Food Chemistry, 237*, 1163–1171.
52. De Araujo, S. M. C. P., Nogueira, R. I., Paim, D. R. S. F., Gouvêa, A. C. M. S., De Oliveira, G. R. L., Peixoto, F. M., & Freitas, S. P., (2016). Effects of encapsulating

agents on anthocyanin retention in pomegranate powder obtained by the spray drying process. *LWT, 73*, 551–556.
53. De Kruif, C. G., Weinbreck, F., & De Vries, R., (2004). Complex coacervation of proteins and anionic polysaccharides. *Current Opinion in Colloid & Interface Science, 9*(5), 340–349.
54. De Moura, S. C. S. R., Berling, C. L., Germer, S. P. M., Alvim, I. D., & Hubinger, M. D., (2018). Encapsulating anthocyanins from *hibiscus sabdariffa* L. calyces by ionic gelation: Pigment stability during storage of microparticles. *Food Chemistry, 241*, 317–327.
55. De Souza, V. B., Thomazini, M., Barrientos, M. A. E., Nalin, C. M., Ferro-Furtado, R., Genovese, M. I., & Favaro-Trindade, C. S., (2018). Functional properties and encapsulation of a proanthocyanidin-rich cinnamon extract (*Cinnamomum zeylanicum*) by complex coacervation using gelatin and different polysaccharides. *Food Hydrocolloids, 77*, 297–306.
56. Del Valle, E. M., (2004). Cyclodextrins and their uses: A review. *Process Biochemistry, 39*(9), 1033–1046.
57. Deldar, Y., Zarghami, F., Pilehvar-Soltanahmadi, Y., Dadashpour, M., & Zarghami, N., (2017). Antioxidant effects of chrysin-loaded electrospun nanofibrous mats on proliferation and stemness preservation of human adipose-derived stem cells. *Cell and Tissue Banking, 18*(4), 475–487.
58. Deng, L., Kang, X., Liu, Y., Feng, F., & Zhang, H., (2017). Effects of surfactants on the formation of gelatin nanofibres for controlled release of curcumin. *Food Chemistry, 231*, 70–77.
59. Ding, H., Chao, J., Zhang, G., Shuang, S., & Pan, J., (2003). Preparation and spectral investigation on inclusion complex of β-cyclodextrin with rutin. *Spectrochimica Acta Part A: Molecular and Biomolecular Spectroscopy, 59*(14), 3421–3429.
60. Do Vale, M. A. R., Do Nascimento, A. É., Júnior, F. H. X., De Oliveira, C. M., Marcelino, H. R., Barratt, G., & Elaissari, A., (2016). Freeze-drying of emulsified systems: A review. *International Journal of Pharmaceutics, 503*(1, 2), 102–114.
61. Draget, K. I., & Taylor, C., (2011). Chemical, physical and biological properties of alginates and their biomedical implications. *Food Hydrocolloids, 25*(2), 251–256.
62. Eskitoros-Togay, Ş. M., Bulbul, Y. E., & Dilsiz, N., (2018). Quercetin-loaded and unloaded electrospun membranes: Synthesis, characterization and *in vitro* release study. *Journal of Drug Delivery Science and Technology, 47*, 22–30.
63. Fang, Z., & Bhandari, B., (2010). Encapsulation of polyphenols: A review. *Trends in Food Science & Technology, 21*(10), 510–523.
64. Feng, W., Yue, C., Wusigale, Ni, Y., & Liang, L., (2018). Preparation and characterization of emulsion-filled gel beads for the encapsulation and protection of resveratrol and α-tocopherol. *Food Research International, 108*, 161–171.
65. Fissore, D., Pisano, R., & Barresi, A. A., (2011). Advanced approach to build the design space for the primary drying of a pharmaceutical freeze-drying process. *Journal of Pharmaceutical Sciences, 100*(11), 4922–4933.
66. Gibbs, B. F., Kermasha, S., Alli, I., & Mulligan, C. N., (1999). Encapsulation in the food industry: A review. *International Journal of Food Sciences and Nutrition, 50*, 213–224.
67. Goëlo, V., Chaumun, M., Gonçalves, A., Estevinho, B. N., & Rocha, F., (2020). Polysaccharide-based delivery systems for curcumin and turmeric powder encapsulation using a spray-drying process. *Powder Technology, 370*, 137–146.
68. Gómez-Estaca, J., Gavara, R., & Hernández-Muñoz, P., (2015). Encapsulation of curcumin in electrosprayed gelatin microspheres enhances its bioaccessibility and

widens its uses in food applications. *Innovative Food Science & Emerging Technologies, 29*, 302–307.
69. Gómez-Mascaraque, G. L., Martínez-Sanz, M., Fabra, M. J., & López-Rubio, A., (2019). Development of gelatin-coated ι-carrageenan hydrogel capsules by electric field-aided extrusion. Impact of phenolic compounds on their performance. *Food Hydrocolloids, 90*, 523–533.
70. González, E., Gómez-Caravaca, A. M., Giménez, B., Cebrián, R., Maqueda, M., Martínez-Férez, A., & Robert, P., (2019). Evolution of the phenolic compounds profile of olive leaf extract encapsulated by spray-drying during *in vitro* gastrointestinal digestion. *Food Chemistry, 279*, 40–48.
71. González-Ortega, R., Faieta, M., Di Mattia, C. D., Valbonetti, L., & Pittia, P., (2020). Microencapsulation of olive leaf extract by freeze-drying: Effect of carrier composition on process efficiency and technological properties of the powders. *Journal of Food Engineering, 285*, 110089.
72. Gorbunova, N., Bannikova, A., Evteev, A., Evdokimov, I., & Kasapis, S., (2018). Alginate-based encapsulation of extracts from beta vulgaris cv. beet greens: Stability and controlled release under simulated gastrointestinal conditions. *LWT, 93*, 442–449.
73. Grant, G. T., Morris, E. R., Rees, D. A., Smith, P. J., & Thom, D., (1973). Biological interactions between polysaccharides and divalent cations: The egg-box model. *FEBS Letters, 32*(1), 195–198.
74. Guamán-Balcázar, M. C., Montes, A., Pereyra, C., & De La Ossa, E. M., (2019). Production of submicron particles of the antioxidants of mango leaves/PVP by supercritical antisolvent extraction process. *The Journal of Supercritical Fluids, 143*, 294–304.
75. Haslam, E., (1996). Natural polyphenols (vegetable tannins) as drugs: Possible modes of action. *Journal of Natural Product, 59*, 205–215.
76. He, X., Liu, Y., Li, H., & Li, H., (2016). Single-stranded structure of alginate and its conformation evolvement after an interaction with calcium ions as revealed by electron microscopy. *RSC Advances, 6*(115), 114779–114782.
77. Hee, Y. Y., Rahman, R. A., Adzahan, N. M., & Chonga, G. H., (2016). Effect of pressure on the microencapsulation of virgin coconut oil by supercritical anti-solvent (SAS). In: *3rd International Conference on Agricultural and Food Engineering* (pp. 416–421). Kuala Lumpur–Malaysia: Universiti Putra Malaysia.
78. Himanshu, A., Sitasharan, P., & Singhai, A. K., (2011). Liposomes as drug carriers. *International Journal of Pharmacy and Life Sciences, 2*(7), 945–951.
79. Huynh, U. T. D., Assifaoui, A., & Chambin, O., (2017). Pellets based on polyuronates: Relationship between gelation and release properties. *Journal Food Engineering, 199*, 27–35.
80. Ingvarsson, P. T., Yang, M., Nielsen, H. M., Rantanen, J., & Foged, C., (2011). Stabilization of liposomes during drying. *Expert Opinion on Drug Delivery, 8*(3), 375–388.
81. Isik, B. S., Altay, F., & Capanoglu, E., (2018). The uniaxial and coaxial encapsulations of sour cherry (*Prunus cerasus* L.) concentrate by electrospinning and their *in vitro* bioaccessibility. *Food Chemistry, 265*, 260–273.
82. Jafari, S. M., Ghalenoei, M. G., & Dehnad, D., (2017). Influence of spray drying on water solubility index, apparent density, and anthocyanin content of pomegranate juice powder. *Powder Technology, 311*, 59–65.
83. Jafari, S. M., Mahdavi-Khazaei, K., & Hemmati-Kakhki, A., (2016). Microencapsulation of saffron petal anthocyanins with cress seed gum compared with Arabic gum through freeze drying. *Carbohydrate Polymers, 140*, 20–25.

84. Jovanović, A. A., Balanč, B. D., Djordjević, V. B., Ota, A., Skrt, M., Šavikin, K. P., & Ulrih, N. P., (2019). Effect of gentisic acid on the structural-functional properties of liposomes incorporating β-sitosterol. *Colloids and Surfaces B: Biointerfaces, 183*, 110422.
85. Joye, I. J., & McClements, D. J., (2013). Production of nanoparticles by anti-solvent precipitation for use in food systems. *Trends in Food Science & Technology, 34*, 109–123.
86. Kalani, M., & Yunus, R., (2011). Application of supercritical antisolvent method in drug encapsulation: A review. *International Journal of Nanomedicine, 6*, 1429–1442.
87. Kalogeropoulos, N., Yannakopoulou, K., Gioxari, A., Chiou, A., & Makris, D. P., (2010). Polyphenol characterization and encapsulation in β-cyclodextrin of a flavonoid-rich *Hypericum perforatum* (St John's Wort) extract. *LWT-Food Science and Technology, 43*, 882–889.
88. Khazaei, K. M., Jafari, S. M., Ghorbani, M., & Kakhki, A. H., (2014). Application of maltodextrin and gum Arabic in microencapsulation of saffron petal's anthocyanins and evaluating their storage stability and color. *Carbohydrate Polymers, 105*, 57–62.
89. Kirby, C. J., & Gregoriadis, G., (1984). A simple procedure for preparing liposomes capable of high encapsulation efficiency under mild conditions. *Liposome Technology, 1*, 19–27.
90. Kurozawa, L. E., & Hubinger, M. D., (2017). Hydrophilic food compounds encapsulation by ionic gelation. *Current Opinion in Food Science, 15*, 50–55.
91. Lee, C. G., Ahmed, M., Jiang, G. H., & Eun, J. B., (2017). Color, bioactive compounds and morphological characteristics of encapsulated Asian pear juice powder during spray drying. *Journal of Food Science and Technology, 54*(9), 2717–2727.
92. Lee, K. Y., & Mooney, D. J., (2012). Alginate: Properties and biomedical applications. *Progress in Polymer Science, 37*, 106–126.
93. Lemetter, C. Y. G., Meeuse, F. M., & Zuidam, N. J., (2009). Control of the morphology and the size of complex coacervate microcapsules during scale-up. *AIChE Journal, 55*(6), 1487–1496.
94. Li, C., Wang, Z. H., & Yu, D. G., (2014). Higher quality quercetin sustained release ethyl cellulose nanofibers fabricated using a spinneret with a Teflon nozzle. *Colloids and Surfaces B: Biointerfaces, 114*, 404–409.
95. Li, J., Kim, S. Y., Chen, X., & Park, H. J., (2016). Calcium-alginate beads loaded with gallic acid: Preparation and characterization. *LWT-Food Science and Technology, 68*, 667–673.
96. Liu, F., Antoniou, J., Li, Y., Majeed, H., Liang, R., Ma, Y., Ma, J., & Zhong, F., (2016). Chitosan/sulfobutylether-β-cyclodextrin nanoparticles as a potential approach for tea polyphenol encapsulation. *Food Hydrocolloids, 57*, 291–300.
97. Liu, W., Chen, X. D., Cheng, Z., & Selomulya, C., (2016). On enhancing the solubility of curcumin by microencapsulation in whey protein isolate via spray drying. *Journal of Food Engineering, 169*, 189–195.
98. Lootens, D., Capel, F., Durand, D., Nicolai, T.,, Boulenguer, P., & Langendorff, V., (2003). Influence of pH, Ca concentration, temperature and amidation on the gelation of low methoxyl pectin. *Food Hydrocolloids, 17*, 237–244.
99. Lupo, B., Maestro, A., Gutiérrez, J. M., & González, C., (2015). Characterization of alginate beads with encapsulated cocoa extract to prepare functional food: Comparison of two gelation mechanisms. *Food Hydrocolloids, 49*, 25–34.

100. Machado, A. P. D., Rezende, C. A., Rodrigues, R. A., Barbero, G. F., Rosa, P. D. V. E., & Martínez, J., (2018). Encapsulation of anthocyanin-rich extract from blackberry residues by spray-drying, freeze-drying and supercritical antisolvent. *Powder Technology, 340*, 553–562.
101. Madan, P. L., (1978). Microencapsulation I. phase separation or coacervation. *Drug Development and Industrial Pharmacy, 4*(1), 95–116.
102. Magnin, D., & Dumitriu, S., (2004). Interactions between polysaccharides and polypeptides; Chapter 12. In: Dumitriu, S., (ed.), *Polysaccharides: Structural Diversity and Functional Versatility* (2nd edn., pp. 305–356). Boca Raton–FL: CRC Press.
103. Mai, Z., Chen, J., He, T., Hu, Y., Dong, X., Zhang, H., & Zhou, W., (2017). Electrospray biodegradable microcapsules loaded with curcumin for drug delivery systems with high bioactivity. *RSC Advances, 7*(3), 1724–1734.
104. Mangolim, C. S., Moriwaki, C., Nogueira, A. C., Sato, F., Baesso, M. L., & Neto, A. M., (2014). Curcumineb-cyclodextrin inclusion complex: Stability, solubility, characterization by FT-IR, FT-Raman, x-ray diffraction and photoacoustic spectroscopy, and food application. *Food Chemistry, 153*, 361–370.
105. Mantegna, S., Binello, A., Boffa, L., Giorgis, M., Cena, C., & Cravotto, G., (2012). A one-pot ultrasound-assisted water extraction/cyclodextrin encapsulation of resveratrol from *Polygonum cuspidatum*. *Food Chemistry, 130*, 746–750.
106. Marisa, R. A., Estevinho, B. N., & Rocha, F., (2019). Microencapsulation of polyphenols-the specific case of the microencapsulation of *Sambucus nigra* L. extracts: A review. *Trends in Food Science & Technology, 105*, 454–467.
107. Medina-Torres, L., Núñez-Ramírez, D. M., Calderas, F., Bernad-Bernad, M. J., Gracia-Mora, J., Rodríguez-Ramírez, J., González-Laredo, R., et al., (2019). Curcumin encapsulation by spray drying using *Aloe vera* mucilage as encapsulating agent. *Journal of Food Process Engineering, 42*(2), e12972.
108. Meister, E., & Gieseler, H., (2009). Freeze-dry microscopy of protein/sugar mixtures: Drying behavior, interpretation of collapse temperatures and a comparison to corresponding glass transition data. *Journal of Pharmaceutical Sciences, 98*(9), 3072–3087.
109. Mercader-Ros, M. T., Lucas-Abellán, C., Fortea, M. I., Gabaldón, J. A., & Núñez-Delicado, E., (2010). Effect of HP-β-cyclodextrins complexation on the antioxidant activity of flavonols. *Food Chemistry, 118*, 769–773.
110. Mezzomo, N., Oliveira, D. A., Comim, S. R. R., & Ferreira, S. R. S., (2016). Encapsulation of extract from winery industry residue using the supercritical anti-solvent technique. *Brazilian Journal of Chemical Engineering, 33*(3), 589–598.
111. Mourtzinos, I., Salta, F., Yannakopoulou, K., Chiou, A., & Karathanos, V. T., (2017). Encapsulation of olive leaf extract in β-cyclodextrin. *Journal of Agricultural and Food Chemistry, 55*, 8088–8094.
112. Muangrat, R., Ravichai, K., & Jirarattanarangsri, W., (2019). Encapsulation of polyphenols from fermented wastewater of miang processing by freeze drying using a maltodextrin/gum Arabic mixture as coating material. *Journal of Food Processing and Preservation, 43*(4), e13908.
113. Murali, S., Kar, A., Mohapatra, D., & Kalia, P., (2015). Encapsulation of black carrot juice using spray and freeze drying. *Food Science and Technology International, 21*(8), 604–612.
114. Nairm, J. G., (1995). Coacervation-phase separation technology; Chapter 3. In: Ganderton, D., Jones, T., & McGinity, J., (eds.), *Advances in Pharmaceutical Sciences* (Vol. 7; pp. 93–219). London–UK: Academic Press.

115. Narayanan, R. P., Melman, G., Letourneau, N. J., Mendelson, N. L., & Melman, A., (2012). Photodegradable iron (III) cross-linked alginate gels. *Biomacromolecules, 13*, 2465–2471.
116. Nedovic, V., Kalusevic, A., Manojlovic, V., & Levic, S., (2011). An overview of encapsulation technologies for food applications. *Italian Oral Surgery, 1*, 1806–1815.
117. Neves, M. I. L., Desobry-Banon, S., Perrone, I. T., Desobry, S., & Petit, J., (2019). Encapsulation of curcumin in milk powders by spray-drying: Physicochemistry, rehydration properties, and stability during storage. *Powder Technology, 345*, 601–607.
118. Nguyen, D. N., Clasen, C., & Van, D. M. G., (2016). Pharmaceutical applications of electrospraying. *Journal of Pharmaceutical Sciences, 105*(9), 2601–2620.
119. Nogueira, G. F., Soares, C. T., Martin, L. G. P., Fakhouri, F. M., & De Oliveira, R. A., (2020). Influence of spray drying on bioactive compounds of blackberry pulp microencapsulated with arrowroot starch and gum Arabic mixture. *Journal of Microencapsulation, 37*(1), 65–76.
120. Norkaew, O., Thitisut, P., Mahatheeranont, S., Pawin, B., Sookwong, P., Yodpitak, S., & Lungkaphin, A., (2019). Effect of wall materials on some physicochemical properties and release characteristics of encapsulated black rice anthocyanin microcapsules. *Food Chemistry, 294*, 493–502.
121. Ong, Y. X. J., Lee, L. Y., Davoodi, P., & Wang, C. H., (2018). Production of drug-releasing biodegradable microporous scaffold using a two-step micro-encapsulation/supercritical foaming process. *The Journal of Supercritical Fluids, 133*, 263–269.
122. Orozco-Villafuerte, J., Escobar-Rojas, A., Buendía-González, L., García-Morales, C., Hernandez-Jaimes, C., & Alvarez-Ramirez, J., (2019). Evaluation of the protection and release rate of bougainvillea (*Bougainvillea spectabilis*) extracts encapsulated in alginate beads. *Journal of Dispersion Science and Technology, 40*, 1065–1074.
123. Ortega-Rivas, E., Juliano, P., & Yan, H., (2005). *Food Powders: Physical Properties, Processing, and Functionality* (p. 372). Verlag–US: Springer.
124. Ozkan, G., Franco, P., De Marco, I., Xiao, J., & Capanoglu, E., (2019). A review of microencapsulation methods for food antioxidants: Principles, advantages, drawbacks and applications. *Food Chemistry, 272*, 494–506.
125. Paini, M., Aliakbarian, B., Casazza, A. A., Lagazzo, A., Botter, R., & Perego, P., (2015). Microencapsulation of phenolic compounds from olive pomace using spray drying: A study of operative parameters. *LWT-Food Science and Technology, 62*(1), 177–186.
126. Pan, R., Zou, Y., Wang, J., Wan, Z., Guo, J., Yang, J., & Yang, X., (2018). Gamma/alpha-zein hydrolysates as oral delivery vehicles: Enhanced physicochemical stability and *in vitro* bioaccessibility of curcumin. *International Journal Food Science and Technology, 53*, 1622–1630.
127. Papoutsis, K., Golding, J. B., Vuong, Q., Pristijono, P., Stathopoulos, C. E., Scarlett, C. J., & Bowyer, M., (2018). Encapsulation of citrus by-product extracts by spray-drying and freeze-drying using combinations of maltodextrin with soybean protein and ι-carrageenan. *Foods, 7*(7), 115.
128. Pasukamonset, P., Kwon, O., & Adisakwattana, S., (2016). Alginate-based encapsulation of polyphenols from *Clitoria ternatea* petal flower extract enhances stability and biological activity under simulated gastrointestinal conditions. *Food Hydrocolloids, 61*, 772–779.
129. Patel, A. S., Lakshmibalasubramaniam, S., & Nayak, B., (2020). Steric stabilization of phycobiliprotein loaded liposome through polyethylene glycol adsorbed cellulose

nanocrystals and their impact on the gastrointestinal tract. *Food Hydrocolloids, 98*, 105252.

130. Patras, A., Brunton, N. P., O'Donnell, C., & Tiwari, B. K., (2010). Effect of thermal processing on anthocyanin stability in foods; mechanisms and kinetics of degradation. *Trends in Food Science & Technology, 21*(1), 3–11.
131. Pilehvar-Soltanahmadi, Y., Dadashpour, M., Mohajeri, A., Fattahi, A., Sheervalilou, R., & Zarghami, N., (2018). An overview on application of natural substances incorporated with electrospun nanofibrous scaffolds to development of innovative wound dressings. *Mini Reviews in Medicinal Chemistry, 18*(5), 414–427.
132. Pravilović, R. N., Balanč, B. D., Djordjević, V. B., Bošković-Vragolović, N. M., Bugarski, B. M., & Pjanović, R. V., (2019). Diffusion of polyphenols from alginate, alginate/chitosan, and alginate/inulin particles. *Journal of Food Process Engineering, 42*, 1–6.
133. Quideau, S., & Feldman, K. S., (1996). Ellagitannin chemistry. *Chemical Reviews, 96*, 475–503.
134. Radünz, M., Hackbart, H. C. D. S., Bona, N. P., Pedra, N. S., Hoffmann, J. F., Stefanello, F. M., & Zavareze, E. D. R., (2020). Glucosinolates and phenolic compounds rich broccoli extract: Encapsulation by eectrospraying and antitumor activity against glial tumor cells. *Colloids and Surfaces B: Biointerfaces, 192*, 111020.
135. Rafiee, Z., Nejatian, M., Daeihamed, M., & Jafari, S. M., (2019). Application of curcumin-loaded nanocarriers for food, drug and cosmetic purposes. *Trends in Food Science & Technology, 88*, 445–458.
136. Rafiee, Z., Nejatian, M., Daeihamed, M., & Jafari, S. M., (2019). Application of different nanocarriers for encapsulation of curcumin. *Critical Reviews in Food Science and Nutrition, 59*(21), 3468–3497.
137. Ravichai, K., & Muangrat, R., (2019). Effect of different coating materials on freeze-drying encapsulation of bioactive compounds from fermented tea leaf wastewater. *Journal of Food Processing and Preservation, 43*(10), e14145.
138. Ray, S., Raychaudhuri, U., & Chakraborty, R., (2016). An overview of encapsulation of active compounds used in food products by drying technology. *Food Bioscience, 13*, 76–83.
139. Rayleigh, L., (1882). On the equilibrium of liquid conducting masses charged with electricity. *The London, Edinburgh, and Dublin Philosophical Magazine and Journal of Science, 14*(87), 184–186.
140. Reineccius, T. A., Reineccius, G. A., & Peppard, T. L., (2002). Encapsulation of flavors using cyclodextrins: Comparison of flavor retention in alpha, beta, and gamma types. *Journal of Food Science, 67*, 3271–3279.
141. Rezende, Y. R. R. S., Nogueira, J. P., & Narain, N., (2018). Microencapsulation of extracts of bioactive compounds obtained from acerola (*Malpighia emarginata* DC) pulp and residue by spray and freeze drying: Chemical, morphological and chemometric characterization. *Food Chemistry, 254*, 281–291.
142. Riaz, M., (1996). Liposomes preparation methods. *Pakistan Journal of Pharmaceutical Sciences, 9*(1), 65–77.
143. Ribeiro, A. M., Estevinho, B. N., & Rocha, F., (2019). Spray drying encapsulation of elderberry extract and evaluating the release and stability of phenolic compounds in encapsulated powders. *Food and Bioprocess Technology, 12*(8), 1381–1394.
144. Rigon, R. T., & Noreña, C. P. Z., (2016). Microencapsulation by spray-drying of bioactive compounds extracted from blackberry (*Rubus fruticosus*). *Journal of Food Science and Technology, 53*(3), 1515–1524.

145. Rijo, P., Falé, P. L., Serralheiro, M. L., Simões, M. F., Gomes, A., & Reis, C., (2014). Optimization of medicinal plant extraction methods and their encapsulation through extrusion technology. *Measurement, 58*, 249–255.
146. Ru, Q., Wang, Y., Lee, J., Ding, Y., & Huang, Q., (2012). Turbidity and rheological properties of bovine serum albumin/pectin coacervates: Effect of salt concentration and initial protein/polysaccharide ratio. *Carbohydrate Polymers, 88*(3), 838–846.
147. Saénz, C., Tapia, S., Chávez, J., & Robert, P., (2009). Microencapsulation by spray drying of bioactive compounds from cactus pear (*Opuntia ficus*-indica). *Food Chemistry, 114*(2), 616–622.
148. Scalbert, A., Johnson, I. T., & Saltmarsh, M., (2005). Polyphenols: Antioxidants and beyond. *American Journal of Clinical Nutrition, 81*, 215S–217S.
149. Scalbert, A., Manach, C., Morand, C., Reémésy, C., & Jiménez, L., (2005). Dietary polyphenols and the prevention of diseases. *Critical Reviews in Food Science and Nutrition, 45*(20), 287–306.
150. Schmitt, C., & Turgeon, S. L., (2011). Protein/polysaccharide complexes and coacervates in food systems. *Advances in Colloid and Interface Science, 167*(1, 2), 63–70.
151. Secolin, V. A., Souza, C. R., & Oliveira, W. P., (2017). Spray drying of lipid-based systems loaded with *Camellia sinensis* polyphenols. *Journal of Liposome Research, 27*(1), 11–20.
152. Šeregelj, V., Tumbas, Š. V., Lević, S., Kalušević, A., Ćetković, G., Čanadanović-Brunet, J., & Vidaković, A., (2019). Application of encapsulated natural bioactive compounds from red pepper waste in yogurt. *Journal of Microencapsulation, 36*(8), 704–714.
153. Sharif, N., Fabra, M. J., & López-Rubio, A., (2019). Nanostructures of zein for encapsulation of food ingredients; Chapter 9. In: Jafari, S. M., (ed.), *Biopolymer Nanostructures for Food Encapsulation Purposes* (1st edn., Vol. 1, pp. 217–245). London–United Kingdom; Academic Press.
154. Sharif, N., Golmakani, M. T., Niakousari, M., Ghorani, B., & Lopez-Rubio, A., (2019). Food-grade gliadin microstructures obtained by electrohydrodynamic processing. *Food Research International, 116*, 1366–1373.
155. Silva, C. M., Ribeiro, J., & Vit, I., (2006). Alginate microspheres prepared by internal gelation: Development and effect on insulin stability. *International Journal of Pharmaceutics, 311*, 1–10.
156. Simon-Brown, K., Solval, K. M., Chotiko, A., Alfaro, L., Reyes, V., Liu, C., & Xu, Z., (2016). Microencapsulation of ginger (*Zingiber officinale*) extract by spray drying technology. *LWT, 70*, 119–125.
157. Siow, L. F., & Ong, C. S., (2013). Effect of pH on garlic oil encapsulation by complex coacervation. *Journal of Food Processing and Technology, 4*, 199–204.
158. Soleimanifar, M., Jafari, S. M., & Assadpour, E., (2020). Encapsulation of olive leaf phenolics within electrosprayed whey protein nanoparticles: Production and characterization. *Food Hydrocolloids, 101*, 105572.
159. Sonego, J. M., Santagapita, P. R., Perullini, M., & Jobbágy, M., (2016). Ca(II) and Ce(III) homogeneous alginate hydrogels from the parent alginic acid precursor: A structural study. *Dalton Transactions, 45*(24), 10050–10057.
160. Stojanović, R., Belščak-Cvitanović, A., Manojlović, V., Komes, D., Nedović, V., & Bugarski, B., (2012). Encapsulation of thyme (*Thymus serpyllum* L.) aqueous extract in calcium-alginate beads. *Journal of the Science of Food and Agriculture, 92*, 685–696.

161. Stokke, B. T., Draget, K. I., Smidsrød, O., Yuguchi, Y., Urakawa, H., & Kajiwara, K., (2000). Small-angle x-ray scattering and rheological characterization of alginate gels. 1. Ca-alginate gels. *Macromolecules, 33*(5), 1853–1863.
162. Surh, Y. J., (2003). Cancer chemoprevention with dietary phytochemicals. *Nature Reviews Cancer, 3*, 768–780.
163. Tai, K., Rappolt, M., He, X., Wei, Y., Zhu, S., Zhang, J., & Yuan, F., (2019). Effect of β-sitosterol on the curcumin-loaded liposomes: Vesicle characteristics, physicochemical stability, *in vitro* release and bioavailability. *Food Chemistry, 293*, 92–102.
164. Tan, S. P., Tuyen, C. K., Parks, S. E., Stathopoulos, C. E., & Roach, P. D., (2015). Effects of the spray-drying temperatures on the physiochemical properties of an encapsulated bitter melon aqueous extract powder. *Powder Technology, 281*, 65–75.
165. Tang, X. C., & Pikal, M. J., (2004). Design of freeze-drying processes for pharmaceuticals: Practical advice. *Pharmaceutical Research, 21*(2), 191–200.
166. Tapia-Hernández, J. A., Torres-Chávez, P. I., Ramírez-Wong, B., Rascón-Chu, A., Plascencia-Jatomea, M., Barreras-Urbina, C. G., Rangel-Vázquez, N. A., & Rodríguez-Félix, F., (2015). Micro-and nanoparticles by electrospray: Advances and applications in foods. *Journal of Agricultural and Food Chemistry, 63*(19), 4699–4707.
167. Taylor, G. I., (1964). Disintegration of water drops in an electric field. *Proceedings of the Royal Society of London. Series A. Mathematical and Physical Sciences, 280*(1382), 383–397.
168. Tengse, D. D., Priya, B., & Kumar, P. A. R., (2017). Optimization for encapsulation of green tea (*Camellia sinensis* L.) extract by spray drying technology. *Journal of Food Measurement and Characterization, 11*(1), 85–92.
169. Thorat, A. A., & Dalvi, S. V., (2012). Liquid antisolvent precipitation and stabilization of nanoparticles of poorly water-soluble drugs in aqueous suspensions: Recent developments and future perspective. *Chemical Engineering Journal, 181*, 1–34.
170. Timilsena, Y. P., Akanbi, T. O., Khalid, N., Adhikari, B., & Barrow, C. J., (2019). Complex coacervation: Principles, mechanisms and applications in microencapsulation. *International Journal of Biological Macromolecules, 121*, 1276–1286.
171. Tommasini, S., Calabro, M. L., Stancanelli, R., Donato, P., Costa, C., Catania, S., Villari, V., Ficarra, P., & Ficarra, R., (2005). The inclusion complexes of hesperetin and its 7-rhamnoglucoside with (2-hydroxypropyl)-β-cyclodextrin. *Journal of Pharmaceutical and Biomedical Analysis, 39*(3–4), 572–580.
172. Trucillo, P., Campardelli, R., Aliakbarian, B., Perego, P., & Reverchon, E., (2018). Supercritical assisted process for the encapsulation of olive pomace extract into liposomes. *Journal of Supercritical Fluids, 135*, 152–159.
173. Tsung, M., & Burgess, D. J., (1997). Preparation and stabilization of heparin/gelatin complex coacervate microcapsules. *Journal of Pharmaceutical Sciences, 86*(5), 603–607.
174. Turgeon, S. L., Schmitt, C., & Sanchez, C., (2007). Protein–polysaccharide complexes and coacervates. *Current Opinion in Colloid & Interface Science, 12*(4, 5), 166–178.
175. Tuyen, C. K., Phan-Tai, H., & Nguyen, M. H., (2014). Effects of pre-treatments on the yield and carotenoid content of Gac oil using supercritical carbon dioxide extraction. *Journal of Food Engineering, 120*, 44–49.
176. Vinceković, M., Viskić, M., Jurić, S., Giacometti, J., Kovačević, D. B., Putnik, P., Donsì, F., et al., (2017). Innovative technologies for encapsulation of Mediterranean plants extracts. *Trends in Food Science & Technology, 69*, 1–12.

177. Wang, H., Gong, X., Guo, X., Liu, C., Fan, Y. Y., Zhang, J., Niu, B., & Li, W., (2019). Characterization, release, and antioxidant activity of curcumin-loaded sodium alginate/ZnO hydrogel beads. *International Journal of Biological Macromolecules, 121*, 1118–1125.
178. Wang, S., Marcone, M. F., Barbut, S., & Lim, L. T., (2013). Electrospun soy protein isolate-based fiber fortified with anthocyanin-rich red raspberry (*Rubus strigosus*) extracts. *Food Research International, 52*(2), 467–472.
179. Wang, W., (2000). Lyophilization and development of solid protein pharmaceuticals. *International Journal of Pharmaceutics, 203*(1, 2), 1–60.
180. Wang, W., Chen, M., & Chen, G., (2012). Issues in freeze drying of aqueous solutions. *Chinese Journal of Chemical Engineering, 20*(3), 551–559.
181. Weinbreck, F., De Vries, R., Schrooyen, P., & De Kruif, C. G., (2003). Complex coacervation of whey proteins and gum Arabic. *Biomacromolecules, 4*(2), 293–303.
182. Whelehan, M., & Marison, I., (2011). Microencapsulation by dripping and jet break up. In: *Bioencapsulation Research Group* (pp. 4–10). France; September.
183. Wilkowska, A., Ambroziak, W., Czyżowska, A., & Adamiec, J., (2016). Effect of microencapsulation by spray-drying and freeze-drying technique on the antioxidant properties of blueberry (*Vaccinium myrtillus*) juice polyphenolic compounds. *Polish Journal of Food and Nutrition Sciences, 66*(1), 11–16.
184. Xiao, Z., Liu, W., Zhu, G., Zhou, R., & Niu, Y., (2014). A review of the preparation and application of flavor and essential oils microcapsules based on complex coacervation technology. *Journal of the Science of Food and Agriculture, 94*(8), 1482–1494.
185. Yamashita, C., Chung, M. M. S., Dos Santos, C., Mayer, C. R. M., Moraes, I. C. F., & Branco, I. G., (2017). Microencapsulation of an anthocyanin-rich blackberry (*Rubus* spp.) by-product extract by freeze-drying. *LWT, 84*, 256–262.
186. Yao, K., Chen, W., Song, F., McClements, D. J., & Hu, K., (2018). Tailoring zein nanoparticle functionality using biopolymer coatings: Impact on curcumin bioaccessibility and antioxidant capacity under simulated gastrointestinal conditions. *Food Hydrocolloids, 79*, 262–272.
187. Yeo, L. Y., Gagnon, Z., & Chang, H. C., (2005). AC electrospray biomaterials synthesis. *Biomaterials, 26*(31), 6122–6128.
188. Zazzali, I., Aguirre, C. T. R., Pizones Ruíz-Henestrosa, V. M., Santagapita, P. R., & Perullini, M., (2019). Effects of pH, extrusion tip size and storage protocol on the structural properties of Ca(II)-alginate beads. *Carbohydrate Polymer, 206*, 749–756.
189. Zeeb, B., Saberi, A. H., Weiss, J., & McClements, D. J., (2015). Formation and characterization of filled hydrogel beads based on calcium alginate: Factors influencing nanoemulsion retention and release. *Food Hydrocolloids, 50*, 27–36.
190. Zeleny, J., (1914). The electrical discharge from liquid points, and a hydrostatic method of measuring the electric intensity at their surfaces. *Physical Review, 3*(2), 69.
191. Zhang, L., Liu, L., Qian, Y., & Chen, Y., (2008). The effects of cryoprotectants on the freeze-drying of ibuprofen-loaded solid lipid microparticles (SLM). *European Journal of Pharmaceutics and Biopharmaceutics, 69*(2), 750–759.
192. Zhao, R., Sun, J., Torley, P., Wang, D., & Niu, S., (2008). Measurement of particle diameter of *lactobacillus acidophilus* microcapsule by spray drying and analysis on its microstructure. *World Journal of Microbiology Biotechnology and Biotechnology, 24*(8), 1349–1354.

193. Zheng, B., Zhang, Z., Chen, F., Luo, X., & McClements, D. J., (2017). Impact of delivery system type on curcumin stability: Comparison of curcumin degradation in aqueous solutions, emulsions, and hydrogel beads. *Food Hydrocolloids, 71*, 187–197.
194. Zuidam, N. J., & Shimoni, E., (2010). Overview of microencapsulates for use in food products or processes and methods to make them; Chapter 2. In: Zuidam, N. J., & Nedovik, V., (eds.), *Encapsulation Technologies for Active Food Ingredients and Food Processing* (1st edn., pp. 3–29). Verlag, New York: Springer.

CHAPTER 7

MICROENCAPSULATION OF NATURAL PIGMENTS

R. C. RANVEER, BHAGWAN K. SAKHALE, and U. S. ANNAPURE

ABSTRACT

The plants parts (flowers, fruits, and leaves), animals, and microorganisms appear in different colors due to the presence of different pigments. These colors are mainly occurred owing to the presence of different coloring compounds such as chlorophylls, carotenoids, anthocyanins, etc. The natural pigments are used as a colorant in food, pharma, and textile industry. Many of these pigments possess antimicrobial and antioxidant properties. These colorants have been also accounted for many pharmacological properties as used for curing of cancer and cardiovascular diseases (CVDs). These pigments are vulnerable to degrade in the presence of high temperature, sunlight, and oxygen, which affect the utilization of these pigments in processed product. The microencapsulation is the technique where bioactive constituents are enclosed in the carrier material to extend its stability. The various techniques like freeze drying, spray drying, extrusion, emulsification, etc., may be used for microencapsulation of pigments. The encapsulation gives better stability of active components against sunlight, oxidation, and temperature.

7.1 INTRODUCTION

Natural pigments found everywhere in life. It appears in leaves, flowers, fruits, and vegetables, also in the animal skin, blood, eyes, and other tissues, even it found in microorganisms such as fungi and bacteria. These pigments

Novel Processing Methods for Plant-Based Health Foods: Extraction, Encapsulation, and Health Benefits of Bioactive Compounds. Megh R. Goyal, N. Veena & Ritesh B. Watharkar (Eds.)

not only add to the beauty of the material but also improve the esthetic value of it. The pigments have important functions, such as, oxygen carrier, photosynthesis, and protection.

Pigments are substances which absorb illumination from the visible region wavelength [15]. The color is appeared to eyes, because non-absorbed light is reflected, and the impression of these is arrested by the eyes. The natural pigments are basically divided into four groups, such as: chlorophylls, carotenoids, anthocyanins, and betalains [15, 23]. Chlorophylls are represented by green color, carotenoids are represented by yellow, orange, and red color, anthocyanins are accountable for orange, reddish, pinkish, purple, and bluish, and betalains are represented by the red color.

Many researchers have reported pharmacological activities, such as: antioxidants, anticancer, antimicrobial, etc. [22]. The stability of these pigments is affected in the presence of sunlight, oxygen, and temperature. These pigments convert into isomers due to sunlight, oxygen, and higher temperature, which affect the pharmacological and nutraceutical activities. Microencapsulation is one of the ways to stabilize the pigments. Broadly in food products, encapsulation comprises coating of the tiny bioactive/food components as well as food ingredient with micro coating/capsules. These capsules have the ability to prevent oxidation, and protect from sunrays and high temperature. The microencapsulation process has various benefits, such as [18]: (i) microencapsulation can protect bioactive from degradation by external environment; (ii) vaporization of volatile components can be prevented; (iii) the physical state can be modified for easier handling; (iv) the product can be solely released over time at specific point; (v) the active material can be diluted when required in minute quantity and can achieve uniform dispersion; and (vi) two different materials can be separated within the mixture by means of encapsulation. The present chapter highlights the encapsulation of different plant pigments to enhance their stability.

7.2 NATURAL PIGMENTS (Table 7.1)

7.2.1 CHLOROPHYLLS

Chlorophyll is responsible for the typical green color of plants. It is oil soluble pigments. Five different types of chlorophylls appear in plants and animal those are capable of photosynthesis, whereas it is found majorly in the form of Chlorophyll *a*, which consist of $-CH_3$ and Chlorophyll *b* which consist of $-CHO$ in plant (Figure 7.1). Chlorophyll *a* represented by bluish-green while

Chlorophyll *b* represented by yellowish-green color [65]. The major function of chlorophyll in photosynthesis. Also, various pharmaceutical studies have proven that it can be useful for skin conditions, odorants for the body and against some types of cancer.

TABLE 7.1 Some of Important Natural Pigments and Its Sources

Pigment	Source	Image
Anthocyanin	Purple carrot	
	Red cabbage	
	Grapes	
	Black current	
	Red radish	
	Elderberry	
Betalain	Beetroot	

TABLE 7.1 *(Continued)*

Pigment	Source	Image
	Swiss chard	
	Amaranthus	
	Cactus fruits	
Carotenoids	Mango	
	Pumpkin	
	Palm fruit	
	Cassava	
Chlorophylls	Green peas	
	Spinach	

TABLE 7.1 *(Continued)*

Pigment	Source	Image
	Cucumber	
	Celery	

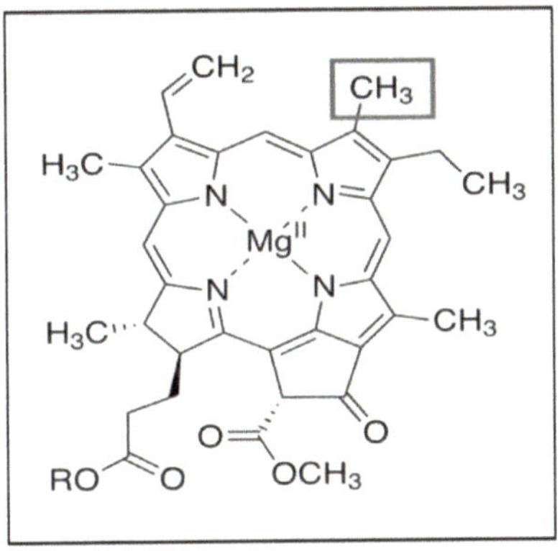

FIGURE 7.1 Structure of chlorophyll.

Chlorophylls degradation can occur in the presence of high temperature, sunlight, oxygen, acids, and enzymes [49]. Mg-dechelatase appears in the plants and algae, and acids responsible for replacement of central magnesium atom with hydrogen ions of chlorophylls leading to transformation into pheophytin, which is responsible for formation of olive-brown color [16]. Oxidative enzymes and chlorophyllase are also responsible for degradation in chlorophylls [44]. Higher temperature is responsible for the formation of olive-green color due to degradation of chlorophylls [40].

7.2.2 CAROTENOIDS

Carotenoids are the derivatives of lycopene obtained after reactions such as hydrogenation and dehydrogenation, cyclization, oxidation, migration of double bond, methylation, and chain shortening [20]. Carotenoids are classified as primary and secondary.

- Primary carotenoids include β-carotene, violaxanthin, and neoxanthin, and those are essential for plant photosynthesis; and
- Secondary carotenoids consist of α-carotene, β-cryptoxanthin, zeaxanthin, antheraxanthin, capsanthin, capsorubin, and gives attractive colors to fruits and vegetables. These appear in higher amounts in flora and microorganisms, such as: bacteria, algae, and fungi. It possesses bright yellow, orange, and red color [20, 22].

Carotenoid possesses important functions, such as: (i) it works as accessory pigment in photosynthesis; (ii) it gives protection against oxidative degradation; (iii) antheraxanthin and zeaxanthin protect plants from damage occurred due to high intensity of light; (iv) it serves as antioxidant; and (v) these pigments possess pharmacological properties including anticancer and antimicrobial.

Carotenoids degradation is caused due to oxidation and isomerization reaction leading to reduction in redness and yellowness [52]. Oxidation occurs in the presence of oxygen, the rate of oxidation increases in the presence of sunlight, high temperature, peroxide, metal ions and enzymes [8]. In food processing, *trans*-isomers convert into the *cis*-isomers by the isomerization reaction [34]. The isomerization reaction is also facilitated by high temperature, sunlight, and acidic condition of substance [52].

7.2.3 ANTHOCYANINS

Anthocyanins are accountable for striking colors of fruits, vegetables, and flowers [21]. The anthocyanin produces infinite colors by combining with glucosides/acyl groups and by their reactions with other molecules and/or media environment [10]. The anthocyanin structure is presented in Figure 7.2.

The anthocyanins are water-soluble colors and found in higher plants, whereas these are not found in lower plants. It can be found in leaves of many ornamental plants as a complex mixture of anthocyanins. In apple, cherry, fig, and peach, a single anthocyanin (i.e., Cyanidin) is present; in eggplants and pomegranate, two major anthocyanins are present, and in grapes several anthocyanins can be accumulated.

The anthocyanins give attractive colors to the plants thus helping them in pollination, seed dispersion and anti-feedant. It can also be used to find out adulterations in pigmented food. Anthocyanins exhibit antibacterial, antiviral, and antifungal properties. It possesses strong antioxidant activities. It lowers down the risk of coronary heart diseases and cancer [9].

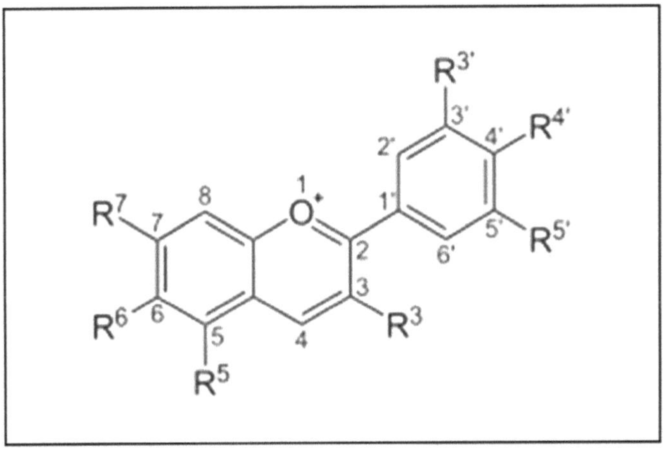

FIGURE 7.2 Basic structure of anthocyanins.

The stability of anthocyanins was hampered due to its structural features and due to hydrogen ion concentration (pH), heat, sunlight, isomers, enzymes, O_2 and sugars [59]. Reversible anthocyanin transformation can be due to alteration in pH, which may be responsible for change in the color. Most of the pigments are susceptible to the high temperature. Anthocyanin degradation starts at around 100°C [27].

7.2.4 BETALAINS

Betalains are the ammonium derivatives of betalamic acid [67]. It is categorized into betaxanthins, which possesses yellow tint, and betacyanins with reddish-purple tint. Structure of betacyanin shows change in acyl group and sugars whereas the betaxanthins illustrate bonding with amines and amino acids in their structures. Betalains gives red, yellow, pink, and orange shades to the higher plants (Figure 7.3).

The betalains have been considered for taxonomical identification of plants and it was known that 11 families of order *Caryophyllales* showed the presence of betalains [61]. Even though it is alkaloids in nature, yet it is non-toxic to human beings so that it is considered as alternative to the synthetic color. It shows antimicrobial and antiviral activities.

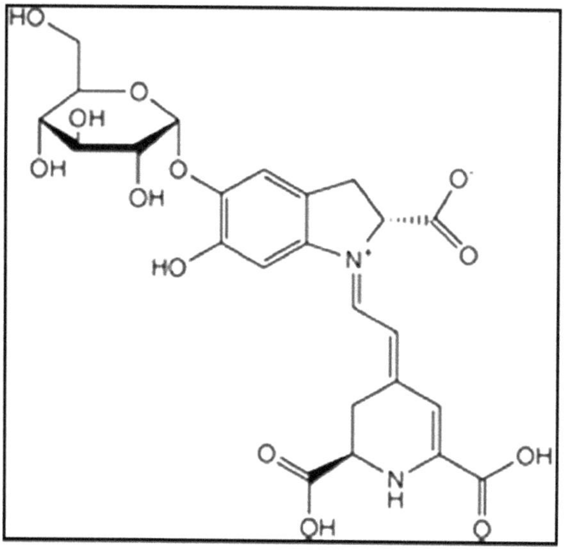

FIGURE 7.3 General structure of the Betalain.

Betalains are highly unstable to high temperature, sunlight, alkalinity, oxygen, and ions of metals, which confine their food applications [33]. High temperature is considered as a major factor responsible for degradation of betalains [56]. Oxidation is induced in the presence of heat resulting in decarboxylation, and ultimately changes in color from orange to yellow [19]. These pigments are stable pigment at pH from 3 to 7 [66]. In the presence of metal ions, the oxidation process can be accelerated [66].

The natural pigments are used as colorants in food, pharmaceuticals, and textile industries. The main constraint for utilization of the natural pigment is their sensitivity against heat, light, and oxygen (Table 7.2). There are various methods that can be employed to overcome this problem. The microencapsulation can be used to improve the stability of natural pigment.

7.3 MICROENCAPSULATION

It is a process of encapsulation of solids, liquids, or gaseous active ingredients in minute, enacted microcapsules which can discharge active ingredients at controlled rates under definite circumstances [13, 37].

Properties of capsules may alter to costume definite active material for applications including composition, release mechanism, size of the particles, physical structure, and price. The structural design of capsules is normally classified into several random and overlapping classifications (Figure 7.4) and matrix encapsulation is classification among them. The simplest arrangement includes wall material of uniform thickness consist of core material in the center.

TABLE 7.2 Sensitivity of Natural Pigments

Pigment	Sensitivity			
	High Temperature	Sunlight	O_2	Change in pH
Anthocyanins	High to moderate	Low	Low	Moderate
Betalains	Moderate to low	Low	Low	High
Carotenoids	Moderate	High	High	Low
Chlorophylls	Moderate	Low	Low	High

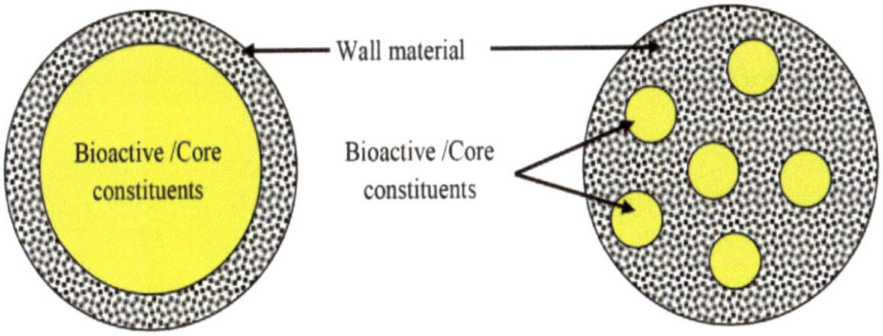

FIGURE 7.4 Schematic diagrams of types of capsules.

7.3.1 MICROENCAPSULATION TECHNIQUES

The microencapsulation techniques are classified into physical and chemical methods [41, 71]. In case physical method, polymers are used so that there will not be any chemical reaction and structure will be fabricated. While in case of chemical method, chemical reaction takes place for formation of microcapsule. The commonly used microencapsulation techniques for the encapsulation of pigments are shown in Table 7.3.

7.3.2 EMULSION POLYMERIZATION

In this method, the monomer (for example, alkyl acrylate) is mixed in active material (core material) drop by drop with continuous stirring to form stable emulsion [71]. In the process of polymerization, precipitation of polymer occurred in an aqueous medium to develop primary nuclei structure, further this nuclei structure develops steadily to entrap bioactive material to form capsules. Lipophilic materials are considered as appropriate material for this method.

TABLE 7.3 Microencapsulation Techniques for the Pigments

Microencapsulation Technique	Name of Pigment	References
Emulsion method	Anthocyanin	[6]
Freeze-drying	β-carotene, saffron	[14, 43]
Microwave drying	Anthocyanin	[62]
Solvent evaporation	Astaxanthin	[24]
Spray drying	Anthocyanin, astaxanthin-oleoresin, betacyanin, betalain, bixin, chlorophyll, β-carotene, curcumin	[3, 11, 28, 58, 63, 64]

7.3.3 INTERFACIAL POLY-CONDENSATION

The capsules are prepared by mixing of two-phase system under controlled condition to develop minute droplet of dispersed phase in continuous phase [29]. The core/bioactive material should disperse into the droplets. In the microencapsulation by poly-condensation process, coalescence of droplets/particle coagulation is prevented by addition of a suitable stabilizer.

7.3.4 SUSPENSION CROSS-LINKING

The capsules of protein and polysaccharide may be developed by the suspension cross-linking method [31]. In this method, minute droplets are developed by dispersion of aqueous polymer which contain bioactive/core material. The hardening of droplets is carried out by cross-linking which directly convert into capsules. Cross-linking is performed using either by higher temperature (at >500°C) or by using cross-linking agents.

7.3.5 SOLVENT EVAPORATION/SOLVENT EXTRACTION

The process for capsule development by evaporation/solvent extraction is similar to the cross-linking method [74] except in the case of hydrophobic polymers. These polymers are suspended in water-immiscible organic solvent of volatile nature along with core/bioactive material. The resulted mixture is then added drop by drop in an aqueous stabilizer solution for formation of micro droplets of capsules. The hardening of these capsules is carried out by removing solvent by evaporation or solvent extraction.

7.3.6 COACERVATION/PHASE SEPARATION

Coacervation is broadly applicable for the formation of capsules of gelatin, cellulose, and its derivatives, and synthetic polymers [48]. The coacervation process is classified into simple and complex coacervation method. In the case of simple coacervation method, only one polymer is used, while in the complex coacervation method, two opposite charged polymers soluble in water are used. In encapsulation by coacervation method is performed by dissolving polymer in water, which also consists of the core/bioactive components.

Suitable coacervating agent and stabilizer may be used for formation of partially dissolved polymers and maintain individuality, respectively. Then hardening of this mixture is performed by cooling at about 5–50°C followed by adding cross-linking agent.

7.3.7 SPRAY DRYING

It is a low-cost technique and is commercially used in microencapsulation of food ingredients [63]. It is most frequently used for microencapsulation of food ingredients since it provides protection against oxidation and water vapors. In this method, emulsion of bioactive component in a concentrated wall material is prepared. This mixture is sprayed in the form of fine droplets inside hot chamber. The water portion gets evaporated, resulting in micro-capsules of bioactive materials. Spray drying was used for encapsulation of lycopene inside gelatin capsules [53].

7.3.8 FLUIDIZED BED COATING

It is utilized for microencapsulation of the solid or liquid material which can be absorbed into porous solid material [36]. The fluidized bed coating is more popular in the pharmaceutical industry. Solid constitutes of bioactive materials are dissolved in a jet of air and further coated by spray of liquid encapsulating materials. Afterwards cooling of this material is carried out. In this method, coating is performed till capsules of required thickness are obtained.

7.3.9 LYOPHILIZATION

The microencapsulation of bioactive materials sensitive to high temperature is performed by the lyophilization or freeze-drying. Here, the process is performed in two steps: (i) the emulsion of carrier material and bioactive material is prepared and subjected to the freezing at – 28°C; and (ii) the freeze-drying is carried out which works on the sublimation principle. It is a time-consuming method and takes around 20 h. This technique is generally preferred for the encapsulation of aromatic compounds [38]. The β-carotene encapsulation is carried out by freeze-drying using hydrolyzed starch as a carrier material [14].

7.3.10 EXTRUSION

Extrusion is generally followed for encapsulation of flavoring components; comparatively low temperature is used for encapsulation. The process is same as that of extrusion cooking of cereal-based snacks. In this process, the core/bioactive material is enforced in a liquid mass of carbohydrate through the series of dies with higher pressure (more than 100 psi) [55]. When coating material comes into contact with liquid become hard and form capsulation matrix over core material.

7.3.11 CENTRIFUGAL SUSPENSION SEPARATION

In this encapsulation process, the core and wall material are mixed together and subjected to centrifugation. After centrifugation, the core material is coated and excess liquid is separated. Afterward, these capsules are dried/chilled. This process is very fast and coating can be performed within a few minutes. It is a continuous, high-speed method and most suitable for encapsulation of food ingredients.

7.3.12 CO-CRYSTALLIZATION

The sucrose syrup is converted in supersaturated stage by concentration and its temperature is maintained high enough to avoid crystallization. Known quantity of core material is added into the concentrated syrup and agitated vigorously to afford nucleation for sucrose-core material mixture to crystallize. When the transformation temperature has reached, crystallization starts with elimination of a considerable amount of heat. Nonstop starring is performed to endorse and extend crystallization until agglomerates are discharged from the vessel. Then these encapsulates are dried to a desired moisture content.

7.4 CARRIER/WALL MATERIALS FOR ENCAPSULATION

Generally, polymers are used as carrier materials. The bioactive components are normally referred to as internal phase or core material. The microencapsulation material is considered as external phase, the shell, coating, or membrane [1]. The polymers of polysaccharides (such as: cellulose and its derivatives, chitin, and chitosan, starches, agar, alginate, carrageenan, gums, and pectin), proteins (such as: gelatin, zein, gluten, soy protein and whey protein) and lipids (such as: waxes and paraffin, aceto-glycerides, and shellac resins) are generally used as carrier materials in encapsulations [34] as shown in Table 7.4.

TABLE 7.4 Carrier Materials for Microencapsulation of Natural Pigments

Carrier Material/Wall Material	Pigment	References
Chitosan, gum Arabic, whey protein, maltodextrin, and inulin	Astaxanthin	[11, 24]
Glucose, maltodextrin, native starch, modified starch, gum Arabic, maltodextrin	Betacyanin	[2, 12, 50]
Gum Arabic	Bixin	[5]
Maltodextrin	Betalain	[28]
Maltodextrin, octenyl succinic anhydride (OSA)-modified starch	Chlorophyll	[51]
Modified starch, gelatin, sucrose, gum Arabic, soy protein and modified tapioca starch	Lycopene	[53, 58, 63]
Native tapioca starch, maltodextrin, native, and hydrolyzed starches and furcellaran	β-carotene	[14, 39, 42]
Polyvinylpyrrolidone, gum Arabic and lecithin	Curcumin	[45, 70]

7.4.1 CARBOHYDRATES BASED CARRIER MATERIALS

Carbohydrates with higher molecular weight are generally utilized for encapsulation of pigments. Starches and its derivates are commonly used encapsulation of sensitive bioactive constituents [7]. The carbohydrates have the ability to produce gels and glassy matrices for the encapsulation [32]. They have good water solubility, low viscous in nature and contain high solids. Polysaccharides are hygroscopic in nature and nearly colorless and tasteless [75]. Microencapsulation is carried out using maltodextrin as carrier material protect bioactive components from oxidation and also improve its stability [57]. Starch obtained from tapioca is utilized for the production of powders of natural pigments [73]. Tapioca starch modified with acid shown improvement in encapsulation properties than native starch in encapsulation of carotenoid pigments [47]. Carbohydrate obtained from sea buckthorn used for carotenoids encapsulation, showed 95% encapsulation efficiency (EE) [69]. Besides this, polysaccharides are very hydrophilic in nature resulting in low barrier properties against water and gas. Even though, polysaccharide encapsulation did not show good barrier properties against water, yet these can act as sacrificing agents to lower down moisture loss [4].

7.4.2 PROTEIN-BASED CARRIER MATERIALS

Proteins are made of chains of amino acids. Protein possesses very good binding and functional properties, and these serve good carrier material for the encapsulation of colorants. Capsules prepared from the proteins are effortlessly rehydrated and soluble in water resulting in the release of core immediately. Gelatin and whey protein are usually utilized for microencapsulation of natural pigments by spray drying. Whey protein along with alginate was used to encapsulate paprika to alter the release mechanism [60]. EE can be enhanced by the replacement of whey protein with lactose. In their native states, proteins generally exist as either fibrous proteins, which are water-insoluble and serve as the main structural materials of animal tissues, or globular proteins, which are soluble in water or aqueous solutions of acids, bases or salts and function widely in living systems. Fibrous proteins are fully extended and associated closely with each other in parallel structures.

7.4.3 LIPID-BASED CARRIER MATERIALS

Emulsion formulation can be prepared with lipids, which can develop a matrix surrounding to the bioactive components. Stability of encapsulated bioactive components is affected by the glass transition temperature of polymers. Lower water vapor transmission and gas transmission rate in matrix of lipids can increase the stability of core materials. The presence of high temperature lactoglobulin chitosan can convert into hydrogels, which are used as carrier material for microencapsulation of functional foods [25]. Generally, water vapor permeability decreases when the concentration of hydrophobicity phase increases. Lipid based films are often supported on a polymer structure matrix, usually a polysaccharide, to provide mechanical strength.

7.5 MICROENCAPSULATION OF NATURAL COLORANTS

Several techniques and carrier materials are being used for the encapsulation of natural pigments. Microencapsulation takes care of the stability of bioactive components and their handlings. As per the commercial application, the encapsulation technique and wall materials are selected.

7.5.1 MICROENCAPSULATION OF CAROTENOIDS

The microencapsulation of β-carotene was carried out using native tapioca starch, maltodextrin, and acid modified starch as a wall material by the spray drying technique. The result showed that when acid modified tapioca starch is used for encapsulation of β-carotene, the retention was better than other wall materials [42]. However, when freeze-drying technique used for microencapsulation then maltodextrin (12 dextrose equivalent (DE)) showed higher stability than native starch [14].

The carotene derived from sea buckthorn (*Hippophaë rhamnoides* L.) was encapsulated using furcellaran and reported 97% EE [39]. Modified starches can be used for encapsulation of lycopene. The results reported that EE decreases with increase in the lycopene content. Also, it was reported that the lycopene stability was more at 10°C than 25°C during storage [58]. The gelatin–sucrose combination was used as wall material for the microencapsulation of lycopene by spray drying [63]. The results suggested that the gelatin/sucrose ratio (3/7), core to wall material ratio 1/4, inlet temperature 190°C and homogenization pressure 40 MPa were optimal conditions for the encapsulation of lycopene.

Solvent evaporation technique is used for encapsulation of astaxanthin using chitosan as carrier material. Microencapsulated astaxanthin did not report any degradation and isomerization during storage of 8 weeks at 25, 35 and 45°C temperature [24]. Higher temperature and sunlight may exhilarate isomerization in the carotenoids [68]. Soy protein isolate and gum Arabic are used for encapsulation of paprika oleoresin which consists of paprika yellow and red fractions [46]. Various encapsulants (i.e., gum Arabic, whey protein, inulin, and maltodextrin) were used for encapsulation of astaxanthin and reported higher encapsulation yield (EY) (61.2–70.1%) when whey protein and gum Arabic were used in combination [11]. Higher stability was recorded in bixin microencapsulated with gum Arabic than maltodextrin [5].

7.5.2 MICROENCAPSULATION OF ANTHOCYANINS

Anthocyanin obtained from black carrot was encapsulated using maltodextrins having different DE (i.e., 10, 20, 23, 28 and 31) by spray drying. The highest yield of anthocyanin was recorded when 20 and 23 DE maltodextrin was used as carrier material for microencapsulation than others [17]. Contrary, Tonon et al. [72] reported that maltodextrin with 10 DE gave better retention and higher antioxidant activity than other carrier material using spray drying. Various encapsulants have been reported for encapsulation of anthocyanins for uniform-sized particles of microcapsules when maltodextrin and gum Arabic are used as carrier agents [64]. When the inlet temperature was increased from 140 to 160°C, then more loss of anthocyanins was recorded in the capsules. The half-life of encapsulated anthocyanin was increased by 3 times when it is stored at 4°C instead of 25°C.

7.5.3 MICROENCAPSULATION OF CHLOROPHYLLS

There is not enough literature available on encapsulation of chlorophyll. Gum Arabic, maltodextrin, and OSA-modified starch was used as carrier agent for microencapsulation of chlorophylls. When OSA-modified starch combination was used as a carrier agent, then highest greenness value and total chlorophyll was recorded with highest antioxidant activity. Also, this combination also recorded longer half-life of microencapsulated chlorophyll [51]. Different blends of gum Arabic and maltodextrin were used for encapsulation of chlorophylls by the spray drying method. An increasing amount of maltodextrin in wall material is associated with lower moisture content (0.56%), higher EE (77.19%), and chlorophyll retentions [30]. Chlorophyll becomes unstable when exposed to oxygen, high temperature, or light environments. Its stability enhanced by microencapsulation by polymer encapsulation. Polycaprolactone (PCL) is used as a carrier agent for microencapsulation of chlorophyll, and the particles size of the composites was controlled through droplet microfluidics [26].

7.5.4 MICROENCAPSULATION OF BETALAINS

Betacyanins and Betaxanthins are two major types of the betalains. The heat, sunlight, water activity (a_w), enzymes, oxygen, and metals may affect the stability of betalains [23]. The encapsulated betalains were stored at different water activity levels and the highest stability was observed when the a_w was <0.521 [50]. Low crystallized maltodextrin was used as encapsulant for microencapsulation of betalains obtained from beetroots by spray drying method [28]. The higher loss of betacyanin was occurred with higher inlet temperature [12].

7.6 SUMMARY

The fruits, leaves, flowers, etc., consist of different pigments, such as: chlorophyll, carotenoids, and anthocyanins. These pigments show different pharmaceutical activities and health benefits. The stability of these pigments against temperature, light, and oxygen is very low. The microencapsulation can improve the stability of the pigments during storage. Various encapsulating materials and methods are employed for the encapsulation of natural pigments. However, the maltodextrin as encapsulating material and spray drying methods is generally applied for the encapsulation of natural pigments.

KEYWORDS

- **anthocyanin**
- **carotenoids**
- **dextrose equivalent**
- **encapsulation yield**
- **microencapsulation**
- **pigments**
- **supercritical anti-solvent**

REFERENCES

1. Atmane, M., Muriel, J., Joe, S., & Stephane, D., (2006). Flavor encapsulation and controlled release: A review. *International Journal of Food Science and Technology, 41*(1), 1–21.
2. Azeredo, H., Santos, A., & Souza, A., (2007). Betacyanin stability during processing and storage of a microencapsulated red beetroot extract. *American Journal of Food Technology, 4*(2), 307–312.
3. Bakowska-Barczak, A. M., & Kolodziejczyk, P. P., (2011). Black currant polyphenols: Their storage stability and microencapsulation. *Industrial Crops and Production, 34*(2), 1301–1309.
4. Balasubramanian, V. M., Chinnan, M. S., Mallikarjunan, P., & Phillips, R. D., (2007). The effect of edible film on oil uptake and moisture retention of a deep-fried poultry product. *Journal of Food Processing and Engineering, 20*(1), 17–29.
5. Barbosa, A., Borsarelli, C., & Mercadante, A., (2005). Light stability of spray-dried bixin encapsulated with different edible polysaccharide preparations. *Food Research International, 38*(8, 9), 989–994.
6. Betz, M., & Kulozik, U., (2011). Microencapsulation of bioactive bilberry anthocyanins by means of whey protein gels. *Procedia Food Science, 1,* 2047–2056.
7. Bhosale, R., & Singhal, R., (2006). Process optimization for the synthesis of octenyl succinyl derivative of waxy corn and amaranth starches. *Carbohydrate Polymers, 66*(4), 521–527.
8. Boon, C. S., McClements, D. J., Weiss, J., & Decker, E. A., (2010). Factors influencing the chemical stability of carotenoids in foods. *Critical Reviews in Food Science and Nutrition, 50*(6), 515–532.
9. Bridle, P., & Timberlake, C. F., (1997). Anthocyanins as natural food color-selected aspects. *Food Chemistry, 58*(1, 2), 103–109.
10. Brouillard, R., Dangles, O., Jay, M., Biolley, J. P., & Chirol, N., (1993). Polyphenols and pigmentation in plants. In: Scalbert, A., (ed.), *Polyphenolic Phenomena: Science Update* (pp. 41–47). Paris: INRA.
11. Bustos-Garza, C., Yáñez-Fernández, J., & Barragán-Huerta, B., (2013). Thermal and pH stability of spray-dried encapsulated astaxanthin oleoresin from *Haematococcus pluvialis* using several encapsulation wall material. *Food Research International, 54*(1), 641–649.
12. Cai, Y., & Corke, H., (2000). Production and properties of spray-dried *amaranthus betacyanin* pigments. *Journal of Food Science, 65*(6), 1248–1252.
13. Chen, X. G., Lee, C. M., & Park, H. J., (2003). O/W emulsification for the self-aggregation and nanoparticle formation of linolenic acid-modified chitosan in the aqueous system. *Journal of Agricultural and Food Chemistry, 51*(10), 3135–3139.
14. Corralo, S., Zapata, N. C., Ferreira, M. L., & Cristina, T. I., (2012). study on the stability of β-carotene microencapsulated with pinhão (*araucaria angustifolia* seeds) starch. *Carbohydrate Polymers, 89*(4), 1166–1173.
15. Delgado-Vargas, F., Jimenez, A. R., & Paredes-Lopez, O., (2000). Natural pigments: Carotenoids, anthocyanins, and betalains–characteristics, biosynthesis, processing, and stability. *Critical Reviews in Food Science and Nutrition, 40*(3), 173–289.
16. Delgado-Vargas, F., & Paredes-Lopez, O., (2002). *Natural Colorants for Food and Nutraceutical Uses* (1st edn., p. 344). Boca Raton: CRC Press.

17. Ersus, S., & Yurdagel, U., (2007). Microencapsulation of anthocyanin pigments of black carrot (*Daucus carota L.*) by spray dryer. *Journal of Food Engineering, 80*(3), 805–812.
18. Gibbs, B. F., Kermasha, S., Alli, I., & Mulligan, C. N., (1999). Encapsulation in food industry: A review. *International Journal of Food Science and Food Nutrition, 50*(3), 213–234.
19. Goncalves, L. C. P., Genova, B. M. D., Dorr, F. A., Pinto, E., & Bastos, E. L., (2013). Effect of dielectric microwave heating on the color and antiradical capacity of betanine. *Journal of Food Engineering, 118*(1), 49–55.
20. Goodwin, T. W., (1980). *Biochemistry of the Carotenoids: Volume I Plants* (2nd edn., Vol. 1; pp. 1–95). Netherlands: Springer.
21. Harborne, J. B., (1988). The flavonoids: Recent advances; Chapter 7. In: Goodwin, T. W., (ed.), *Plant Pigments* (pp. 299–343). London–United Kingdom: Academic Press.
22. Hari, R. K., Patel, T. R., & Martin, A. M., (1994). An overview of pigment production in biological systems: Functions, biosynthesis, and applications in food industry. *Food Reviews International, 10*(1), 49–70.
23. Hendry, G. A. F., & Houghton, J. D., (1996). *Natural Food Colorants* (2nd edn., p. 348). London–United Kingdom: Chapman & Hall.
24. Higuera-Ciapara, I., Felix-Valenzuela, L., Goycoolea, F., & Arguelles-Monal, W., (2003). Microencapsulation of astaxanthin in a chitosan matrix. *Carbohydrate Polymers, 56*(1), 41–45.
25. Hong, Y. H., & McClements, D. J., (2007). Formation of hydrogel particles by thermal treatment of lactoglobulin-chitosan complexes. *Journal of Agriculture and Food Chemistry, 55*(14), 5653–5660.
26. Hsiao, C. J., Lin, J. F., Wen, H. Y., Lin, Y. M., Yang, C. H., Huang, K. S., & Shaw, J. F., (2020). Enhancement of the stability of chlorophyll using chlorophyll-encapsulated polycaprolactone microparticles based on droplet microfluidics. *Food Chemistry, 306*, 125300.
27. Jackman, R. L., & Smith, J. L., (1996). Anthocyanins and betalains. In: Hendry, G. A. F., & Houghton, J. D., (eds.), *Natural Food Colorants* (2nd edn., pp. 244–309). London–United Kingdom: Chapman & Hall.
28. Janiszewska, E., & Wlodarczyk, J., (2013). Influence of spray drying conditions on beetroot pigments retention after microencapsulation process. *Acta Agrophysica, 20*(2), 343–356.
29. Janssen, L. J. J. M., Boersma, A., & Te Nijenhuis, K., (1993). Encapsulation by interfacial polycondensation III. Microencapsulation; the influence of process conditions on wall permeability. *Journal of Membrane Science, 79*(1), 11–26.
30. Kang, Y. R., Lee, Y. K., Kim, Y. J., & Chang, Y. H., (2019). Characterization and storage stability of chlorophylls microencapsulated in different combination of gum Arabic and maltodextrin. *Food Chemistry, 272*, 337–346.
31. Katti, D., & Krishnamurti, N., (1999). Preparation of albumin microspheres by an improved process. *Journal of Microencapsulation, 16*(2), 231–242.
32. Kenyon, M. M., (1995). Modified starch, maltodextrin, and corn syrup solids as wall materials for food encapsulation. In: Risch, S. J., & Reineccius, G. A., (eds.), *Encapsulation and Controlled Release of Food Ingredients, ACS Symposium Series No. 590* (pp. 42–50). Washington–DC: American Chemical Society.
33. Khan, M. I., & Giridhar, P., (2014). Enhanced chemical stability, chromatic properties and regeneration of betalains in *Rivina humilis* L. berry juice. *LWT–Food Science and Technology, 58*(2), 649–657.

34. Khanvilkar, A. M., Ranveer, R. C., & Sahoo, A. K., (2016). Carrier materials for encapsulation of bioactive components of food: A review. *International Journal of Pharmaceutical Sciences Review and Research, 40*(1), 62–73.
35. Khazaei, K. M., Jafari, S., Ghorbani, M., & Kakhki, A. H., (2014). Application of maltodextrin and gum Arabic in microencapsulation of saffron petals's anthocyanins and evaluating their storage stability and color. *Carbohydrate Polymers, 105*, 57–62.
36. Khoo, H. E., Prasad, K. N., Kong, K. W., Jiang, Y., & Ismail, A., (2011). Carotenoids and their isomers: color pigments in fruits and vegetables. *Molecules, 16*(2), 1710–1738.
37. Knezevic, Z., Gosak, D., Hraste, M., & Jalsenjako, I., (1998). Fluid bed microencapsulation of ascorbic acid. *Journal of Microencapsulation, 15*(2), 237–252.
38. Ko, J. A., Park, H. J., Hwang, S. J., Park, J. B., & Lee, J. S., (2002). Preparation and characterization of chitosan microparticles intended for controlled drug delivery. *International Journal of Pharmaceutics, 249*(1, 2), 165–174.
39. Kopelman, I. J., Meydav, S., & Wwilmersdorf, P., (1977). Storage studies of freeze dried lemon crystals. *Journal of Food Technology, 12*(4), 65–69.
40. Laos, K., Lougas, T., Mandmets, A., & Vokk, R., (2007). Encapsulation of β-carotene from sea buckthorn (*Hippophaë rhamnoides L.*) juice in furcellaran beads. *Innovative Food Science and Emerging Technologies, 8*(3), 395–398.
41. Lau, M. H., Tang, J., & Swanson, B. G., (2000). Kinetics of textural and color changes in green asparagus during thermal treatments. *Journal of Food Engineering, 45*(4), 231–236.
42. Ley, S. V., Ramarao, C., Lee, A. L., Ostergaard, N., Smith, S. C., & Shirley, I. M., (2003). Microencapsulation of osmium etroxide in polyurea. *Organic Letters, 5*(2), 185–187.
43. Loksuwan, J., (2007). Characteristics of microencapsulated β-carotene formed by spray drying with modified tapioca starch, native tapioca starch and maltodextrin. *Food Hydrocolloids, 21*(5, 6), 928–935.
44. Marquez, U. M. L., & Sinnecker, P., (2008). Chlorophylls in foods: Sources and stability. In: Socaciu, C., (ed.), *Food Colorants: Chemical and Functional Properties* (1st edn., pp. 195–212).Boca Raton: CRC Press.
45. Martins, R., Pereira, S., Siqueira, S., Salomão, W., & Freitas, L., (2013). Curcuminoid content and antioxidant activity in spray-dried microparticles containing turmeric extract. *Food Research International, 50*(2), 657–663.
46. Mínguez-Mosquera, M. I., Jaren-Galán, M., & Garrido-Fernández, J., (1992). Color quality in paprika. *Journal of Agriculture and Food Chemistry, 40*(12), 2384–2388.
47. Nilsson, L., & Bergenstahl, B., (2007). Emulsification and adsorption properties of hydrophobically modified potato and barley starch. *Journal of Agricultural and Food Chemistry, 55*(4), 1469–1474.
48. Nimmannit, U., & Suwanpatra, N., (1996). Microencapsulation of drugs by the coacervation technique using ethyl cellulose and acrylate methacrylate copolymer as wall materials. *Journal of Microencapsulation, 13*(6), 643–649.
49. Ozkan, G., & Bilek, S. E., (2015). Enzyme-assisted extraction of stabilized chlorophyll from spinach. *Food Chemistry, 176,* 152–157.
50. Pitalua, E., Jimenez, M., Vernon-Carter, E., & Beristain, C., (2010). Antioxidative Activity of microcapsules with beetroot juice using gum Arabic as wall material. *Food and Bioprocess Processing, 88*(2, 3), 253–258.
51. Porrarud, S., & Pranee, A., (2010). Microencapsulation of Zn chlorophyll pigment from pandan leaf by spray drying and its characteristic. *International Food Research Journal, 17*(4), 1031–1042.

52. Provesi, J. G., & Amante, E. R., (2015). Carotenoids in pumpkin and impact of processing treatments and storage, Chapter 9. In: Preedy, V. R., (ed.), *Processing and Impact on Active Components in Food* (1st edn., pp. 71–80). London–UK: Elsevier.
53. Ranveer, R. C., Gatade, A. A., Kamble, H. A., & Sahoo, A. K., (2015). Effect of different storage conditions on storage stability of microencapsulated lycopene. *Brazilian Archives of Biology and Technology, 58*(6), 953–960.
54. Rascon, M. P., Beristain, C., Garcia, H., & Salgado, M., (2011). Carotenoid retention and storage stability of spray-dried encapsulated paprika oleoresin using gum Arabic and soy protein isolate as wall materials. *LWT-Food Science and Technology, 44*(2), 549–557.
55. Reineccius, G. A., (1989). Flavor encapsulation. *Food Reviews International, 5*(2), 147–150.
56. Reshmi, S. K., Aravindhan, K. M., & Devi, P. S., (2012). The effect of light, temperature, pH on stability of betacyanin pigment in *Basella alba* fruit. *Asian Journal of Pharmaceutical and Clinical Research, 5*(4), 107–110.
57. Righetto, A. M., & Netto, F. M., (2005). Effect of encapsulation materials on water sorption, glass transition, and stability of juice from immature acerole. *International Journal of Food Properties, 8*(2), 337–346.
58. Rocha, G., Favaro-Trindade, C., & Ferreira, G. C., (2012). Microencapsulation of lycopene by spray drying: Characterization, stability and application of microcapsules. *Food and Bioproducts Processing, 90*(1), 37–42.
59. Rodriguez-Saona, L. E., Giusti, M. M., & Wrolstad, R. E., (1999). Color and pigment stability of red radish and red-fleshed potato anthocyanins in juice model systems. *Journal of Food Science, 64*(3), 451–456.
60. Rosenberg, M., & Lee, S. J., (2004). Water-insoluble whey protein-based microsphere prepared by an all aqueous process. *Journal of Food Science, 69*(1), 50–58.
61. Rosendal-Jensen, S., Juhl-Nielsen, B., & Dahlgren, R., (1989). Use of chemistry in plant classification. *Revista Latinoamericana de Química,* (Suppl 1), 66–89.
62. Sahat, N. S., Norulfairuz, D., Zaidel, A., Muhamad, I. I., & Zainal, A. M. N. H., (2014). Stability study of water-in-oil emulsion containing anthocyanins from red cabbage. *Jurnal Teknologi, 69*(4), 1–5.
63. Shu, B., Yu, W., Zhao, Y., & Liu, X., (2006). Study on microencapsulation of lycopene by spray-drying. *Journal of Food Engineering, 76*(4), 664–669.
64. Silva, I., Stringheta, C., Teofilo, F., & Nolasco De, O. I., (2013). Parameter optimization for spray-drying microencapsulation of jaboticaba (*Myrciaria jaboticaba*) peel extracts using simultaneous analysis of responses. *Journal of Food Engineering, 117*(4), 538–544.
65. Steet, J. A., & Tong, C. H., (1996). Degradation kinetics of green color and chlorophylls in peas by colorimetry and HPLC. *Journal of Food Science, 61*(5), 924–928.
66. Stintzing, F. C., & Carle, R., (2008). Betalains in food: Occurrence, stability, and postharvest modifications. In: Socaciu, C., (ed.), *Food Colorants: Chemical and Functional Properties* (pp. 277–290). Boca Raton: CRC Press.
67. Strack, D., Steglich, W., & Wray, V., (1993). Betalains. In: Waterman, P. G., Dey, P. M., & Harborne, J. B., (eds.), *Methods in Plant Biochemistry: Alkaloids and Sulphur Compounds* (Vol. 8, pp. 421–450). London, UK: Academic Press.
68. Sweeney, J. P., & Marsh, A. C., (1973). Liver storage of vitamin A in rats fed carotene stereoisomers. *Journal of Nutrition, 103*(1), 20–25.
69. Szente, L., & Szejtli, J., (2004). Cyclodextrins as food ingredients. *Trends in Food Science & Technology, 15*(3, 4), 137–142.

70. Takahashi, M., Kei-Ichiroinafuku, Takashi, M., Hirosuke, O., Koji, W., Tomohiro, I., & Dai, K., (2007). Effect of preparation of liposomes encapsulating food material using lecithins by a mechanochemical method. *Journal of Oleo Science, 56*(1), 35–42.
71. Tiarks, F., Landfester, K., & Antonietti, M., (2001). Preparation of polymeric nanocapsules by miniemulsion polymerization. *Langmuir, 17*(3), 908–918.
72. Tonon, R. V., Grosso, C. R. F., & Hubinger, M. D., (2011). Influence of emulsion composition and inlet air temperature on the microencapsulation of flaxseed oil by spray drying. *Food Research International, 44*(1), 282–289.
73. Tonon, R., Brabet, C., & Hubinger, D., (2010). Anthocyanin stability and antioxidant activity of spray-dried açai (*Euterpe oleracea* Mart.) juice produced with different carrier agents. *Food Research International, 43*(3), 907–914.
74. Yang, Y. Y., Chung, T. S., & Ngee, N. P., (2001). Morphology, drug distribution, and *in vitro* release profiles of biodegradable polymeric microspheres containing protein fabricated by double emulsion solvent extraction/ evaporation method. *Biomaterials, 22*(3), 231–241.
75. Zuidam, N. J., & Nedovic, V., (2010). *Encapsulation Technologies for Active Food Ingredients and Food Processing* (1st edn., p. 400). Verlag, New York: Springer.

PART III

HEALTH PROMOTING ACTIVITIES OF BIOACTIVE COMPOUNDS

CHAPTER 8

ROLE OF PHYTOCHEMICALS IN HUMAN HEALTH

BHAGWAN K. SAKHALE, NAMRATA A. GIRI, and B. B. BORSE

ABSTRACT

In the wake of awareness of disease prevention, foods that provide potential health benefits have become a new opportunity as well as a challenge for the food industry. "Let food be thy medicine and medicine be thy food," well phrased by Hippocrates, has gained more attention by the food scientists and the consumers in these days. Hence, the term functional food represents the connection between the importance of health and nutrition, which not only promotes health but also reduces the risk of diseases. Based on clinical trials, data showed that the consumption of plant-based diet having phytochemicals and phytonutrients helps in reduction of the risk of chronic diseases. Nowadays, the food industry is exploring the use of these phytochemicals in functional foods for the enhancement of human health. The consumers are more conscious about their health and demanding the specialized foods (not as the source of calories but) for the excellent delivery of some nutraceuticals and bioactive compounds. In general, phytochemicals in the development of new food products have a role to provide benefits beyond traditional caloric nutrition along with health security to the consumers. The major food commodities such as fruits, vegetables, and tuber crops possess pharmacologically active principles and important medicinal properties. Over the past few decades, there has been observed significant revival of interest among the people on various natural products as powerful antioxidants, bioactive compounds and natural drugs, therefore current research on

Novel Processing Methods for Plant-Based Health Foods: Extraction, Encapsulation, and Health Benefits of Bioactive Compounds. Megh R. Goyal, N. Veena & Ritesh B. Watharkar (Eds.)

natural products remains the main means of discovering potential principle bioactive health compounds in the existing unexplored crop species. The nutritionally sound and nutraceuticals enriched foods are superimposing the existing trade to underline their demand in national and international markets (the online marketing has made available flurry of nutraceutical and functional foods).

8.1 INTRODUCTION

Phytochemical is the word derived from Greek *'phyto'* means 'plant' and consists of two parts, 'phyto' and 'chemical,' which means chemicals of plant origin. The chemicals found in plant having potential to prevent diseases and used as therapeutics for human being. It may be the part of plant or secreted from plant which is considered as good for human health in order to prevent the occurrence of chronic disease.

Non-nutritive compounds associated with defense are plant-based bioactive compounds or 'phytochemicals.' They have been used against chronic degenerative disorders in alternative medicine and have been used in traditional remedies programs because patients are more interested in treating any illness using natural sources rather than allopathic medications [79]. Now a days the application of phytochemicals expanded particularly in nutraceuticals and functional foods [23].

The change in life style and faulty food habits leads to different life style disease such as obesity, cancer, diabetics, etc. Consumers prefer more healthy diet alternative which provide health benefits along with nutrition. The phytochemicals naturally present in plants could be acting to fulfill nutritional requirements along with health benefiting components. The increasing variety of cases of cardiovascular diseases, cancers, polygenic disorder, and lots of alternative chronic diseases, are pertained to below consumption of the fruits and vegetables in diet that conjointly contain bioactive compounds called plant chemicals or phytochemicals. Numerous epidemiological studies have shown that phytochemical-rich diets not only protect against chronic diseases [84], but also help to avoid oxidative damage to cellular systems [49].

This chapter highlights the potential of phytochemicals in human health, various sources of phytochemicals, consumers' acceptance, challenges associated, and use of phytochemicals in the development of functional foods.

8.2 CLASSIFICATION OF PHYTOCHEMICALS

Till date, different types of phytochemicals (Table 8.1) have been identified and extracted by researchers. These are classified based on various biological activity, physical characteristics, and chemical characteristics [26, 67, 79, 80].

TABLE 8.1 Classification of Major Phytochemicals

Phytochemical Class	Major Phytochemicals
Alkaloids and nitrogen-containing compounds	Quinolone, Pepperdine, pyrrolidine,
Organosulfur compounds	Indoles, glucosinolates, isothiocyanates
Phenolics	Lignans, phenolic acids, tannins, flavonoids
Terpenoids	Limonoids, carotenoids, saponins

8.3 SOURCES OF PHYTOCHEMICALS

8.3.1 FRUITS AND VEGETABLES

Fruits and vegetables are the store house of vital vitamins, minerals along with beneficial phytochemicals. These are preferred by consumers as fresh or processed forms. The known phytochemicals from these commodities are about 200,000 [57]. The nutritional health advantages and organic process importance of the potent phytochemicals is advantageous in fruits and vegetables area unit well studied and documented and principally below prescription for the person tormented by chronic diseases and diabetes, cancers, cardiovascular diseases [9]. These phytochemicals present in minute quantity in fruits and vegetables, which makes their ability to prevent the diseases [68].

The phytochemicals include antioxidants, and phenolic compounds (such as: flavonoids, carotenoids, and tocopherols). The antioxidant capacity of phytochemicals helps to fight against free radicals, responsible for the incidence of chronic disease. The researchers studied the impact of intake of fruits and vegetables in patient's diet tormented by cancers of the respiratory organ, breast, esophagus, colon, cervix, Rima, oris, pancreas, stomach, bladder, and ovary [10].

Phytochemicals recognized in some of vegetables, their chemical composition and biological activity unit of measurement stratified per their helpful activity. Fruits and vegetables inclusion in the diet can decrease the danger of certain chronic disorders [17].

8.3.1.1 CAROTENOIDS

Carotenoids having pro-vitamin A activity is after all naturally occurring plant pigment that's accountable for the colors in fruits and vegetables [31]. Keep with study, until date 600 totally different carotenoids were known and among that 50 are often born-again to axerophthol. The dietary demand of significant amine A in adult is consummated from fertilizer supply [70]. The potential health edges of carotenoids unit having high inhibitor activity that helps to cut back the possibility of cardiovascular diseases (CVDs), cancers, degeneration, and conjointly cataracts, and also helps in boosting the immune system [47]. Among carotenoids, β-carotene that's generally found associated with fruits poses the highest pro-vitamin A activity [75]. Studies showed that, β-carotene reduces the injury of skin caused because of UV [12] and conjointly reduce the danger of cancer or cardiopathy [92].

The highest inhibitor activity in carotenoids family is carotenoid. It represents the red watermelon, tomato, pink grapefruit, and alternative red fruits. The carotenoid is effective to forestall the prevalence of liver, brain, colon, breast, cervix, and prostate carcinogens, therefore eliminating or delaying bound sorts of carcinogens [15, 19]. The coronary disorder is additionally reduced by consumption of fruits and vegetables enriched in carotenoids [46]. Xanthophyll and carotenoid are also carotenoids notably found in inexperienced and yellow leaf like vegetables having potential health edges to chop back the age connected downside [72].

8.3.1.2 FLAVONOIDS

Anthocyanins, flavanones, catechins, isoflavones, and flavones are a unit resin compounds composed of flavonoids. Naturally, these flavonoids are a unit gift in fruits and vegetables like bananas, citrus fruits, broccoli, cabbage, peppers, etc. Quercetin might be a plant flavonol within the main found in onion, apple, broccoli, etc. [2, 34]. Whereas kaempferol, that's additionally a kind of flavanol found in radish, horseradish, etc. [33]. Flavonoids in a regular diet will facilitate to chop back the rationale behind varied diseases like urinary tract infections (UTI), vessel disorders, cancers, and alternative chronic diseases [20, 42], and this health benefit is because of higher inhibitor property [41, 59].

Anthocyanins are red cherry, strawberry, and blueberry pigments. In biological systems, the potent inhibitor property of anthocyanins is principally because of the scavenging activity of free radicals. The free radical scavenging (FRS) activity of anthocyanin has significance in the prevention of many diseases [3, 88]. It is well accepted that the free radicals are responsible for obscurant elementary cell parts, for instance, polymer, and thus the cell layer; nonetheless, concentrates on whether or not cancer preventing agent supplementation might limit aerobic pressure instigated observe are till now dubious [64]. In addition, anthocyanins are concerned within the treatment of capillary fragility [22].

In berries, particularly cranberries, blueberries, blackberries, and dark raspberries, pro-anthocyanidins are found. It is been demonstrated that they need an undertaking to finish in decreasing the peril of UTI that make when microorganisms are brought into the urinary plot and stick with body tissues. Pro-anthocyanidins and a few different flavonoids are fit official to cell dividers, along these lines forestalling bacterial attachment. Some studies also reported that cranberry found effective in reduction of cavity so as to reinforce dental hygiene, the gathering of certain substances that cause cavity are derived from the berries [86, 87].

The importance of quercetin is anticarcinogenic nature and prevents the formation of bad cholesterol [21]. It helps to scale back the allergic symptoms by associating with carcinogens [74]. The inclusion of quercetin in diet indicated the lowering bad cholesterol [18, 48] and protect from CVD [50].

8.3.1.3 GLUCOSINOLATES

This is the assembly of organosulfur aggravates that can be changed over into indoles and isothiocyanates. The majority of glucosinolates intake is accounted for by vegetables, for example, broccoli, kale, cabbage, cauliflower, and brussels sprout. Indoles, found in broccoli, cabbage, and other cruciferous vegetables, are known to be anti-carcinogenic as well as a potential compound for detoxification [14]. It can increase the ability of carcinogen metabolism [52]. Indole-3-carbinol can, moreover, activate cytochrome P450 enzymes that have been shown to metabolize estrogen [13] and suppose to prevent the breast and uterine cancers. Isothiocyanates, in addition, are widely distributed in watercress, broccoli, and radish, for example, are cruciferous vegetables. Numerous investigations have indicated that isothiocyanates are equipped for setting off stage II detoxification catalysts and smother stage I malignant growth advancing proteins, activities that may add to the hindrance of tumorigenesis [91].

Isothiocyanates have moreover been seemed to have a guarded effect against tumors in various tissues including the mammary organ, liver, bladder, throat, pancreas, and colon, notwithstanding diminishing the danger of cellular breakdown in the lungs [32, 91].

It was reported that the regular consumption of foods enriched in polyphenols may reduce the chance of CVD, cancer but may disturb urinary bladder dysfunctions and Alzheimer's disease [24]. The health benefits of phytochemicals are well explained by food scientists when to consume on a regular basis in diet [81]. The remaining phenolic compounds, i.e., because of their antibacterial, antiviral, antioxidant, and anticancer effects, flavonoids, isoflavonoids, anthocyanin are also important in the diet [45]. Nowadays, polyphenols are gaining a lot of importance due to its anti-stressing property [65]. Polyphenols are also having antiviral, antifungal properties, especially flavonoids [37].

Fruits and vegetables are the commodities with high phenolic compounds, naturally coloring pigments with high antioxidant activity, which prevents the chronic diseases [63]. Phenolics also have antiviral, antimicrobial, and antifungal properties. An antioxidant property of fruits and vegetables are mainly classified as vitamins, phenolics, and carotenoids [78].

8.3.2 SPICES, CONDIMENTS, AND PLANTATION CROPS

India is 'a Land of Spices' and through scientifically recorded phytochemicals and therapeutic properties, the origin, utility, taste, color, and functionality of Indian spices (major, minor, and herbs) is recognized around the world [5]. Spices and condiments are classified as "vegetable products or mixtures free of foreign matter, used in foods to flavor, season, and impart aroma" [36]. The word "applies equally to the product in the whole form or in the ground form." Total of 109 spices have been mentioned and enlisted [56].

The addition of spices into food or beverages, enhance the taste and its nutritional value and also helps to fight against diseases [7]. Now a days, demand for spices has been increased globally due to immune-boosting ability of spices which is specially required during pandemic situations. Whole spices or its extract are used in medicine, food, beverages, and other industries too [58]. The application of spices and its extract not only enhance the taste but also color and functional value of food. The compound responsible for characteristic color, flavor, and nutraceutical value is essential oil or oleoresins. These are extracted from different spices, herbs, and condiments. These are having high cost and demand in the market. These can be used to make herbal tea and for other medical applications and food flavoring [56].

8.3.3 TROPICAL TUBER CROPS

Many tropical tuber crops like cassava, sweet potato, taro, yam species, elephant foot yam, costus, coleus, typhonium species, canna species, tacca species, giant taro and aroids are grown in many tropical regions of the world. In addition to the food value the edible tuber crops possess several health benefits and medicinal properties. The tuber crops are not only storehouse of carbohydrates, but also rich sources of various antioxidant compounds like carotenoids, polyphenols, phenolic acids, flavonoids, and triterpenoids. Apart from the edible tuber crop species, which are important for their food value, a large number of tuber crops possess physiological and pharmacologically active principles and are important for their medicinal value are listed in Table 8.2, these crops are well studied extensively for their therapeutic potency. The important medicinal tuber crop species comprise of *Acorus calamus*, *Safed musli*, Asparagus, Alocasia, Alpinia, Amorphophallus (wild species), *Coleus forskohlii*, canna, ceropegia, costus, and Ipomea [66].

8.4 ROLE OF SPECIFIC PHYTOCHEMICALS

8.4.1 PHENOLICS

This is the largest category of phytochemicals consists of flavonoids, phenolic acid, and tannins. Phenolics exhibit many useful effects with their inhibitor properties as most. Phenolics exhibit many useful effects with their inhibitor properties as most vital attributable to their role in suppressing free radical-mediated malady processes [69].

The components of phenolics which protect from oxidative stress are known as phenolic acids. Chlorogenic acid is among the phenolics present at large extent naturally in fruits and vegetables. Its importance is recorded for weight loss and inhibition of fatty acid biosynthesis [8]. It is also richly present in coffee. Ferulic acid, a phenolic compound found in rice bran oil is advantageous in lowering obesity linked to a high fat diet [73].

TABLE 8.2 Tropical Tuber Crops: Potential Sources of Antioxidants

Tuber Crop	Parts Used	Phytochemical Present	Biological Activities	References
Cassava	Starch	Phenolic compounds	Antioxidant and free radical scavenging activity	[4]
	Rind	Proanthocyanidin and phenolics content	Antioxidant activity	[71]
	Stem	Phenolic compounds	Antioxidant activity	[90]
Taro	Tubers	Phenolics compounds and peptides	Anticancerous activity (antimetastatic activity)	[53]
	Tubers, stem, and leaves	Octadecadienoic acid and hexadecanoic acid	Anticancer activity	[85]
Elephant foot yam	Tubers	Flavonoids	Antioxidant activity	[38]
Sweet potato	Leaves	Carotenoid	Anticancer activity	[27]
	Leaves	Phenolic and flavonoids content	Antioxidant activity	[35]
	Tubers	Anthocyanins and phenolics content	Radical scavenging activity	[11]
	Purple sweet potato tubers	Anthocyanins	Radical scavenging effects	[39]
	Storage roots	Phenolic compounds	Antioxidant activity	[77]
Giant swamp Taro	Tubers	Mucilage content	Antioxidant activity	[60]
West Indian arrowroot	Whole plants (tubers, leaves, and flowers)	Phenols and flavonoids	Anticancer activity	[55]
Curcuma zedoaria, Curcuma angustifolia, Curcuma caesia	Rhizomes	Phenolic compounds	Antioxidant activity and free radical quenching ability	[25]

Tannins are small molecular weight phenols and commercially utilized in aroma therapy products and dietary supplements. It can be of two forms, such as condensed tannins and hydrolysable tannins. Because of ecological pressure, tainting, and openness to extreme bright light, they are fabricated in plants and furthermore diminish the damage caused. The largest group under phenols is flavonoids which are classified as anthocyanins and anthoxanthins. The presence of anthocyanins that have high antioxidant activity and function as anti-inflammatory and anti-obese properties is due to the purple, blue, and red color of vegetables and fruits. White molecule is nothing but anthoxanthins which is further categorized into flavonols, flavanone, flavones, and isoflavones [82].

8.4.2 TERPENOIDS

Phytochemical responsible for growth, metabolism, and development of plant known as terpenoids [69]. The commercial utilization of terpenoids is in preparation of functional foods, flavoring agents in ice creams, non-alcoholic beverages, chewing gum, candy, bakery products, etc., and biocolorant in pharmaceuticals, cosmetics, etc. The main class of terpenoids is carotenoids which are responsible for yellow and orange color in different fruits and vegetables. It not only gives color but also acts as an antioxidant [44]. Carotenoids are further classified as carotene and xanthophyll. Carotene is reported to help protect against cancers of the uterus, prostate, breast, colorectal, and lungs. Xanthophylls, on the other hand, also act as antioxidants. Zeaxanthin, cryptoxanthin, and astaxanthin are significant types of xanthophyll used in different preparations as functional food ingredients [69].

8.4.3 GLUCOSINOLATES

Sulphur containing phytochemicals mostly found in cruciferous vegetables such as cabbage, broccoli, and also in mustard seed as well as rape seed which have anticarcinogenic properties are known as glucosinolates [62].

8.4.4 POLYACETYLENES

Polyacetylenes are natural chemically reactive metabolites derived from various flora. In the Apiaceae (fennel, celery, and carrot), Araliaceae (hedraspp and ginseng) and Asteraceae (sunflower, lettuce, chicory, and artichoke) gatherings, their event is most noteworthy. The natural pesticides having three compounds viz. falcarinol, falcarindiol, and falcarindiol-3-acetate are released as a natural defense against pest attack in carrots and have a strong functional profile as a functional ingredient that has recently received considerable scientific attention [1].

8.4.5 *PHYTOSTEROLS AND PHYTOSTANOLS*

Plant sterols are called phytosterols. Phytosterols are fundamentally derived from vegetable oils, cereals, and organic products, while phytosterols are bountiful in maize, wheat, rye, and rice and are available in acceptable sums. The function of cholesterol in the human body and phytosterol in plant are the same. The different forms of phytosterols are free alcohol, fatty acid esters, steryl glycosides, acylatedsteryl glycosides and phytosteryl-hydroxycinnamic-acid esters. Both phytosterols and phytostanols majorly found in maize oil, rapeseed oil, sunflower oil, soybean oil, nuts, beans, and grains [30]. These are essential for reducing cholesterol, cancer prevention, immunomodulation, and skin protection [16].

8.4.6 *NON-DIGESTIBLE CARBOHYDRATES (NDC)*

The complex carbohydrates which resist to digestion are known as non-digestible carbohydrates (NDC). Non-digestible oligosaccharides, resistant starch, non-starch polysaccharides are graded. These are important in the human diet, which improves the gastrointestinal (GI) health by preventing constipation, diverticular disease, irritable bowel syndrome, and colon cancer [53]. Complex, heterogeneous dietary substances derived mainly from plants are NDC. These are important components of the diet and GI disorders can often result from insufficient intake.

The closely resembling carbs that fall into two gatherings as indicated by their solvency are dietary filaments, for example, water-dissolvable strands. The potential health benefits of resistant starch, which either undergo slow digestion or doesn't undergo digestion, passes to GI tract, and fermented in the colon and converted to short-chain fatty acids, which help to lowers cholesterol level and acts as substrate for probiotic microflora.

8.5 FUNCTIONAL FOODS BASED ON PHYTOCHEMICALS

Exposure to individual phytochemicals is less convincing, and epidemiological studies, natural, and exploratory assessments, and preliminary clinical trials [29, 51] revealed that plant-based foods reduce the risk of degenerative infections, especially fatal neoplasms; [40, 43, 89]. This is because a plant-based diet prevents almost half of all cancers [75, 89]. In addition, nutritional proposals to combat all forms of life-threatening neoplasms and other persistent diseases have continually emphasized the need to use a range of plant-based food sources to validate wellness.

The single compound methodology has offered route to the idea that general assurance against illness is given by a scope of phytochemicals contained in food sources and flavors, for example, cinnamon, ginger, pepper, tulsi – *Ocimum sanctum* [83] even some antagonistic reports accessible in specific species and physiological conditions (like in pregnancy, liver injury and so on) for well-being and alert [76].

Plant-based foods and drinks (coarse grains/millet, legumes, fenugreek, quinoa, soybeans, fruits, berries, all main/minor spices including herbs, grapes, and wine, citrus fruits, tomatoes, flax seeds, oats, cruciferous vegetables, tea, coffee, cocoa, herbal teas and their compounds and phytochemicals, etc.), are proposed or are created as functional foods, each with series of logical insights into positive effects of dietary supplements and phytochemicals on well-being. The majority of phytochemicals tend to deteriorate in handling and capacity.

Therefore, the creation of food has to be combined with an ideal handling and capacity of food, which can significantly affect the welfare-promoting capacity of food. As mentioned earlier, manufacturing is important to modify some compounds and their bioavailability by improving an appearance of bioactive mixtures from the food structure, an underlying cycle stage of the human stomach [28, 61].

Thereafter, the ultimate aim of the handling conditions should be to prevent the calculable problems of the phytochemical while improving its bioavailability.

8.6 CURRENT TRENDS AND CHALLENGES

Over the past few decades, the number of known physically active phytochemicals has increased significantly. The phytochemicals and their beneficial effects on human health have been revealed by many researchers. A clear correlation is observed between antioxidant intake and lower risk of various lifestyle diseases like heart cancer, diabetes, hypertension, and other medical conditions [21, 78].

In the 1990s, foods with specific functions and nutritional value emerged as a major trend for the food industry internationally. Nutritionists and scientists have often documented the nutritional significance and health benefits of food and are legally supported by public policy and legislative mandates for diets and dietary supplements. The developments have received considerable attention from agriculture, biotechnology, and life science-based industries for developing the raw materials for nutrition and food manufacturers to design new products. The new foods are clearly prepared for the 21st century of the sunrise industry. They promise value-added and new marketing opportunities in food industries.

They offer advances in public health as the marketing messages of health claims enable consumers to choose healthier foods. Making the newly designed and repositioned food products available to consumers is seen as a challenge and an opportunity for added value to the industry. Healthcare professionals are gradually realizing the role of phytochemicals in improving health. As antithrombotic, anti-inflammatory, and carcinogenic agents, phytochemicals such as phenolic compounds are of considerable importance. In both cases, companies in the food industry tend to focus on normal businesses due to the expected adverse effects of phytochemicals fortifications and can be used as substances added to food or as drug additives. Particular attention to regulatory issues and careful consideration of conveying meaningful messages to consumers about the benefits of products are necessary for marketing and positioning of health claims products in the functional food sector. An elaboration of the product positioning is proposed, taking into account that consumer's receptivity is often dependent on perception of taste, acceptance, quality, and well-being rather than the stated specifications for product strength and clinical benefit.

The regulations govern the language and scientific benefits that can be expressed in product labels, literature of marketing and advertisement of the products. However, consumers are often more satisfied with the subtleties, associations, and promises of fitness than with scientific claims and digital literature. In order to target genetic weapons in plants, food regulators must agree that increasing levels will provide health benefits to all segments of the population or, conversely, will not pose a threat to certain groups of the population. This will be a difficult challenge. Growing traditional plants can lead to major changes in the composition of food plants that have not been well documented, but there is no evidence of adverse health effects. Moreover, an additional data is needed to understand the harmonic effects of phytochemicals found in foods and how they can protect humans from chronic disease.

8.7 SUMMARY

Developed countries like Japan, the USA, Russia, Australia, and European countries are the main exporters of the Indian food products, extracts, phytochemicals, and spices (suggested by Spice Council, India). Such types of developed nations have their own stringent regulations and laws for foods as we now have unified food law, food standards and safety measures for Indian foods through FSSAI. The major objectives of such stringent laws are to safeguard the health of the people. The developed nations like USA and Russia only permits the import of raw commodities such as foods, spices, plant products (tea, coffee, cocoa, etc.), only if they comply with their mandates regarding food safety rules and converts these commodities into high value extracts and phytochemicals for a very high export value [6].

Globally major spices produced include chili/paprika (capsaicin), turmeric (curcumin), ginger (gingerol and shogol), black pepper (piperine), cardamom (sineol and alpha terpenyl acetate), garlic (allicin) and coriander (decanol), cumin (cumin aldehyde), celery (epigenin), fennel (anethole), fenugreek (galactomannan), azoan (thymol), dead seed (dilapiol), tamarind, cloves (eugenol), nutmeg, *Mentha arvensis* (menthol), basil (ascamine), rosemary (rosamarinol), thyme, marjoram, etc. The commercial platform for these commodities is not organized and the specialty segment share is nearly 15%. The players like MTR, Badshah, Catch, Everest, Ramdev, etc., dominate organized Indian spice market.

To promote the export of food, spices, and phytochemicals and to better manage the industry, food business operators have high-quality food, beverage, phytochemicals, and spices manufactured, sold, stored, and distributed. There are standards that must be maintained in the market. For countries around the world, this COVID-19 challenge includes food safety and hygiene (HACCP).

KEYWORDS

- **cardiovascular diseases**
- **chronic disease**
- **functional food**
- **human health**
- **lycopene**
- **phytochemicals**
- **phytonutrients**

REFERENCES

1. Acworth, I., Plante, M., Bailey, B., Crafts, C., & Waraska, J., (2011). Simple and direct analysis of falcarinol and other polyacetylene oxylipins in carrots by reversed-phase HPLC and charged aerosol detection. *Planta Medica, 77*(12), 1263–1270.
2. Aherne, S. A., & O'Brien, N. M., (2002). Dietary flavonols: Chemistry, food content and metabolism. *Nutrition, 18*(1), 75–81.
3. Ames, B. N., Shigenga, M., & Hagen, T. M., (1993). Oxidants, antioxidants and degenerative diseases of aging. *Proceeding of National Academy of Sciences USA, 90*(17), 7915–7922.
4. Amornpitak, T. Y., Koguchi, M., & Teramoto, Y., (2012). Antioxidant activity of herbal wine made from cassava starch. *World Applied Sciences Journal, 16*(6), 874–878.
5. Annual Report, (2017–2018). *Spices Board, Ministry of Commerce and Industry* (p. 63). Government of India, Cochin. http://indianspices.com/sites/default/files/Spices%20Board%20Annual%20Report%202017-18.pdf (accessed on 30 May 2022).
6. ASTA. (2022). *American Spice Trade Association: The Expert Voice of the U.S. Spice Industry in the Global Market*. Washington, DC, USA; https://www.astaspice.org (accessed on 30 May 2022).
7. Ayush Ministry, Government of India. https://www.ayush.gov.in (accessed on 30 May 2022).
8. Baile, C. A., Yang, J. Y., Rayalam, S., Hartzell, D. L., Lai, C. Y., Andersen, C., & Della-Fera, M. A., (2011). Effect of resveratrol on fat mobilization. *Annals of the New York Academy of Sciences, 1215*(1), 40–47.
9. Bureau of Indian Standards (BIS). *Ministry of Consumer Affairs, Food and Public Distribution, Government of India, Manak Bhawan, Old Delhi*. https://bis.gov.in (accessed on 30 May 2022).
10. Block, G., Patterson, B., & Subar, A., (1992). Fruit, vegetables and cancer prevention: A review of the epidemiological evidence. *Nutrition and Cancer, 18*(1), 1–29.
11. Boliavar, A., Cevallos, C., & Zevallos, L. C., (2003). Stoichiometric and kinetic studies of phenolic antioxidants from Andean purple corn and red-fleshed sweet potato. *Journal of Agricultural and Food Chemistry, 51*(11), 3313–3319.

12. Bose, B., & Chatterjee, S. N., (1993). Effect of UVA irradiation on linoleic acid micelles. *Radiation Research, 133*(3), 340–344.
13. Bradfield, C. A., & Bjeldanes, L. F., (1984). Effect of dietary indole-3-carbinol on intestinal and hepatic monooxygenase, glutathione S-transferase and epoxide hydrolase activities in the rat. *Food and Chemical Toxicology, 22*(12), 977–982.
14. Bradfield, C. A., & Bjeldanes, L. F., (1987). Structure-activity relationships of dietary indoles. A proposed mechanism of action modifiers of xenobiotic metabolism. *Journal of Toxicology and Environmental Health, 21*(3), 311–323.
15. Bramley, P. M., (2000). Is lycopene beneficial to human health? *Phytochemistry, 54*(3), 233–236.
16. Bruce, J., & Grattan, J., (2013). Plant sterols as anticancer nutrients: Evidence for their role in breast cancer. *Nutrients, 5*(2), 359–387.
17. Chen, L., Vigneault, C., Vijaya, R. G. S., & Kubow, S., (2007). Importance of the phytochemical content of fruits and vegetables to human health. *Stewart Postharvest Review, 3*(2), 1–5.
18. Chopra, M., Fitzsimons, P. E., Strain, J. J., Thurnham, D. I., & Howard, A. N., (2000). Non-alcoholic red wine extract and quercetin inhibit LDL oxidation without affecting plasma antioxidant vitamin and carotenoid concentrations. *Clinical Chemistry, 46*(8), 1162–1170.
19. Clinton, S. K., (1998). Lycopene: Chemistry, biology, and implications for human health and disease. *Nutrition Reviews, 56*(2), 35–51.
20. Cox, B. D., Whichelow, M. J., & Prevost, A. T., (2000). Seasonal consumption of salad vegetables and fresh fruit in relation to the development of cardiovascular disease and cancer. *Public Health Nutrition, 3*(1), 19–29.
21. Craig, W. J., (1997). Phytochemicals: Guardians of our health. *Journal of the American Dietetic Association, 97*(10), S199–S204.
22. Cristoni, A., & Magistretti, M. J., (1987). Antiulcer and healing activity of *Vaccinium myrtillus* anthocyanosides. *IL Farmaco, 42*(2), 29–43.
23. Das, L., Bhaumik, E., Raychaudhuri, U., & Chakraborty, R., (2011). Role of nutraceuticals in human health. *Journal of Food Science and Technology, 49*(2), 173–183.
24. De Pascual-Teresa, S., Moreno, D. A., & Garcia-Viguera, C., (2010). Flavanols and anthocyanins in cardiovascular health: A review of current evidence. *International. Journal of Molecular Sciences, 11*(4), 1679–1703.
25. Dhal, Y., Deo, B., & Sahu, R. K., (2012). Comparative antioxidant activity of non-enzymatic and enzymatic extracts of *Curcuma zedoaria, Curcuma angustifolia* and *Curcuma caesia. International Journal of Plant, Animal and Environmental Science, 2*(4), 232–239.
26. Dillard, C. J., & German, J. B., (2000). Phytochemicals: Nutraceuticals and human health. *Journal of the Science of Food and Agriculture, 80*(12), 1744–1756.
27. Fidrianny, I., Windyaswari, A. S., & Wirasutisna, K. R., (2013). DPPH scavenging activity of various extracts of sweet potato leaves with varying tubers colors. *International Journal of Research in Pharmacy and Science, 3*(2), 133–145.
28. Gartner, C., Stahl, W., & Sies, H., (1997). Lycopene is more bioavailable from tomato paste than from fresh tomatoes. *American Journal of Clinical Nutrition, 66*(1), 116–122.
29. Gholap, S., & Kar, A., (2004). Hypoglycaemic effects of some plant extracts are possibly mediated through inhibition in corticosteroid concentration. *Die Pharmazie, 59*(11), 876–878.

30. Harbourne, N., Marete, E., Jacquire, J. C., & O'Riordan, D., (2013). Stability of phytochemicals as sources of anti-inflammatory nutraceuticals in beverages: A review. *Food Research International, 50*(2), 480–486.
31. Heber, D., (2004). Vegetables, fruits and phytoestrogens in the prevention of diseases. *Journal of Postgraduate Medicine, 50*(2), 145–149.
32. Hecht, S. S., (1995). Chemoprevention by isothiocyanates. *Journal of Cellular Biochemistry, 59*(22), 195–209.
33. Hertog, M. G. L., Feskens, E. J. M., Hollman, P. C. H., Katan, M. B., & Kromhout, D., (1993). Dietary antioxidant flavonoids and risk of coronary heart disease: The Zutphen elderly study. *Lancet, 342*(8878), 1007–1011.
34. Hollman, P. C. H., & Arts, I. C. W., (2000). Flavonols, flavones and flavanols–nature, occurrence and dietary burden. *Journal of the Science of Food and Agriculture, 80*(7), 1081–1093.
35. Hue, S. M., Boyce, A. N., & Somasundram, C., (2012). Antioxidant activity, phenolic and flavonoid contents in the leaves of different varieties of sweet potato (*Ipomoea batatas*). *Australian Journal of Crop Science, 6*(3), 375–380.
36. ISO, (1995). *ISO 676:1995, Spices and Condiments–Botanical Nomenclature* (p. 25).
37. Ippolito, A., & Nigro, F., (2005). Natural antimicrobials for preserving fresh fruit and vegetables. In: Wim, J., (ed.), *Improving the Safety of Fresh Fruit and Vegetables* (pp. 201–234). Boca Raton, DC–USA: CRC Press.
38. Jagatheesh, K., Arumugam, V., Elangovan, N., & PavanKumar, P., (2010). Evaluation of the anti-tumor and antioxidant activity of *Amorphophallus paeonifolius* on DMBA induced mammary carcinoma. *International Journal of Chemistry and Pharmaceutical Science, 1*(2), 40–50.
39. Jiao, Y., Jiang, Y., Zhai, W., & Yang, Z., (2012). Studies on antioxidant capacity of anthocyanin extract from purple sweet potato (*Ipomoea batatas* L.). *African Journal of Biotechnology, 11*(27), 7046–7054.
40. Kaur, C., & Kapoor, H. C., (2001). Antioxidants in fruits and vegetables–the millennium's health. *International Journal of Food Science and Technology, 36*(7), 703–725.
41. Kinsella, J. E., Frankel, E., German, B., & Kanner, J., (1993). Possible mechanisms for the protective role of antioxidants in wine and plant foods. *Food Technology, 47*(4), 85–89.
42. Knekt, P., Jarvinen, R., Reppanen, R., Heliovaara, M., Teppo, L., Pukkala, E., & Aroma, A., (1997). Dietary flavonoids and the risk of lung cancer and other malignant neoplasms. *American Journal of Epidemiology, 146*(3), 223–230.
43. Law, M. R., & Morris, J. K., (1998). By how much does fruit and vegetable consumption reduce the risk of ischaemic heart disease? *European Journal of Clinical Nutrition, 52*(8), 549–556.
44. Liebler, D. C., (1996). Antioxidant reactions of carotenoids. *Annals of the New York Academy of Sciences, 691*(1), 20–31.
45. Makkar, H. P. S., Sidhuraju, P., & Becker, K., (2007). *Plant Secondary Metabolites* (pp. 130). Totowa–New Jersey: Humana Press.
46. Manson, J. E., Gaziano, J. M., Jonas, M. A., & Hennekens, C. H., (1993). Antioxidants and cardiovascular disease: A review. *Journal of the American College of Nutrition, 12*(4), 426–432.
47. Mathews-Roth, M. M., (1991). Recent progress in the medical applications of carotenoids. *Pure and Applied Chemistry, 63*(1), 147–156.

48. McAnlis, G. T., McEneny, J., Pearce, J., & Young, I. S., (1999). Absorption and antioxidant effects of quercetin from onions in man. *European Journal of Nutrition, 53*(2), 92–96.
49. McClements, D. J., (2020). Advances in nanoparticle and microparticle delivery systems for increasing the dispersibility, stability, and bioactivity of phytochemicals. *Biotechnology Advances, 38*, 107287.
50. Mirsafaei, L., Reiner, Ž., Shafabakhsh, R., & Asemi, Z., (2020). Molecular and biological functions of quercetin as a natural solution for cardiovascular disease prevention and treatment. *Plant Foods for Human Nutrition, 75*(3), 307–315.
51. Mondal, S., Varma, S., Bamola, V. D., Naik, S. N., Mirdha, B. R., Padhi, M. M., & Mahapatra, S. C., (2011). Double-blinded randomized controlled trial for immunomodulatory effects of tulsi (*Ocimum sanctum* Linn.) leaf extract on healthy volunteers. *Journal of Ethnopharmacology, 136*(3), 452–456.
52. Morse, M. A., LaGreca, S. D., Amin, S. G., & Chung, F. L., (1990). Effects of indole-3-carbinol on lung tumorigenesis and DNA methylation induced by 4-(methylnitrosamino)-1-(3-pyridyl)-1-butanone (NNK) and on the metabolism and disposition of NNK in A/J mice. *Cancer Research, 50*(9), 2613–2617.
53. Mussatto, S. I., & Mancilha, I. M., (2007). Non-digestible oligosaccharides: A review. *Carbohydrate Polymers, 68*(3), 587–597.
54. Namita, K., Patricia, C., Brian, H., Lina, C., Xinrong, M., Nicholas, A., Frank, Z., et al., (2012). Antimetastatic activity isolated from *Colocasia esculenta* (Taro). *Anticancer Drugs, 23*(2), 200–211.
55. Nishaa, S., Vishnupriya, M., Sasikumar, J. M., Christabel, H. P., & Gopalakrishnan, V. K., (2012). Antioxidant activity of ethanolic extract of *Maranta arundinacea* tuberous rhizomes. *Asian Journal of Pharmaceutical and Clinical Research, 5*(4), 85–88.
56. Parthasarathy, V. A., & Kandiannan, K., (2007). *Spices and Condiments* (p. 171). Indian Institute of Spices Research (ICAR), Calicut.
57. Patra, A. K., (2012). *Dietary Phytochemicals and Microbes* (p. 400). New York: Springer Science & Business Media.
58. Rajashri, K., Mudhol, S., Peddha, M. S., & Borse, B. B., (2020). Neuroprotective effect of spice oleoresins on memory and cognitive impairment associated with scopolamine-induced Alzheimer's disease in rats. *ACS Omega, 5*(48), 30898–30905.
59. Rice-Evans, C. A., Miller, N. J., & Paganga, G., (1997). Antioxidant properties of phenolic compounds. *Trends in Plant Science, 2*(4), 152–159.
60. Richard, M. N., Thaddee, B., Nicolas, Y. N., Makhlouf, H., Joel, S., & Carl, M. F. M., (2014). Mucilage chemical profile and antioxidant properties of giant swamp taro tubers. *Journal of Food Science and Technology, 51*(12), 3559–3567.
61. Rock, C. L., Lovalvo, J. L., Emenhiser, C., Ruffin, M. T., Flatt, S. W., & Schwart, S. J., (1998). Bioavailability of β-carotene is lower in raw than in processed carrots and spinach in women. *The Journal of Nutrition, 128*(5), 913–916.
62. Rodríguez-Fragoso, L., Osuna-Martínez, U., Gonzaga-Morales, A. I., & Reyes-Esparza, J., (2013). Potential antioxidant benefits of commonly used fruits and vegetables around the world; Chapter 2. In: Pereira, D. A. M., (ed.), *Medicinal Plants: Antioxidant Properties, Traditional Uses and Conservation Strategies* (pp. 41–90). Hauppauge, NY–USA: Nova Science Publishers Inc.
63. Sacheck, J., & Blumberg, J., (2001). Role of vitamin E and oxidative stress in exercise. *Nutrition, 17*(10), 809–814.

64. Sakakibara, H., & Shimoi, K., (2020). Anti-stress effects of polyphenols: Animal models and human trials. *Food and Function, 11*(7), 5702–5717.
65. Sakhale, B. K., & Giri, N. A., (2020). Nutritional values and processing of tropical tuber crops; Chapter 9. In: Deka, S. C., Seth, D., & Swami, H. N. R., (eds.), *Technologies for Value Addition in Food Products and Processes* (pp. 191–233). Boca Raton: Apple Academic Press.
66. Saxena, M., Saxena, J., Nema, R., Singh, D., & Gupta, A., (2013). Phytochemistry of medicinal plants. *Journal of Pharmacognosy and Phytochemistry, 1*(6), 168–182.
67. Schreiner, M., & Huyskens-Keil, S., (2006). Phytochemicals in fruit and vegetables. *Critical Reviews in Plant Sciences, 25*(3), 267–278.
68. Sharma, R., Kumar, S., Kumar, V., & Thakur, A., (2019). Comprehensive review on nutraceutical significance of phytochemicals as functional food ingredients for human health management. *Journal of Pharmacognosy and Phytochemistry, 8*(5), 385–395.
69. Silva, I. D., Rodrigues, A., Gaspar, J., Maia, R., Laires, A., & Rueff, J., (2006). Mutagenicity of kaempferol in V79 cells: The role of cytochromes P450. *Teratogenesis, Carcinogenesis, and Mutagenesis, 16*(4), 229–241.
70. Simpson, K. L., (1982). Relative value of carotenoids as precursors of vitamin A. *Proceedings of the Nutrition Society, 42*(1), 7–17.
71. Siyumbwa, S., Nomishan, N., Benny, & Okechukwu, P. N., (2014). Antioxidant and gastro-protective activity of ethanolic rind extract of *Manihot esculenta* Crantz. *International Journal of Pharmaceutical Science and Research Scholars, 3*(1), 798–808.
72. Sommerburg, O., Keunen, J. E., Bird, A. C., & Kuijk, F., (1998). Fruits and vegetables that are sources for lutein and zeaxanthin: The macular pigment in human eyes. *British Journal of Ophthalmology, 82*(8), 907–910.
73. Son, M. J., Rico, C. W., Nam, S. H., & Kang, M. Y., (2010). Influence of oryzanol and ferulic acid on the lipid metabolism and antioxidative status in high fat-fed mice. *Journal of Clinical Biochemistry and Nutrition, 46*(2), 150–156.
74. Starvic, B., Matula, T. I., Klassen, R., Downie, R. H., & Wood, R. J., (1992). Effect of flavonoids on mutagenicity and bioavailability of xenobiotics in food. In: Huang, H. M. T., Ho, C. T., & Lee, C. Y., (eds.), *Phenolic Compounds in Food and Their Effects on Health. II. Antioxidants and Cancer Prevention* (*ACS Symposium Series 507)* (pp. 239–249). Washington–DC: American Chemical Society.
75. Steinmetz, K. A., & Potter, J. D., (1996). Vegetables, fruit, and cancer prevention: A review. *Journal of the American Dietetic Association, 96*(10), 1027c1039.
76. Suzuki, Y., Yamazaki, Y., Hashizume, H., Oyama, T., Horiguchi, N., Sato, K., Kakizaki, S., & Yamada, M., (2015). Drug-induced liver injury caused by a dietary supplement (kin-toki shoga (®)) made from ginger. *Nihon Shokakibyo Gakkai Zasshi, 112*(1), 108–114.
77. Teow, C. C., Truong, V. D., McFeeters, R. F., Thompson, R. L., Pecota, K. V., & Yencho, G. C., (2007). Antioxidant activities, phenolic and β-carotene contents of sweet potato genotypes with varying flesh colors. *Food Chemistry, 103*(3), 829–838.
78. Thaiponga, K., Boonprakoba, U., Crosbyb, K., Cisneros-Zevallosc, L., & Byrnec, D. H., (2006). Comparison of ABTS, DPPH, FRAP and ORAC assays for estimating antioxidant activity from guava fruit extracts. *Journal of Food Composition and Analysis, 19*(7), 669–675.
79. Thakur, A., & Sharma, R., (2018). Health-promoting phytochemicals in vegetables: A mini-review. *International Journal of Food and Fermentation Technology, 8*(2), 107–117.

80. Tharanathan, R. N., (2002). Food-derived carbohydrates: Structural complexity and functional diversity. *Critical Reviews in Biotechnology, 22*(1), 65–84.
81. Tlili, N., Mejri, H., Yahia, Y., Saadaoui, E., Rejeb, S., Khaldi, A., & Nasri, N., (2014). Phytochemicals and antioxidant activities of *Rhus tripartitum* (ucria) fruits depending on locality and different stages of maturity. *Food Chemistry, 160*, 98–103.
82. Tsuda, T., (2008). Regulation of adipocyte function by anthocyanins; possibility of preventing the metabolic syndrome. *Journal of Agricultural and Food Chemistry, 56*(3), 642–646.
83. Ulbricht, C., Seamon, E., Windsor, R. C., Armbruester, N., Bryan, J. K., Costa, D., Giese, N., et al., (2011). An evidence-based systematic review of cinnamon (*Cinnamomum* spp.) by the natural standard research collaboration. *Journal of Dietary Supplements, 8*(4), 378–454.
84. Vinson, J. A., Su, X., Zubik, L., & Bose, P., (2001). Phenol antioxidant quantity and quality in foods: Fruits. *Journal of Agricultural and Food Chemistry, 49*(11), 5315–5321.
85. Wei, L. S., Wee, W., Siong, J. Y. F., & Syamsumir, D. F., (2011). Antimicrobial, antioxidant, anticancerous property and chemical composition of different parts (corm, stem, leaves) of *Colocasia esculenta* extract. *Annales-Medical University Lubin-Polonia, 24*(3), 9–16.
86. Weiss, E. L., Lev-Dor, R., Sharon, N., & Ofek, I., (2002). Inhibitory effect of a high-molecular-weight constituent of cranberry on adhesion of oral bacteria. *Critical Reviews in Food Science and Nutrition, 42*(3), 285–292.
87. Wilson, T., & Marley, J. C., (2001). Effects of cranberry juice consumption on platelet aggregation. *FASEB Journal, 15*(4), A286.
88. Witt, E. H., Reznick, A. Z., Viguie, C. A., Starke-Reed, P., & Packer, L., (1992). Exercise, oxidative damage and effects of antioxidant manipulation (review). *The Journal of Nutrition, 122*(3), 766–773.
89. World Cancer Research Fund, (1997). *Food, Nutrition and the Prevention of Cancer: A Global Perspective* (Vol. 11, pp. 9–23). Washington-DC: American Institute for Cancer Research.
90. Yi, B., Hu, L., Mei, W., Zhou, K., Wang, H., Luo, Y., Wei, X., & Dai, H., (2010). Antioxidant phenolic compounds of cassava (*Manihot esculenta*) from Hainan. *Molecules, 16*(12), 10157–10167.
91. Zhang, Y., & Talalay, P., (1994). Anticarcinogenic activities of organic isothiocyanates: Chemistry and mechanisms. *Cancer Research, 54*(7), 1976–1981.
92. Ziegler, R. G., (1991). Vegetables, fruits, and carotenoids and the risk of cancer. *American Journal of Clinical Nutrition, 53*(Suppl. 1), 251S–259S.

CHAPTER 9

NUTRACEUTICALS WITH HEALTH-PROMOTING ACTIVITIES

SANGEETA SAIKIA, NIKHIL K. MAHNOT, and KULDEEP GUPTA

ABSTRACT

For consumers, nutraceuticals are delivered as any dietary supplement with naturally derived bioactive components with health benefits. Nutraceuticals in disease management is a new alternative to modern medicine and has gained preference in the present society. The bioactive components exhibit many health-promoting properties, such as antioxidant, anticancer, anti-inflammatory, antibacterial, antidiabetic, antihypertensive, etc. The nutraceuticals are generally categorized based on their sources and chemical properties and their mode of action. The bioactive compounds like phenolics and flavonoids are commonly derived from plant-based sources, although other components like unsaturated fatty acids (MUFA and PUFA), biopeptides are from animals. In recent times, researchers have explored the possibility of extraction and application of novel bioactives obtained from medicinal plants and marine sources, which are not conventional food sources for all. This will promote the development of newer nutraceuticals with enhanced therapeutic properties to cater to the needs of the ever-changing consumers.

9.1 INTRODUCTION

The current society is well informed regarding the correlation of a balanced diet and health. There is an increased awareness to maintain a healthy life. Consequently, the demand for healthy and nutritious food as well as

Novel Processing Methods for Plant-Based Health Foods: Extraction, Encapsulation, and Health Benefits of Bioactive Compounds. Megh R. Goyal, N. Veena & Ritesh B. Watharkar (Eds.)

products with added health-promoting properties has gone up. This has fueled an increase in research to determine the health-promoting properties of bioactive compounds derived from plant-based food sources as well as the production of newer products with enhanced health benefits. Generally, such products grouped under functional foods and nutraceuticals. Functional foods are those food products which impart health-promoting effects apart from fulfilling basic nutritional needs. On the other hand, nutraceuticals are isolated or purified bioactive rich components obtained from natural sources and aimed for prevention and management of chronic diseases and are delivered in high concentrations as form of tablets or pills.

The term 'nutraceutical' was coined by Dr. Stephen L. DeFelice. It refers to the combining of the field of both 'nutrition' and 'pharmaceutics.' Nutraceuticals are derived from both herbal and other natural sources. At present, nutraceuticals are the new additions in management of chronic degenerative diseases. They consist of biologically active components, termed as bioactive compounds, which has many health-promoting properties like antioxidant, anticancer, anti-inflammatory, antimicrobial, anticancer, etc.

The present chapter gives an overview on the bioactive compounds, their role in health management and as viable source of nutraceuticals.

9.2 BIOACTIVE COMPOUNDS

Secondary plant metabolites are the prime sources of bioactive compounds. They are widely regarded to boost human health and have proven to be beneficial in the management and avoidance of varied chronic disorders [43, 65]. They are classified into polyphenols, vitamins, minerals, natural pigments, dietary fiber, phytosterols, bioactive lipids, bioactive peptides (BP), etc. Bioactive constituents are abundantly present in varied food sources, including fresh vegetables and raw fruits, nuts, and seeds, marine algae, meat, dairy, marine fish, medicinal plants, etc.

9.2.1 POLYPHENOLS

The polyphenols comprise of one or more hydroxyl-substituted benzene ring, i.e., having a polyphenolic structure. Till date > 8,000 polyphenolic structures have been identified and studied [65]. Polyphenols can be broadly classified into flavonoid and non-flavonoid groups. The flavonoid group is further divided into sub-classes of flavones, flavonol, flavanone, isoflavones, dihydroflavanols, flavan-3-ols, anthocyanidins, and proanthocyanidins [22]. The non-flavonoid group consists of tannins, phenolic acids, xanthones, acetophenones, chalcone, stilbenes, lignans, secoiridoids, and phenylacetic acid.

Further, the phenolic acids are sub-classed into hydoxycinnamic acid and hydroxybenzoic acid derivatives. Hydroxybenzoic acids are derivatives of benzoic acids having a C6–C1 type general, while hydroxycinnamic acids are cinnamic acid derivatives having C6–C3 type structures.

The common examples of hydroxybenzoic acid derivatives are gallic, gentisic, salicylic, ellagic and vanillic acids. Whereas ferulic, caffeic, chlorogenic, sinapic, coumaric acids, etc., belong to hydroxycinnamic acid derivatives.

Flavonoids are phenolics having a diphenylpropane skeleton (C6–C3–C6) structure, consisting of two aromatic rings linked through a heterocyclic closed pyrene ring (containing oxygen) [31]. Fruits, vegetables, and medicinal plants are major sources of polyphenols.

9.2.2 NATURAL PIGMENTS

Pigments are present in almost all living organisms, but the plants are the leading source of pigments. Pigments are found naturally in leaves, flowers, fruits, stems, etc. Some pigments are also present in bacteria and fungi. The pigments find their applications in medicines, foods, textile, cosmetics sector, etc. Based on color, pigments are broadly classified into chlorophyll, carotenoids, anthocyanins, and betalains. Chlorophyll is the most abundant pigment and is responsible for photosynthesis in plants.

Carotenoids are fat-soluble color pigments ranging from yellow to red through orange. Chemically, carotenoids consist of 40-carbon isoprene units covalently linked with multiple conjugated double bonds [49]. Broadly, they have been divided into carotenes and xanthophylls. Carotenes are constituted of only carbon and hydrogen atoms whereas, xanthophylls also contains oxygen. The carotenes are further grouped into phytoene, phytofluene, lycopene, and β-carotene. Similarly, xanthophylls are sub-grouped as β-cryptoxanthin, zeaxanthin, lutein, astaxanthin, and fucoxanthin [5]. Some of these carotenoid pigments are precursors for vitamin A synthesis and are needed to maintain visual health in human.

Carotenoid pigments play a predominant role as an antioxidant molecule in a lipid rich medium. Lycopene is one such carotene, it is associated with reducing blood pressure and has a protective effect on the cardiovascular system. A diet rich in lycopene has been associated with lowering the risk of damage to the heart muscle fibers, breast, and uterine cancer [59]. Lycopene is found in high amounts in guavas, tomatoes, grapefruit, watermelon, papaya, sweet red peppers, persimmons, etc. Considerable amounts of carotenoids are present in pumpkin, carrot, grapefruit, orange, and apricots.

Anthocyanins are water-soluble pigments belonging to the flavonoid class. The glycosylated forms of anthocyanins are blue, red, or purple in color depending upon surrounding pH. In acidic pH, anthocyanins appear red while in basic pH it appears blue in color. The sugar free counterpart of anthocyanins is known as anthocyanidins and classified into cyanidin, delphinidin, pelargonidin, peonidin, petunidin, and malvidin. They belong to the flavonol subgroup of the phenolic group [38]. Light and temperature are the known factors affecting the stability of the anthocyanin pigments.

In the plant kingdom, glycosylated form of cyanidin (cyanidin-3-glucoside) is the major anthocyanin pigment and is abundant in different flowers as well as fruits (such as: berries, currants, grapes). Leafy vegetables like black carrot, red cabbage, brinjal, black, and red rice varieties, black corn, sweet potato, etc., are also high in anthocyanin. Anthocyanins are generally used as a source of natural colorant in foods and textile industries, but they also confer many health benefits. They exhibit antioxidant, antidiabetic, anticancer, anti-inflammatory, antimicrobial properties, etc.

Betalains are also types of pigments and are predominant in the vacuoles of plants belonging to the families under the order Caryophyllales [37]. The betacyanin (red-violet colored) and betaxanthin (yellow colored) are good examples. Betalains are formed due to condensation of betalamic acid [4-(2-oxoethylidene)-1,2,3,4-tetrahydropyridene-2,6-dicarboxylic acid] with amino compounds (cyclo-DOPA and/or its glucosyl derivatives) to betacyanins. If betalamic acid condenses with an amino group, betaxanthin is obtained. Researchers have reported that betalains under *in vitro* and *in vivo* conditions exhibited anti-inflammatory, radical scavenging, antidiabetic, and anticancer properties [4, 30, 37]. Beetroot, cactus pear, amaranth, red-purple pitaya contain large amounts of betalains, which can be extracted for use as nutraceuticals.

9.2.3 VITAMINS AND MINERALS

Vitamins are organic compounds indispensable for the growth and maintenance of optimal health in humans. Therefore, a diet enriched with vitamins is vital. Vitamins are classified as fat-soluble (vitamin A, D, E, and K) and water-soluble (ascorbic acid and vitamin B complex). The fruits and vegetables, marine fishes, seaweeds, mushrooms, eggs, meat, dairy products, legumes as well as fortified cereals are good sources of different vitamins in diet. Depending on the type of vitamins, the functional role in the body differs. Vitamin A is vital for visual health and vitamin D aids in bone health maintenance. Similarly, vitamin E has antioxidant properties and has a beneficial effect on the nervous and cardiovascular system [18, 56].

Ascorbic acid is well known for its excellent antioxidant activity and helps in prevention of oxidative cell damage. The B-complex vitamins group is essential for the normal growth and development of the body, correct fat and carbohydrate metabolism, proper functioning of the nerves, and red blood cell formation. Importantly, the vitamin B complex acts as cofactors in various intricate biochemical reactions occurring in the body during metabolism [44].

Minerals are also vital for normal functioning of many metabolic reactions. An adequate mineral balance is required for the maintenance of teeth and bone health as well as smooth functioning of nervous and vascular systems [17]. Minerals essential for body function are divided into macro [magnesium (Mg), potassium (K), calcium (Ca), chlorine (Cl), phosphorus (P), sodium (Na), sulfur (S)] and micro minerals [manganese (Mn), chromium (Cr), selenium (Se), cobalt (Co), copper (Cu), iodine (I), iron (Fe), molybdenum (Mo), and zinc (Zn)].

In many biological reactions, minerals (Zn, Cu, K, Mn, Fe, Ca, etc.), act as a co-factor for the enzymes catalyzed reactions. Mg, P, and Mn play a role in energy production in the body while, Cu, Zn, Fe, and Se are also required for a healthy immune response. Additionally, Cu, Se, and Zn are noted for their antioxidant properties, thereby protecting the cells from going under oxidative stress [17, 25]. Nuts, meat, fish, dairy products, fruits, and vegetables are well-known sources of minerals. Lately, marine sources such as seaweed are also listed as major sources of minerals for a healthy diet.

9.2.4 DIETARY FIBERS (DFS)

Byproducts from vegetable and fruit wastes like peel, pomace, seed, and seed coats are naturally rich in dietary fibers (DFs) with health-promoting properties. American Association of Cereal Chemists [1] has defined DF as edible part of plants and analogous carbohydrates that are resistant to digestion and absorption in the human small intestine, undergoes complete or partial fermentation in the human large intestine. DFs are broadly classified as soluble (SDF) and insoluble dietary fibers (IDF) owing to their respective solubility in water. The SDF include gums, mucilage, pectin, and hemicelluloses. While IDF consists of mainly cellulose, other types of hemicelluloses and lignin [65].

DF acts as a bulking agent, which helps in the maintenance of a healthy bowel movement. IDF help to increase fecal bulk and plays a role in decreasing intestinal transit. Additionally, DF is also associated with other health benefits such as lowering of glucose and cholesterol level in blood, maintenance of healthy body weight, prevention of cardiovascular diseases (CVDs), and diabetes. SDF cause an increase in viscosity and help to reduce plasma cholesterol and glycemic levels [2, 55, 63]. The DFs derived from vegetable and fruit wastes contain polyphenols embedded in the fiber matrix [20]. Such type of DF imparts antioxidant effect in addition to the benefits associated with consumption of fiber [65]. Another important role of DF is to increase the fecal mass as well as maintenance of colon health and also aids in preventing colon cancer. This is due to the generation of short-chain fatty acids, as inherent microflora of the large intestine ferments the DFs.

9.2.5 *PHYTOSTEROL*

Sterols or steroid alcohols are lipids that belong to a subgroup of steroids. They are found naturally in plants, fungi, and in animals. Sterols of plant origin are termed as phytosterol and are derived from squalene, a member of triterpene family. β-Sitosterol, stigmasterol, campesterol, brassicasterol are major phytosterols with functional properties [14]. Dietary intake of phytosterol helps in the maintenance of optimal health as they have anti-carcinogenic, antioxidant, anti-inflammatory properties and helps in reduction of LDL (bad cholesterol) level [32, 81, 82]. Nuts, edible seed and oils, whole grains, legumes, vegetables, and fruits are the major sources of phytosterols.

9.2.6 *BIOACTIVE LIPIDS*

Some lipids in addition to being a source of energy for bodily functions also have health-promoting effects and are termed as bioactive lipids or bio-lipids. A bioactive lipid usually imparts its bioactivity either through changing the fatty acid composition of different tissues or by preventing cell signaling pathways [87]. Both polyunsaturated fatty acids (PUFA) and monounsaturated fatty acids (MUFA) are considered as bioactive lipids with an array of health-promoting properties. PUFA is classified into ω-3- and ω-6-fatty acid. Intake of diet rich in PUFAs (such as α-linoleic acid ($C_{18}H_{30}O_2$), eicosapentaenoic acid (EPA, $C_{20}H_{30}O_2$), docosahexaenoic acid (DHA, $C_{22}H_{32}O_2$) may impart beneficial role in metabolic function and also has preventive role in cancer, diabetes, inflammatory bowel disorders, CVDs, neurodegenerative conditions, etc. [21]. DHA is an important molecule required for proper brain development in growing years. EPA intake is shown to slow down cognitive decline and dementia associated with aging [87]. Marine fish, seaweeds, oils derived from sunflower seed, flax seed, corn, soybean, and safflower are rich in PUFA. Marine fish oils are major sources of EPA and DHA.

The ω-7-fatty acid (Palmitoleic acid, C16:1n–7) and ω-9-fatty acid (Oleic acid, C18:1n–9) are MUFAs, which are also beneficial for human health. Palmitoleic acid helps in reduction of hyperglycemia, hypertriglyceridemia conditions and helps in improving insulin sensitivity. Similarly, oleic acid regulates many biological processes and decreases the chance of occurrence of coronary heart diseases and other metabolic disorders [87]. Animal fat, marine fish, and oils derived from macadamia nuts, peanuts, canola, sunflower, sesame, poppy seeds, avocado, etc., are the main sources of MUFAs.

9.2.7 *BIOACTIVE PEPTIDES (BPS)*

The BPs exhibit functional properties that have a beneficial role in metabolic functions and health [11]. They can be used for treatment of many ailments of the digestive, endocrine, cardiovascular, immune, and nervous system [41, 68]. These functional peptides can be derived both from animal and plant sources. Among the animal sources, blood, a by-product of animal slaughterhouses is a promising source of BPs. From the hydrolyzed animal blood proteins, a number of peptides are obtained that showcase bioactivity like inhibiting angiotensin-converting enzyme (ACE), glucose regulation by inhibiting dipeptidyl peptidase-IV (DPP-IV), and antioxidant potential [6, 68]. BPs are also derived from cheese and bovine milk, meat, egg, and fish through chemical and enzymatic hydrolysis [51] Similarly, some sequences of peptides present in rice, wheat, and soy proteins are also used for BP production. Mushrooms [92] and seaweeds [39] derived BP exhibits antihypertensive, antioxidant, and antimicrobial properties.

9.3 FUNCTIONAL ROLE OF BIOACTIVE COMPOUNDS

Bioactive compounds derived from food sources as well as medicinal plants have numerous functions in the prevention and maintenance of different metabolic disorders. Therefore, these compounds play an active role in the improvement of human health; hence, quality of life. Some of the major functional role of bioactive compounds has been discussed briefly in this section.

9.3.1 ANTIOXIDANT ACTIVITY

The metabolic processes in the human body led to the production of free radicals, and these free radicals play a major role in human metabolic processes and diseases. Free radicals are molecular species that contain an unpaired electron which are highly unstable, reactive in nature and behave as an oxidant. Hydroxyl, superoxide anion, peroxide, singlet oxygen, nitric oxide (NO), etc., are some free radical groups that are capable of mutilating biomolecules such as DNA, proteins, carbohydrates, and lipids in the cell which in turn could cause a cascade of unwanted biochemical reactions [42]. However, an optimal quantity of free radicals is required for normal body functions.

An imbalance in the free radical generation can trigger a health crisis due to an increase in oxidative stress. A prolonged oxidative stress could lead to the onset of many degenerative conditions like diabetes, CVDs, neurodegenerative disorders, cancer, etc. Bioactive compounds have antioxidant properties that can neutralize or control the oxidative stress in the body. An antioxidant acts as a barrier that protects the important biological sites inside the cell by scavenging or quenching the free radicals. The human body has an innate defense mechanism to overcome the free radical inflicted oxidative stress during metabolism. Enzymes like superoxide dismutase, glutathione peroxidase, and micronutrients help to remove free radical-induced damages [66].

A diet rich in bioactive compounds will therefore aid the innate antioxidant system to overcome any kind of oxidative stress. Vitamin C, tocopherols, carotenoids, and polyphenols exhibit antioxidant properties in addition to other health-promoting properties [27]. Bioactive compounds present in fruits such as strawberry can inhibit LDL-cholesterol oxidation [76]. Honey also shows antioxidant properties that can be ascribed to the presence of polyphenols, carotenoids, and ascorbic acid in them [23, 46]. Curcumin in *Curcuma longa* and resveratrol in grapes also exhibit antioxidant properties in addition to other functional properties.

9.3.2 *HYPOGLYCEMIC ACTIVITY*

Type-2 diabetes has emerged as one of the most prevalent metabolic disorders in the present society. Change in lifestyle, diet, stress, and other metabolic anomalies often lead to an increased blood glucose level. Plant bioactive compounds like polyphenols and DF have hypoglycemic activity and can bring down the blood glucose levels in diabetic conditions. Polyphenols such as phenolic acids (cinnamic, ferulic, catechin, chlorogenic, rosmarinic, and caffeic acids), tannins, flavonoids (myricetin, apigenin, quercetin, luteolin, and vitexin), stilbenoids (resveratrol, pterostilbene, and polydatin) derived from fruits, vegetables, tea, coffee, etc., have significant hypoglycemic activity [10, 50]. Depending on the type of polyphenols, they help in lowering of glucose in a number of ways such as by protection of the pancreatic islet β-cells, promotion of the β-cells proliferation and reduction in their apoptosis. They also help in stimulation of pancreatic activities and reduction of oxidative stress.

Additionally, polyphenols can inhibit the enzymes responsible for carbohydrate metabolism as well as reduce the generation of advanced glycation end products, thus regulating glucose absorption [78]. Similarly, isoflavones present in soyabean can reduce glucose intolerance and insulin resistance in blood as well as inhibit glucose uptake in the small intestine [12, 78].

DF is another plant-derived bioactive that has hypoglycemic property. They can reduce the glucose absorption by binding to the glucose molecules or alleviate insulin resistance in blood [45]. Shtriker et al. [73] reported that fiber derived from fenugreek and citrus fruits help in reduction of blood glucose level by inhibiting the function of α-amylase enzyme. Additionally, Novel BPs derived from hydrolysis of meat, soy, egg, and other plant proteins also exhibit hypoglycemic effects [62] by inhibiting the activities of enzymes of concern like α-amylase, α-glucosidase, and DPP-IV [89]. Restricting the DPP-IV enzyme helps to stimulate insulin secretion and further inhibition of glucagon release, thus controlling the increase in blood glucose level [34].

9.3.3 *ANTI-INFLAMMATORY ACTIVITY*

Inflammation is a protective response against harmful stimuli (chemical or biological) to a cell. On injury, a varied cascade of reaction starts comprising of overexpression of cell adhesion molecules and interleukin-1, synthesis of proinflammatory cytokines, activation of transcription factor NF-κB, overexpression of phospholipases A2 along with release of reactive oxygen species (ROS). Bioactive components from various foods and medicinal plants are shown to inhibit these reactions thereby subliming inflammation.

Bioactive components of foods and medicinal plant extracts have shown potent application as anti-inflammatory agents. For example, *Albizi myriophylla*-derived ethanolic extracts showed anti-inflammatory property under both *in vitro* and *in vivo* conditions. The major active compounds present in the extracts are flavonones and phytosterols, and they have a significant effect in reduction of skin swelling when tested in rat ear edema model. The proposed mechanism for its action was through inhibiting production NO [9].

Aerial parts methanolic extracts of *Ajuga laxmannii* herb may reduce inflammation in rats by inhibition of phagocytosis through reduced oxidative stress. The major bioactive components isolated from the methanolic extracts were rutin, iridoids, and phytosterols [80]. S-allylcysteine from black garlic, polyphenols from *Uncaria tomentosa, Myrciaria dubia, Harpagophytum procumbens, Ribes nigrum*, and citrus fruits extracts have shown significant anti-inflammatory nature in *in vitro* and *in vivo* studies [70]. More recently, a shift towards marine microalgae and cyanobacteria is seen as important sources of bioactive components which inhibit the inflammatory reaction cascade [79].

9.3.4 LIPID LOWERING ACTIVITY

For a proper absorption of lipid droplets in the body, emulsification of the lipid takes place in the stomach and duodenum due to emulsifying activity of bile salts and phospholipids. The gastric and pancreatic lipases act on these emulsified lipid droplets and lipolysis takes place. The rate of lipolysis is depended on the lipid emulsion stability. Therefore, to manage the lipid absorption rate in the body, the rate of lipid digestion needs to be controlled. DF can interfere with the lipolysis depending on the type, concentration, pH, and ionic strength. They act either by stabilizing or destabilizing the lipid emulsion. In both the cases, the lipid molecules become unavailable to the lipolytic enzymes and thus the rate of lipolysis can be controlled. Therefore, a diet rich in DF can help in prevention and management of atherosclerosis, obesity, and other associated ailments with increased lipid levels in the body [77].

Polyphenols derived from fruits, vegetables, and medicinal plants also have lipid-lowering properties. The lowering of lipids by polyphenols takes place either inhibiting the lipase enzymes or interfering with the emulsion droplets formation. Research has reported that polyphenols in black tea can inhibit the emulsion droplets in the stomach and reduce the surface area hence, making them unavailable for the lipolysis [29]. Black tea polyphenols also can inhibit the activity of pancreatic lipase [57]. Polyphenols such as curcumin, cyanidin-3-glucoside, catechin, chlorogenic acid can suppress the Niemann-Pick C1-Like 1 (NPC1L1) mRNA expression while, luteolin, and epigallocatechin gallate (EGCG) inhibit NPC1L1 by binding directly [33]. NPC1L1 is a protein found on the epithelial cells of the gastrointestinal (GI) tract and hepatocytes of the liver. Hence, inhibition of NPC1L1 by polyphenols results in lowering of the blood cholesterol due to decreased absorption from the intestine [33].

9.3.5 ANTIHYPERTENSIVE ACTIVITY

A holistic approach consisting of simultaneous use of drug, a balanced diet with the richness of bioactive compounds, nutraceuticals, and an active lifestyle may be the key for effective treatment/management of hypertension. Many bioactive compounds have antihypertensive properties. For example, ascorbic acid, tocopherols, BPs, polyphenols, bioactive lipids, phytosterols as well as minerals-like Ca, K, and P have antihypertensive property [7]. BPs derived from milk, bovine blood, fish, seaweeds, and plants exhibit an inhibitory effect on ACE [7, 13]. The main role of ACE is to convert angiotensin I to angiotensin II (an active vasoconstrictor). ACE controls blood pressure by modulating the volume of fluids in the body, and is a central component of the renin-angiotensin system. This enzyme is a metalloprotein with Zn at its catalytic site [3]. Most polyphenols can chelate the Zn at the catalytic site of ACE thus, inhibiting the enzyme's catalytic activity [3, 85]. Garlic's bioactive compounds like allicin and captotril can inhibit the ACE activity [71]. Similarly, polyphenols isolated from *Clerodendrum colebrookianum* also showed ACE inhibitory effects [88].

9.3.6 ANTI-NEURODEGENERATIVE ROLE

Parkinson's disease, Alzheimer's disease, Huntington's disease, etc., are some of the common ages related neurodegenerative disorders. It is a progressive dysfunction and degradation of neurons that leads to neuronal cell damage [91]. Epidemiological studies have identified that certain components in the diet can have a therapeutic role in neurodegenerative conditions [26]. However, development of nutrient based therapeutics for neurodegeneration is often challenged by the blood-brain barrier which hampers the efficacy of any therapy. To overcome this barrier, a proper carrier or vehicle is required for targeted delivery of the bioactive compounds. Use of nano-carriers is one such way for a successful delivery to the targeted site of action. Degeneration of neurons is triggered often due to oxidative stress, neuroinflammation, mitochondrial dysfunction, abnormal protein misfolding, apoptosis, and nerve cells death due to toxicity [52].

Natural products with bioactivity derive from Korean ginseng (*Panax ginseng*), ashwagandha (*Withania somnifera*), honey, propolis, cat's claw herb (*Uncaria rhyncophylla*), seaweeds, turmeric (*Curcuma longa*) rhizomes, fish oil, germinated brown rice are considered to have neuroprotective role [52]. The bioactive compounds present in them have antioxidant, anti-apoptotic, and anti-inflammatory properties, which may prevent or control the undesirable changes in the neuronal structures and cells that leads to neurodegeneration.

9.3.7 ANTIMICROBIAL ACTIVITY

Microorganisms causing infectious diseases are the main cause for creating a huge burden on the healthcare system as well as responsible for increased mortality. Due to increased resistance against commonly available antibiotics, concern to find alternatives of available antibiotics is gaining attention day by day. From ancient time plants are well known to possess medicinal value. Traditionally medicinal plants have been utilized to treat different diseases and even in avoiding food spoilage. The antimicrobial activity of compounds isolated from plants is thought to be an alternative of chemically synthesized antibiotics [28]. These bioactive compounds proved a platform to overcome the concern of developing antibiotics resistance [64]. According to their chemical structures, they are classified into alkaloids, sulfur-containing compounds, terpenoids, and polyphenols.

Bioactive constituents are divided into different groups like polyphenols, alkaloids, sulfur-containing compounds, terpenoids, etc., based on their chemical structure. Piperine, an alkaloid isolated from different piper species have shown growth inhibition of *S. aureus* and many other microorganisms alone or in combination with antibiotics [36].

Berberine (soquinoline alkaloid) can target RNA polymerase and gyrase of bacteria, fungi, protozoa, and viruses thus can inhibit their growth [17]. Other alkaloids such as Ungeremine (such as dictamnine kokusagine, Reserpine, and masculine) have also been found as potent antimicrobial agents against *Staphylococcus spp.*, *E. coli*, *Streptococcus spp.* and *Micrococcus spp.* [35].

Organo-sulfur compounds such as allicin from garlic (*Allium sativum*) is a well-known antimicrobial agent against a wide range of pathogenic microorganisms, i.e., *S. epidermidis*, *P. aeruginosa*, *Streptococcus agalactiae*, etc. [61]. Allicin can act upon alcohol dehydrogenase and RNA polymerase of the microorganisms to kill the microbes. Ajoene an organo-sulfur compound found in garlic extract has also been found to contain broad-spectrum antimicrobial activity and also have antiviral activity.

Polyphenolic bioactive compound like resveratrol is also known for its antimicrobial activity against *Campylobacter jejuni, Arcobacter, and M. smegmatis* [40]. Baicalein has been screened and found potential to restrict the multiplication and growth of varied Gram-positive and Gram-negative microorganisms like *Bacillus cereus*, *E. coli, S. aureus*, *Candida albicans,* and *Pseudomonas aeruginosa.* Kaempferol is a bioactive molecule against different antibiotic resistant microorganisms like fluconazole-resistant *C. albicans* and methicillin resistant *S. aureus* (MRSA) [60, 72]. Curcumin has a promising bactericidal effect against MRSA and uropathogenic *E. coli* by damaging the cell wall of these microorganisms [83].

EGCG has been found to exhibit potent antimicrobial activity against different pathogenic microorganisms. Apart from above mentioned bioactive compounds, other components like tannins, ascorbic acid, gallic acid, and coumarins are also known for their antimicrobial activity. Wu et al. [87] showed that quercetin and apigenin can target D-alanine: D-alanine ligase enzyme in *E. coli* and *Helicobacter pylori* to show their antimicrobial activity.

9.3.8 ANTI-CANCEROUS ACTIVITY

The development of plant-based compounds has been targeted to search some bioactive components against cancer. Several molecules from plants have been used in the cancerous cell lines *in vitro* and have shown the good efficacy, and after animal experiments, some have been sent to the clinical trials also.

Artemisinin a plant active compound from *Artemisia annua* has been found to show liver, breast, and pancreatic anticancer activity [19]. Cabazitaxel, a derivative of natural taxoid have been found to eliminate the prostate cancer in randomized open-label trial [15]. Sinani et al. [74] showed in his study that solamargine, a component of *Solanum nigrum* plant extract has the potential to eradicate human melanoma cancer by activating lysosomal mitochondrial death pathway [67].

Another active component of plant, kaempferol when form a complex with zinc (II) have been evaluated for their anticancer activity and it has been found that it has a great potential in proliferation of lung, breast, and liver cancer cells [84]. Withaferin A, withanolide D, gingerol, colchicine, skimmianine, boswellic acid, and silymarin are some compounds which have also been tested for their anticancer activity. The source plants of these bioactive components have been considered in traditional medicine, but their scientific validation and active component analysis is quite recent, and further approaches are utilized to find out the novel drugs against the cancer.

Curcuma longa contains an active component, i.e., curcumin which has the ability to eliminate colon adenocarcinoma by activating STAT3 and NF-κB (transcription factors in immunity) signaling pathways [86]. Some other compounds which have been found to contain anticancer activity include betulinic acid, asiatic acid, gallic acid, lycopene, plumbagin, allicin, apigenin, calcaelin, and ursolic acid, etc. These components have shown prominent activity against breast, colon, lung, liver, spleen, and skin cancer.

9.4 BIOAVAILABILITY AND DELIVERY OF NUTRACEUTICALS

Bioavailability in general refers to the ingested nutraceuticals fraction that become accessible to absorption in the GI tract, is metabolized and later distributed to organs and tissues. Distinctively, the bioavailability of nutraceuticals is governed by three steps: bioaccessibility, absorption, followed by transformation [47]. The first step being 'bioaccessibility' is defined as the fraction of nutraceutical that is available for absorption through the epithelial membrane of the intestine. This includes liberation of the active nutraceutical molecule from its matrix (food or delivery matrix); solubility in corresponding biological fluid (stomach, pancreatic, intestinal, and bile) where it is available for interacting with other components or systems.

Bioaccessibility is dependent upon the physical (solubility, size, charge, load degree, etc.), and chemical nature of the nutraceutical along with prevailing digestive environment (pH, enzymes, bile salts, FFAs, etc.). 'Absorption' of a biocomponent/nutraceutical takes place at the GI tract epithelial cells marking the second step for bioavailability. Absorption is governed by active or passive transport based upon the nature of the nutraceutical.

Lastly, the "transformation" of biocomponents during digestion and their metabolism in the liver also affects the bioavailability. Transformations such as curcumin degrades in alkaline conditions, cis-trans conversion of carotenoids into inactive forms or oxidations of fatty acids like PUFAs by prooxidants, etc., leads to lowering of bioavailability [16]. Thereby, in general, a lower bioactivity of nutraceuticals is observed in *in vivo* models as compared to *in vitro* models. Considering the above-mentioned points, the limited bioavailability of the most nutraceuticals does not materialize into distinct health benefits on intake. Consequently, to improve bioavailability, researchers are on the pursuit to design delivery matrixes and systems for nutraceuticals.

Delivery matrices for nutraceuticals are designed with objectives, such as [24]: (i) protection against external factors; (ii) easier incorporation into food products; (iii) masking of off-flavors; (iv) controlled release; and (v) maximal retention of functional property in general or till the components reaches its targeted site. Encapsulation of nutraceuticals (such as carotenoids, vitamins, PUFAs, polyphenols, phytosterols, minerals, etc.), has been studied to show such benefits. Common biomaterials for encapsulation are polysaccharides, proteins, lipids, low molecular weight surfactants. Nano-delivery systems increase the surface-to-volume ratio, thus assists in better solubility and felicitate the movement of the nutraceutical through biological barriers/membranes by bypassing transformation steps, thereby increasing bioavailability. Novel nano-delivery systems (such as solid lipid nanoparticles (SLNs), liposomes nano-emulsions, nanostructured lipid carriers (NLCs), self-generating nano-emulsifying drug delivery systems (SNEDDS)) have shown promising applications for increased bioavailability of various nutraceuticals [8, 53].

Researchers have shown the dependence of size on bioavailability. Smaller size of nano-emulsion of vitamin E corresponded to augmented bioavailability [59] and similar results were also observed for β-carotene using SLNs [48]. SNEDDS approach has been most effective to provide higher levels of bioactive component loading, better transport, dissolution, and easier intestinal permeation whilst enhancing bioavailability. SNEDDS formulations have helped to overcome the limited bioavailability for various flavonoids, carotenoids, polyphenols, alkaloids, and vitamins leading to better final results as compared to native compounds [53].

Incorporation of enhancer molecules along with particular nutraceutical has been suggested to increase the absorption of nutraceuticals by increasing membrane permeation. Enhancers (such as: piperine, bile salts, genistein, unsaturated fatty acids, lactose esters, chitosan derivatives) have shown to increase the absorption of curcumin, vitamin D_3, EGCG, ovalbumin, and salvianolic acid, respectively [24]. More recently, use of the Maillard reaction-based protein-polysaccharide conjugates have been potential encapsulant or delivery systems. This is because of their unique characteristics, such as: excellent emulsification capacity, high solubility, and antioxidant property, stability towards a wide range of temperature, pH, and ionic strengths along with better protection of lipophilic bioactive nutraceuticals [54].

Overall, specifically designed delivery systems can substantially increase the efficacy of nutraceuticals for an intended health benefit. The systems can better regulate the bioaccessibility, absorption alongside transformation of the nutraceuticals inside the GI tract. However, further developments related to the selection of other food-grade carrier materials, delivery system, controlled release, and toxicological studies for risk assessments needs to be investigated.

9.5 SUMMARY

Bioactive compounds derived from food and other natural sources has ample scope to be used as nutraceuticals as therapeutics for chronic metabolic disorders viz. diabetes, cancer, hypertension, neurodegenerative diseases, CVDs, etc. Bioactive compounds such as polyphenols, vitamins, minerals, natural pigments, dietary fibers, phytosterols, bioactive lipids, bioactive peptides (BP), etc. have been identified and determined for many therapeutic properties. The bioaccessibility, absorption and transformation are the three important steps to be considered for bioavailability of nutraceuticals. The specially designed delivery system for nutraceuticals has shown a substantial potential to increase the bioavailability and bioaccessibility of varied bioactives. This will help in increasing the efficacy of the intended health benefits of the nutraceuticals. Therefore, nutraceuticals from novel bioactive compounds have immense potential in improving the quality of life by playing an important role in disease management along with conventional medicines.

KEYWORDS

- **bioactive compounds**
- **food sources**
- **medicinal plants**
- **monounsaturated fatty acids**
- **nutraceuticals**
- **therapeutics**

REFERENCES

1. AACC Report, (2001). The definition of dietary fiber. *Cereal Foods World, 46*(3), 112–126.
2. Abdul-Hamid, A., & Luan, Y. S., (2000). Functional properties of dietary fiber from defatted rice bran. *Food Chemistry, 68*(1), 15–19.
3. Adefegha, S. A., & Oboh, G., (2016). Antioxidant and inhibitory properties of *Clerodendrum volubile* leaf extracts on key enzymes relevant to non-insulin-dependent diabetes mellitus and hypertension. *Journal of Taibah University for Science, 10*(4), 521–533.
4. Alexandrino De, O. S. P., Almeida Do, N. H. M., Sampaio, K. B., & Leite De, S. E., (2021). A review on bioactive compounds of beet (*Beta vulgaris* L. subsp. *vulgaris*) with special emphasis on their beneficial effects on gut microbiota and gastrointestinal health. *Critical Reviews in Food Science and Nutrition, 61*(12), 2022–2033.
5. Amengual, J., (2019). Bioactive properties of carotenoids in human health. *Nutrients, 11*(10), 2388–2395.
6. Arrutia, F., Puente, Á., Riera, F. A., Menéndez, C., & González, U. A., (2016). Influence of heat pre-treatment on BSA tryptic hydrolysis and peptide release. *Food Chemistry, 202*(1), 40–48.
7. Aslam, M. Z., Aslam, M. S., Firdos, S., Ghous, G., Firdos, G., Hongfei, Z., & Bolin, Z., (2019). Role of bioactive peptides in reducing the severity of hypertension with the inhibition of ACE. *International Journal of Peptide Research and Therapeutics, 25*, 1639–1649.
8. Ali, A., Ahmad, U., Akhtar, J., Badruddeen, & Khan, M. M., (2019). Engineered nano scale formulation strategies to augment efficiency of nutraceuticals. *Journal of Functional Foods, 62*, 103554.
9. Bakasatae, N., Kunworarath, N., Yupanqui, C. T., Voravuthikunchai, P. V., & Joycharat, N., (2018). Bioactive components, antioxidant, and anti-inflammatory activities of the wood of *Albizia myriophylla. Revista Brasileira de Farmacognosia, 28*(4), 444–450.
10. Bahmanzadeh, M., Goodarzi, M. T., Rezaei, F. A., Fathi, N., & Alizadeh, Z., (2019). Resveratrol supplementation improves DNA integrity and sperm parameters in streptozotocin-nicotinamide-induced type 2 diabetic rats. *Andrologia, 51*(8), e13313.
11. Chakrabarti, S., Snigdha, G. S., & Majumder, K., (2018). Food-derived bioactive peptides in human health: Challenges and opportunities. *Nutrients, 10*(11), 1738.
12. Chen, X., Yu, J., & Shi, J., (2018). Management of diabetes mellitus with puerarin, a natural isoflavone from *Pueraria lobata. American Journal of Chinese Medicine, 46*(8), 1771–1789.
13. Cicero, A. F. G., Fogacci, F., & Colletti, A., (2017). Potential role of bioactive peptides in prevention and treatment of chronic diseases: A narrative review. *Brazilian Journal of Pharmacology, 174*(11), 1378–1394.
14. Da Silva, M. R., Furlan, C. P. B., Marques, A. Y. C., Bicas, J., Pastore, G. M., & Maróstica, Jr. M. R., (2015). Phytosterols: Biological effects and mechanisms of hypercholesterolemic action; Chapter 23. In: Gupta, V. K., Tuohy, M. G., Lohani, M., & O'Donovan, A., (eds.), *Biotechnology of Bioactive Compounds: Sources and Applications* (pp. 565–581). Oxford–UK: John Wiley & Sons Ltd.
15. De Bono, J. S., Oudard, S., Ozguroglu, M., Hansen, S., Machiels, J. P., Kocak, I., Gravis, G., et al., (2010). Prednisone plus cabazitaxel or mitoxantrone for metastatic

castration-resistant prostate cancer progressing after docetaxel treatment: A randomized open-label trial. *Lancet, 376*(9747), 1147–1154.

16. Dima, C., Assadpour, E., Dima, S., & Jafari, S. M., (2020). Bioavailability of nutraceuticals: Role of the food matrix, processing conditions, the gastrointestinal tract, and nano delivery systems. *Comprehensive Reviews in Food Science and Food Safety, 19*(3), 954–994.
17. Domadia, P. N., Bhunia, A., Sivaraman, J., Swarup, S., & Dasgupta, D., (2008). Berberine targets assembly of *Escherichia coli* cell division protein FtsZ. *Biochemistry, 47*(10), 3225–3234.
18. Donati, E., (2018). Vitamins, minerals, and bioactive compounds; Chapter 1.3.1. In: Contò, F., Del Nobile, M. A., Faccia, M., Zambrini, A. V., & Conte, A., (eds.), *Advances in Dairy Products* (pp. 75–97). Oxford–UK: John Wiley & Sons Ltd.
19. Efferth, T., (2017). From ancient herb to modern drug: Artemisia annua and artemisinin for cancer therapy. *Seminars in Cancer Biology, 46*, 65–83.
20. Evans, P., & Halliwell, B., (2001). Micronutrients: Oxidant/antioxidant status. *British Journal of Nutrition, 85*(2), 67–74.
21. Fernandes, I., & Pinto, R., (2019). Fatty acids polyunsaturated as bioactive compounds of microalgae: Contribution to human health. *Global Journal of Nutrition and Food Science, 2*(1), 000526.
22. Francisco, A., Barbetan, T., & Andres-Lacuaeva, C., (2012). Polyphenols and health: Current state and progress. *Journal of Agricultural and Food Chemistry, 60*(36), 8773–8775.
23. Froschle, M., Horn, H., & Spring, O., (2018). Characterization of *Jatropha curcas* honeys originating from the southern highlands of Madagascar. *LWT Food Science Technology, 93*, 525–533.
24. Gonçalves, R. F. S., Martins, J. T., Duarte, C. M. M., Vicente, A. A., & Pinheiro, A. C., (2018). Advances in nutraceutical delivery systems: From formulation design for bioavailability enhancement to efficacy and safety evaluation. *Trends in Food Science & Technology, 78*, 270–291.
25. Goñi, I., & Hernández-Galiot, A., (2019). Intake of nutrient and non-nutrient dietary antioxidants. Contribution of macromolecular antioxidant polyphenols in an elderly Mediterranean population. *Nutrients, 11*(9), 2165.
26. González-Fuentes, J., Selva, J., Moya, C., Castro-Vázquez, L., Lozano, M. V., Marcos, P., Plaza-Oliver, M., et al., (2018). Neuroprotective natural molecules, from food to brain. *Frontiers in Neuroscience, 12*, 721.
27. Gülçin, İ., (2012). Neuroprotective natural molecules, from food to brain. Frontiers in neuroscience antioxidant activity of food constituents: An overview. *Archives of Toxicology, 86*(3), 345–391.
28. Gyawali, R., & Ibrahim, S. A., (2014). Natural products as antimicrobial agents. *Food Control, 46*, 412–429.
29. Hamdaoui, M. H., Snoussi, C., Dhaouadi, K., Fattouch, S., Ducroc, R., Le Gall, M., & Bado, A., (2016). Tea decoctions prevent body weight gain in rats fed high-fat diet: Black tea being more efficient than green tea. *Journal of Nutrition and Intermediary Metabolism, 6*, 33–40.
30. Han, J., Tan, C., Wang, Y., Yang, S., & Tan, D., (2015). Betanin reduces the accumulation and cross-links of collagen in high-fructose-fed rat heart through inhibiting non-enzymatic glycation. *Chemico-Biology Interactions, 227*, 37–44.

31. Harborne, J. B., (1980). Plant phenolics. In: Bell, E. A., & Charlwood, B. V., (eds.), *Encyclopedia of Plant Physiology* (Vol. 8, pp. 329–395). Berlin–Heidelberg: Springer.
32. Jesch, E. D., & Carr, T. P., (2017). Food ingredients that inhibit cholesterol absorption. *Preventive Nutrition and Food Science, 22*(2), 67–80.
33. Kobayashi, P., (2019). The effect of polyphenols on hypercholesterolemia through inhibiting the transport and expression of Niemann-pick C1-like 1. *International Journal of Molecular Sciences, 20*(19), 4939.
34. Kęska, P., Stadnik, J., Bąk, O., & Borowski, P., (2019). Meat proteins as dipeptidyl peptidase IV inhibitors and glucose uptake stimulating peptides for the management of a type 2 diabetes mellitus in silico study. *Nutrients, 11*(10), 2537.
35. Khameneh, B., Iranshahy, M., Soheili, V., & Bazzaz, B. S. F., (2019). Review on plant antimicrobials: A mechanistic viewpoint. *Antimicrobial Resistance and Infection Control, 8*, 118.
36. Khameneh, B., Iranshahy, M., Ghandadi, M., Ghoochi, A. D., Bazzaz, B. S. F., & Iranshahi, M., (2015). Investigation of the antibacterial activity and efflux pump inhibitory effect of co-loaded piperine and gentamicin nanoliposomes in methicillin-resistant *Staphylococcus aureus*. *Drug Development and Industrial Pharmacy, 41*(6), 989–994.
37. Khan, M. I., (2016). Plant betalains: Safety, antioxidant activity, clinical efficacy, and bioavailability. *Comprehensive Reviews in Food Science and Food Safety, 15*(2), 316–330.
38. Khoo, H. E., Azlan, A., Tang, S. T., & Lim, S. M., (2017). Anthocyanidins and anthocyanins: colored pigments as food, pharmaceutical ingredients, and the potential health benefits. *Food & Nutrition Research, 61*(1), 1–21.
39. Lafarga, T., Acién-Fernández, F. G., & Garcia-Vaquero, M., (2020). Bioactive peptides and carbohydrates from seaweed for food applications: Natural occurrence, isolation, purification, and identification. *Algal Research, 48*, 1–10.
40. Lechner, D., Gibbons, S., & Bucar, F., (2008). Plant phenolic compounds as ethidium bromide efflux inhibitors in *Mycobacterium smegmatis*. *Journal of Antimicrobial Chemotherapy, 62*(2), 345–348.
41. Lemes, A. C., Sala, L., Ores, J. D. C., Braga, A. R. C., Egea, M. B., & Fernandes, K. F., (2016). A review of the latest advances in encrypted bioactive peptides from protein-rich waste. *International Journal of Molecular Sciences, 17*(6), 950.
42. Lobo, V., Patil, A., Phatak, A., & Chandra, N., (2010). Free radicals, antioxidants and functional foods: Impact on human health. *Pharmacognosy Reviews, 4*(8), 118–126.
43. Liu, R. H., (2004). Potential synergy of phytochemicals in cancer prevention: Mechanism of action. *The Journal of Nutrition, 134*(12), 3479S–3485S.
44. Lykstad, J., & Sharma, S., (2021). *Biochemistry, Water Soluble Vitamins*. In: StatPearls [Internet]; Treasure Island–FL: StatPearls Publishing. Available from: https://www.ncbi.nlm.nih.gov/books/NBK538510/ (accessed on 30 May 2022).
45. Ma, M., & Mu, T., (2016). Anti-diabetic effects of soluble and insoluble dietary fiber from de-oiled cumin in low-dose streptozotocin and high glucose-fat diet-induced type 2 diabetic rats. *Journal of Functional Foods, 25*, 186–196.
46. McClements, D. J., Li, F., & Xiao, H., (2015). The nutraceutical bioavailability classification scheme: Classifying nutraceuticals according to factors limiting their oral bioavailability. *Annual Reviews of Food Science and Technology, 6*, 299–327.
47. Mehrad, B., Ravanfar, R., Licker, J., Regenstein, J. M., & Abbaspourrad, A., (2018). Enhancing the physicochemical stability of β-carotene solid lipid nanoparticle (SLNP) using whey protein isolate. *Food Research International, 105*, 962–969.

48. Miller, D. D., Li, T., & Liu, R. H., (2014). Antioxidants and phytochemicals. *Reference Module in Biomedical Sciences* (3rd edn., pp. 1–13). Cornell University-ON: Elsevier Inc.
49. Miranda, P. T. F. D., Bonamigo, T. R., Da Silva, J., Vasconcelos, P., Felix, J. M., Cardoso, C. A. L., & Tricheza, V. D. K., (2019). Chemical constituents of *Cochlospermum regium* (Schrank) pilg. Root and its antioxidant, antidiabetic, antiglycation, and anticholinesterase effects in Wistar rats. *Biomedicine & Pharmacotherapy, 111*, 1383–1392.
50. Mohanty, D., Jena, R., Choudhury, P. K., Pattnaik, R., Mohapatra, S., & Saini, M. R., (2016). Milk derived antimicrobial bioactive peptides: A review. *International Journal of Food Properties, 19*(4), 837–846.
51. Mohd, S. N. S., & Sirajudeen, K., (2020). Natural products and their bioactive compounds: Neuroprotective potentials against neurodegenerative diseases. *Evidence-Based Complementary and Alternative Medicine, 2020*, 6565396.
52. Naseem, A., Mohammed, S. A. A., Khan, R. A., Yusuf, M., Singh, V., Mohammed, H. A., Al-Omar, M. S., et al., (2020). Self-generating nano-emulsification techniques for alternatively-routed, bioavailability enhanced delivery, especially for anticancers, anti-diabetics, and miscellaneous drugs of natural, and synthetic origins. *Journal of Drug Delivery Science and Technology, 58*, 101808.
53. Nooshkam, M., & Varidi, M., (2020). Maillard conjugate-based delivery systems for the encapsulation, protection, and controlled release of nutraceuticals and food bioactive ingredients: A review. *Food Hydrocolloids, 100*, 105389.
54. Olsan, A., Gray, G. M., & Chiu, M., (1987). Chemistry and analysis of soluble dietary fiber. *Food Technology, 41*(2), 71–80.
55. Pae, M., Meydani, S. N., & Wu, D., (2012). The role of nutrition in enhancing immunity in aging. *Aging and Disease, 3*(1), 91–129.
56. Pan, H., Gao, Y., & Tu, Y., (2016). Mechanisms of body weight reduction by black tea polyphenols. *Molecules, 21*(12), 1659.
57. Parthasarathi, S., Muthukumar, S. P., & Anandharamakrishnan, C., (2016). The influence of droplet size on the stability, *in vivo* digestion, and oral bioavailability of vitamin E emulsions. *Food & Function, 7*(5), 2294–2302.
58. Przybylska, S., (2020). Lycopene–a bioactive carotenoid offering multiple health benefits: A review. *International Journal of Food Science & Technology, 55*(1), 11–32.
59. Randhawa, H. K., & Hundal, K. K., (2016). Efflux pump inhibitory activity of flavonoids isolated from *Alpinia calcarata* against methicillin-resistant *Staphylococcus aureus*. *Biologia, 71*(5), 484–493.
60. Reiter, J., Levina, N., Van Der, L. M., Gruhlke, M., Martin, C., & Slusarenko, A. J., (2017). Diallylthiosulfinate (allicin), a volatile antimicrobial, from garlic (*Allium sativum*), kills human lung pathogenic bacteria, including MDR strains, as a vapor. *Molecules, 22*(10), 1711.
61. Ren, Y., Liang, K., Jin, Y., Zhang, M., Chen, Y., Wu, H., & Lai, F., (2016). Identification and characterization of two novel α-glucosidase inhibitory oligopeptides from hemp (*Cannabis sativa* L.) seed protein. *Journal of Functional Foods, 26*, 439–450.
62. Roehrig, K. L., (1988). The physiological effects of dietary fiber: A review. *Food Hydrocolloids, 2*(1), 1–18.
63. Rossiter, S. E., Fletcher, M. H., & Wuest, W. M., (2017). Natural products as platforms to overcome antibiotic resistance. *Chemical Reviews, 117*(19), 12415–12474.
64. Saikia, S., & Mahanta, C. L., (2016). *In vitro* physicochemical, phytochemical, and functional properties of fiber rich fractions derived from by-products of six fruits. *Journal of Food Science and Technology, 53*(3), 1496–1504.

65. Saikia, S., (2015). *Studies on the Effects of Processing on the Quality of Fruits and Vegetables and Development of a Functional Beverage Powder* (p. 222). PhD Thesis; Tezpur University, Napaam, India.
66. Samieri, C., Feart, C., Proust-Lima, C., Peuchant, E., Tzourio, C., Stapf, C., Berr, C., & Barberger-Gateau, P., (2011). Olive oil consumption, plasma oleic acid, and stroke incidence the three-city study. *Neurology, 77*(5), 418–425.
67. Sánchez, A., & Vázquez, A., (2017). Bioactive peptides: A review. *Food Quality and Safety, 1*(1), 29–46.
68. Selamassakul, O., Laohakunjit, N., Kerdchoechuen, O., & Ratanakhanokchai, K., (2016). A novel multi-biofunctional protein from brown rice hydrolyzed by endo/endo-exoproteases. *Food & Function, 7*(6), 2635–2644.
69. Serrano, A., Ros, G., & Nieto, G., (2018). Bioactive compounds and extracts from traditional herbs and their potential anti-inflammatory health effects. *Medicines, 5*(3), 76.
70. Shang, A., Cao, S. Y., Xu, X. Y., Gan, R. Y., Tang, G. Y., Corke, H., Mavumengwana, V., & Li, H. B., (2019). Bioactive compounds and biological functions of garlic (*Allium sativum* L.). *Foods, 8*(7), 246.
71. Shao, J., Zhang, M., Wang, T., Li, Y., & Wang, C., (2016). The roles of CDR1, CDR2, and MDR1 in kaempferol-induced suppression with fluconazole-resistant *Candida albicans*. *Pharmaceutical Biology, 54*(6), 984–992.
72. Shtriker, M. G., Hahn, M., Taieb, E., Nyska, A., Moallem, U., Tirosh, O., & Madar, Z., (2018). Fenugreek galactomannan and citrus pectin improve several parameters associated with glucose metabolism and modulate gut microbiota in mice. *Nutrition, 46*, 134–142.e3.
73. Sinani, S. S. A., Eltayeb, E. A., Coomber, B. L., & Adham, S. A., (2016). Solamargine triggers cellular necrosis selectively in different types of human melanoma cancer cells through extrinsic lysosomal mitochondrial death pathway. *Cancer Cell International, 16*, 11–18.
74. Singh, B. P., Vij, S., & Hati, S., (2014). Functional significance of bioactive peptides derived from soybean. *Peptides, 54*, 171–179.
75. Skrovankova, S., Sumczynski, D., Mlcek, J., Jurikova, T., & Sochor, J., (2015). Bioactive compounds and antioxidant activity in different types of berries. *International Journal Molecular Sciences, 16*(10), 24673–24706.
76. Surampudi, P., Enkhmaa, B., Anuurad, E., & Berglund, L., (2016). Lipid-lowering with soluble dietary fiber. *Current Atherosclerosis Reports, 18*(12), 75.
77. Sun, C., Zhao, C., Guven, E. C., Paoli, P., Simal-Gandara, J., Ramkumar, K. M., Wang, S. P., et al., (2020). Dietary polyphenols as antidiabetic agents: Advances and opportunities. *Food Frontiers, 1*(1), 18–44.
78. Suryam, T., Vahideh, A., & Tahereh, H., (2020). Anti-inflammatory activity of bioactive compounds from microalgae and cyanobacteria by focusing on the mechanisms of action. *Molecular Biology Reports, 47*(8), 6193–6205.
79. Toiu, A., Mocan, A., Vlase, L., Pârvu, A. E., Vodnar, D. C., Gheldiu, A. M., Moldovan, C., & Oniga, I., (2018). Phytochemical composition, antioxidant, antimicrobial and *in vivo* anti-inflammatory activity of traditionally used Romanian *Ajuga laxmannii* (Murray) benth. *Frontiers in Pharmacology, 9*, 7.
80. Tolve, R., Cela, N., Condelli, N., Cairano, M. D., Caruso, M. C., & Galgano, F., (2020). Microencapsulation as a tool for the formulation of functional foods: The phytosterols-case study. *Foods, 9*(4), 470.

81. Tolve, R., Condelli, N., Caruso, M. C., Barletta, D., Favati, F., & Galgano, F., (2018). Fortification of dark chocolate with microencapsulated phytosterols: Chemical and sensory evaluation. *Food Function, 9*(2), 1265–1273.
82. Tyagi, P., Singh, M., Kumari, H., Kumari, A., & Mukhopadhyay, K., (2015). Bactericidal activity of curcumin is associated with damaging of bacterial membrane. *PLoS One, 10*(3), e0121313.
83. Tu, L. Y., Pi, J., Jin, H., Cai, J. Y., & Deng, S. P., (2016). Synthesis, characterization and anticancer activity of kaempferol-zinc (II) complex. *Bioorganic & Medicinal Chemistry Letters, 26*(11), 2730–2734.
84. Umamaheswari, M., Ajith, M. P., Asokkumar, K., Sivashanmugam, T., Subhadradevi, V., Jagannath, P., & Madeswaran, A., (2012). *In vitro* angiotensin-converting enzyme inhibitory and antioxidant activities of seed extract of *Apium graveolens* Linn. *Annals of Biological Research, 3*(3), 1274–1282.
85. Vallianou, N. G., Evangelopoulos, A., Schizas, N., & Kazazis, C., (2015). Potential anticancer properties and mechanisms of action of curcumin. *Anticancer Research, 35*(2), 645–651.
86. Vázquez, L., Corzo-Martínez, M., Arranz-Martínez, P., Barroso, E., Reglero, G., & Torres, C., (2019). Bioactive lipids; Chapter 17. In: Mérillon, J. M., & Ramawat, K., (eds.), *Bioactive Molecules in Food* (pp. 467–527). Cambridge: Springer International Publishing.
87. Wu, D., Kong, Y., Han, C., Chen, J., Hu, L., Jiang, H., & Shen, X., (2008). D-alanine: D-alanine ligase as a new target for the flavonoids quercetin and apigenin. *International Journal of Antimicrobial Agents, 32*(5), 421–426.
88. Wang, J. H., Luan, F., He, X. D., Wang, Y., & Li, M. X., (2018). Traditional uses and pharmacological properties of *Clerodendrum* phytochemicals. *Journal of Traditional and Complementary Medicine, 8*(1), 24–38.
89. Yan, J., Zhao, J., & Yang, R., (2019). Bioactive peptides with antidiabetic properties: A review. *International Journal of Food Science & Technology, 54*(6) 1909–1919.
90. Yildiz-Unal, A., Korulu, S., & Karabay, A., (2015). Neuroprotective strategies against calpain-mediated neurodegeneration. *Neuropsychiatric Disease and Treatment, 11*, 297–310.
91. Zhou, J., Chen, M., Wu, S., Liao, X., Wang, J., Wu, Q., Zhuang, M., & Ding, Y., (2020). A review on mushroom-derived bioactive peptides: Preparation and biological activities. *Food Research International, 134*, 109230.

CHAPTER 10

FUNCTIONAL AND NUTRACEUTICAL FOODS: HEALTH AND SAFETY ASPECTS

MONIKA THAKUR, VATSALA SHARMA, and ASHMITA SINGH

ABSTRACT

The philosophy "*Let food be thy medicine and medicine be thy food*" by Hippocrates is re-establishing the interest of the scientific community. Primarily, the consumer's interest is more on health-promoting food with bioactive components having certain disease preventive properties. The phytonutrients play a significant role to withstand various physiological disorders, and therefore, such foods are the necessity in the present-day scenario. Functional foods have been classified at the convergence of drugs and food, and the selection of these bioactive components has been made as per their protective functions, physical, and chemical attributes of the particle. Even though, in that place, evidence is lying that these bioactive components play a focal role in the avoidance of several diseases and promotion of health, but the considerations for lack of side effects have to be of an immense importance and has many challenges as well. Safety concerns have recently been lifted, specially focusing on arbitrary incorporation of bioactive constituents to the foods. The safety issues linked to herbs are complicated and the matter of herb-drug interaction is generating consciousness among consumers and researchers. Food and Drug Administration (FDA) also has issued a Public Health Advisory for various traditional herbs. To date, there are critiques that are focused upon the purpose of functional and nutraceutical components in preventing various health issues; however, no focus has been made on the safety concerns related to such foods. The present review discusses the

Novel Processing Methods for Plant-Based Health Foods: Extraction, Encapsulation, and Health Benefits of Bioactive Compounds. Megh R. Goyal, N. Veena & Ritesh B. Watharkar (Eds.)

health and safety aspects of functional foods and nutraceuticals so that the researchers shall be able to develop an enhanced and robust system to report and examine carefully the health issues linked to consumption and utilization of enriched foods.

10.1 INTRODUCTION

Hippocrates golden words, *"Let food be thy medicine and medicine be thy food"* becomes mandate in the modern world. Food Researchers, producers, and most importantly the food consumers discern that the food consumption is not only limited to perform specific body functions but, it also contributes in the treatment and overall well-being. In the present scenario, food becomes a highly recognized sector; people become very well aware about the food, its bioactive components, and various pharmacological roles other than just being a source of micro and macronutrients [2]. With the advancement in the domain of Food Sciences and Technology, naturally driven health-promoting products received substantial attention for their therapeutic potential and nutritive properties by both the experts as well as the public. The concept of functional food and nutraceutical are expressed by Dr. Stephen Defelice as "Any substance that may be considered as food or part of a food and dispenses medical and health benefits inclusive of the avoidance and treatment of disease."

The term nutraceuticals is basically an amalgam of two words: Nutrient and Pharmaceuticals. As stated by the Association of American Feed Control Officials (AAFCO), 'Nutrient' represents the food constituent in the way at which it facilitates and supports the life of living being while 'nutraceutical' represents the non-toxic or considerably safe component in the food that is proved to possess bioactivity and foster health benefits like avoidance and management of diseases [40]. Functional foods are basically similar to the normal daily routine conventional foods, and are an integral part of our food platter with the nutrients and the bioactive components that can improve biological functions and make them healthier, fitter, and safer [42]. On the other hand, Nutraceuticals are one of the products of food but they are utilized as medicine like a tablet/capsule/powdered form, solutions that are not exactly utilized as a food though consumed as a natural supplement for individuals to treat various chronic diseases and to promote various physiological benefits. They have also been referred as "natural health products" in Canada [46].

Functional foods and nutraceuticals help in providing the means to take the edge off increasing pressure on Medical Health Care System by taking preventive measures. This leads to the increased inquisitiveness towards their continuous use, which is supported by ongoing research attempts to validate the properties and possible administrations of nutraceutical components in food, in combination with communal interest and consumer demand [1, 4, 19, 34, 37, 39, 45].

Consequently, in today's course of events, the usage of functional foods is included in the wish list of every single individual. Indigenously also, the natural and traditional products have been consumed since time immemorial, however the scientific authentication is yet unexplained and inexperienced. Few toxicity studies have also been reported. A lot of *in vitro* work has been done but *in vivo* studies on efficacy, bioavailability, and safety are still in the pipeline. This sector is facing substantial challenges in relation to the safety matters, toxic evaluation, and the regulatory issues globally.

This chapter explores potential and broader overview of functional foods, nutraceuticals, their categories, health, and safety aspects, safety assessments, toxicological studies, and regulatory assessments.

10.2 CATEGORIES OF FUNCTIONAL FOODS

10.2.1 ON THE BASIS OF THEIR ORIGIN

1. **Basic or Natural Functional Foods:** Food products containing natural biological active non-nutritive components, e.g., tomato–lycopene, turmeric–curcumin.
2. **Formulated Foods:** These include the food products especially formulated to have the higher amounts (means they naturally don't have enough amount of that biologically active compounds). They are foods with enhanced functional components, e.g., ω-3 enriched eggs.

10.2.2 ON THE BASIS OF SOURCES (Figure 10.1)

10.2.2.1 PLANT-BASED PHYTOCHEMICALS

Phytochemicals are plant-derived bioactive components. They are a non-essential part of the plant which helps to promote various pharmacological functions like protection against chronic degenerative disorders like cardiovascular diseases (CVDs), cancer, etc. [36]. Major classes of phytochemicals are terpenoids, phenolic metabolites, alkaloids, including phytoestrogens, antioxidants, vitamins, tocopherols, steroids, gamma-linolenic acids (GLA).

Major sources are green-colored foods, such as brassica vegetables, soy plants, grains, etc. [9].

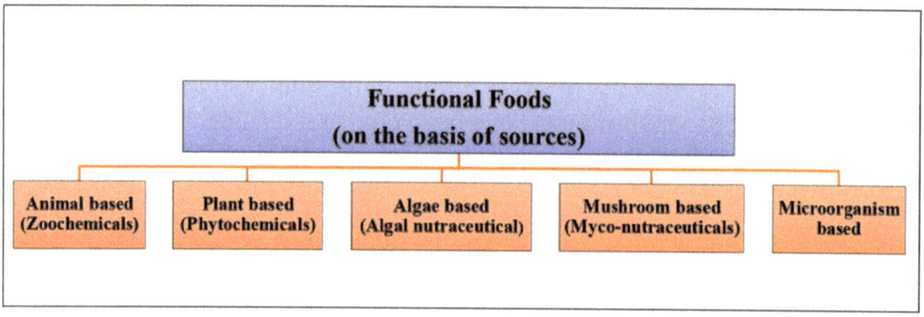

FIGURE 10.1 Categories of functional foods.

10.2.2.2 ANIMAL-BASED ZOOCHEMICALS

Zoochemicals are animal-based bioactive components. Animal-derived functional foods include ω-3 and ω-6 fatty acids, conjugated linoleic acid, small peptides, whey proteins, and casein and glucosamine. The ω-3 fatty acids include alpha-linolenic acid, docosahexaenoic acid (DHA), and eicosapentaenoic acid (EPA).

10.2.2.3 ALGAL-BASED BIOACTIVE COMPONENTS

Bioactive components derived from microalgae reported with many functional properties like antioxidant, anti-aging, anti-cancerous properties, and many more [41, 44]. Various food industries use exo-polysaccharides (EPS) as thickeners and gelling agents produced by several microalgae [17]. They are excellent sources of lipids, specifically of omega-3 fatty acids (DHA and EPA), which can help to facilitate the lactating women and adults [32]. Some examples include *Spirulina*, containing all the essential AAs, whereas *Chlorella* in detoxification and relieves premenstrual syndrome.

10.2.2.4 MUSHROOM-BASED BIOACTIVE COMPONENTS

Mushrooms have a plethora of bioactive components, like polysaccharides, dietary fibers (DFs), oligosaccharides, triterpenoids, peptides, proteins, alcohols, and phenols. These components help in various pharmacological functions like immunomodulatory, anti-tumor, anti-cancerous, anti-hyper-cholesterolemic properties, anti-viral action, and also as aphrodisiacs [28, 29, 42, 45, 46].

10.2.2.5 MICROORGANISM-BASED BIOACTIVE COMPONENTS

Microorganisms are source of several bioactive components. Many strains are reported with many biofunctional effects and have many industrial uses. These strains are utilized in the development of many nutraceutical food products and as dairy starters like lactic acid bacteria (LAB) [22]. The probiotics and prebiotics together as symbiotics are positioned or focused dietary supplements, which help in enhancing the health [30, 43].

Food composed of various non-nutritive components with many pharmacological functions is enlisted in Table 10.1 with their bioactive components and respective health benefits.

10.3 HEALTH AND SAFETY ASPECTS OF FUNCTIONAL FOODS AND NUTRACEUTICALS

The demand of consumers for nutraceuticals is growing rapidly in the market, yet it seems to lie in a gray zone between pharmaceuticals and food. They confront many challenges related to the safety and health claims, the reason being the lack of regulatory shared systems and lack of *in vivo* research with reference to the suggested health claims that avoid the procurement and use of it on the labels of the food items. It is judicious to possess specific defined laws for the use and formulations of nutraceuticals that have all the safety parameters with well-proven clinical data. However, it is appropriate to consider the clinical outlook of nutraceuticals as the "Pharmafoods" or any type of connection between food and drugs assumed together with nutraceuticals.

In the world with nutritional imbalance, these can play a vital role but for that we require the following objectives [37]:

- It is vital to get better nutritional products towards sustainable health, for which we are required to make sure that the assertions made are trustworthy and honest and all are based on the scientific evidence and have verified documentation.

TABLE 10.1 The Bioactive Components with Their Origin and Health Benefits

Origin	Bioactive Compound	Health Benefits	References
		Plant-based Components	
Cocoa powder	Cardiac glycoside, tannin	Potential benefits in renal and hepatic toxicity	[14]
Flax seeds	α-linolenic acid (ALA)	Cancer preventive effects, reduce risk of coronary heart disease	[46]
Red pepper, ginger	Capsaicin	Anti-carcinogenic	[36, 45, 46]
Soybean, flax, maize	Diadzein	Reduce menopause symptoms, improve bone health	[40]
Sweet clover plant	Warfarin	Antimicrobial activities, Sweet-smelling anticoagulant	[36]
		Animal-based Components	
Mackerel and Salmon	ω-3 and ω-6 fatty acids	Precursors of potent lipid mediators, regulation of inflammation, proinflammatory, and immunoactive functions	[49]
Sea cucumbers *Holothuroidea*	Triterpene glycosides (saponins), polysaccharides, sterols, phenolics, peptides, cerebrosides, and lectins	Anti-microbial and anti-cancerous	[3]
		Algal-based Components	
Chorella vulgaris and *Arthrospira platensis* (Microalgae)	Phycobiliproteins	Coloring agent with excellent antioxidant properties and as additives in food and beverages	[32, 41, 44]
Dunaliell salina (Microalgae)	β-Carotene	Coloring additive in food	[32, 44]
Red algae	Ascorbic acid, polyphenols	Antioxidant activity	[41, 44]
Spirulina sp. (Microalgae)	DHA, EPA, ALA, Steridonic acid, vitamins, essential AAs, and phycocyanin	As a human food supplement, natural dyes, fluorescent agents, cosmetics, antioxidant, anti-inflammatory, neuroprotective, or hepatoprotective agent	[17, 33, 44]

TABLE 10.1 *(Continued)*

Origin	Bioactive Compound	Health Benefits	References
	Mushroom-based Components		
Coprinus comatus	Ornithine and γ-amino butyric acid (GABA)	Regulating neuronal excitability throughout the central nervous system in mammals, inhibitory neurotransmitter, anti-anxiety, anticonvulsive effect, and antidiabetic effects	[38, 42]
Cordyceps sinensis and *C. militaris*	Polysaccharide and cordycepin (3'-deoxyadenosine)	Potent anticancer components	[28, 42]
Ganoderma lucidum	Triterpenoids and polysaccharides	Anti-tumor effects, enhance the immune system, reduce blood pressure, and help to reduce blood sugar levels	[28, 42]
Ganoderma lucidum	Ganoderan-bioactive glucan	Heteroglucans and proteoglycans with immune stimulating activity and anticancer activity	[28, 38]
Schizophyllum commune	EPS-Schizophyllan	Use in skin care products as viscosifier and as anti-aging, de-pigmenting, and healing agent	[28, 42]
Volvariella volvacea	Standard and nonstandard AA	Anti-hypertensive and anti-diabetic properties	[28, 42]
	Microorganism-based Components		
Bifidobacterium bifidum BGN4	Probiotic strain	Immune-modulating properties	[1, 27]
Propionibacteria sp.	Vitamin B_{12}	Anti-fungal and anti-bacterial	[1, 22, 23]
Slovenian cheese	*Lactobacillus plantarum* isolates (PCS20, PCS22, PCS25 and PCS26)	High antimicrobial and immunomodulatory capabilities	[1, 4, 30]

- Their presence in the foods must also depict the level of their fortification otherwise toxicity may be reached upon ingestion of an exclusive component rather than micro-doses of multiple components that may otherwise have interaction.
- The food-based approach is always considered easy than the component-based approach.

In the upcoming years, it is anticipated that the nutraceutical market will grow because of the industrial and consumer's interest. The usage of plant wastes and byproducts from agricultural produce targeted the interest in the growth and optimization of bioactive components which shows the possibility for the development of new recipe formulations [8].

10.4 SAFETY ASSESSMENT OF FUNCTIONAL FOODS AND NUTRACEUTICALS

The active components that are possibly utilized on an individual basis or as an additive in food beverage with health-benefiting role and decent safety profile demonstrating safe consumption for individuals can be recognized as functional ingredients. Therefore, the records of advanced and revolutionary technologies are available to attain information and are rapidly scaling up in this age of technology-oriented care for health-related issues. Therefore, with these innovative technologies, it is appropriate to re-visit the analysis of functional components in the context of these emerging technologies [3].

At present, the safety concerns are receiving more importance from industries related to food, health professionals, biomedical communities, policy makers, governments, and lastly by the customers [6, 24, 46]. Functional foods are looked forward to having significant background in the latest nutrition science, to foster health and to mitigate the probability of deadliest diseases. The food product consumed should be developed and validated post investigation of benefits with its long-term risk [3].

Food and Agriculture Organization (FAO) highlighted the importance of Placebo-controlled clinical trials for the food safety aspects and thus for the formulation of such foods with four different phases as: safety, efficiency, effectiveness, and surveillance.

Preference of the trial procedure to evaluate safety should be hinged on a variety of novel food and on the account of whether the similar functional properties are present or not [10]. For computation of safety efficacy of functional food, health experts (dietitian/nutritionist/food science specialists) should be able to answer on following assertions [20]:

- Administration of modulated human clinical mediating trials;
- Amount of ingredient present in one serving;
- Efficacy of single dose of functional ingredient on consumption;
- Frequency of consumption of functional food;
- Functional ingredients present in functional foods;
- Interaction of functional ingredient with drugs;
- Measure of health benefits from functional ingredient;
- Peer-reviewed research studies on functional ingredients;
- Provision of safety information by manufacturer or background of the published research;
- Publication of the studies in peer-reviewed journals;
- Scientific evidence of efficacy of functional food ingredient;
- Standard serving size.

The assessment of safety in functional foods has been based upon 4-step approach: hazard identification, hazard characterization, exposure assessment, and risk characterization.

10.4.1 HAZARD IDENTIFICATION

Gathering pre-existing information about the food and ingredients with a precise biological identification of naturally occurring substance is the initial step in food safety assessment. Description of geographical distribution and details of origin of foods/ingredients with their analysis of variability in composition should be defined [5].

Identification of constituents (such as: protein, AAs, fatty acids, ash, and content of moisture) and chemical hazards (such as: toxicants, presence of antinutrients, mycotoxins, heavy metals, etc.), is necessary. The outcome of the food consumption should be considered to ascertain that no unpropitious effects by the means of malnutrition [11]. In case of scanty data, testing for toxicity should be carried out for safety hazards identification and characterization [3].

10.4.2 HAZARD CHARACTERIZATION

This step elucidates harmful health effects which may result after ingestion of hazards. On the availability of significant data, the characterization of food hazard should manifest information and also the possibility of harmful effects. This can be implemented as a stand-alone process or as a risk.

10.4.3 ASSESSMENT OF COMPONENTS

Analysis of a hazard for a specific pathogen play role as a building block in the assessment of risk of the functional food for a variety of objectives. Characterization of hazard developed for the assessment of water possibly also work for the assessment of food matrix. Generally, hazard characterization of the same pathogen might work in every scenario for the safety assessment [15]. This step for analyzing of water and food follows a six-step approach (Figure 10.2).

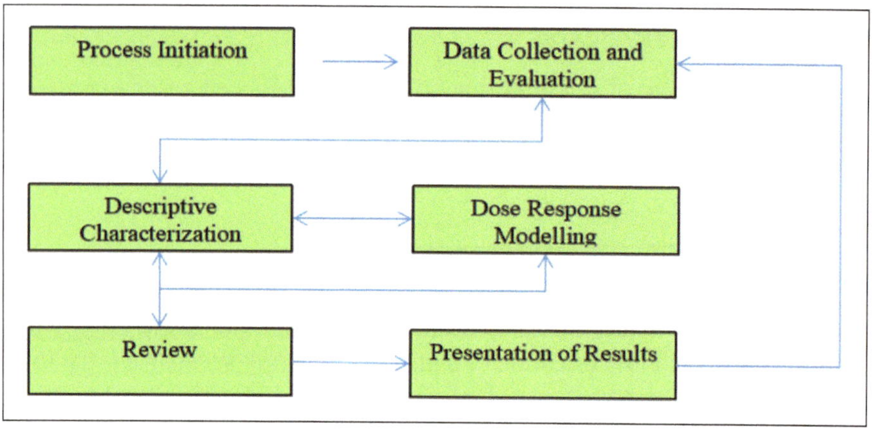

FIGURE 10.2 Process for hazard characterization of pathogens.

10.4.4 EXPOSURE ASSESSMENT

This step takes account of two strands: the amount of the constituent present and consumption amount for an individual population. Information on the quantity of food/food constituent's consumption is procured from the national dietary or nutritional surveys like National Health and Nutrition Examination Survey (NHANES) which are formulated for the assessment of the diet, nutrition intake, and status of nutrition of the broad society including individuals of stage life of over 1.5 years. The intake of food is also quantified by three methods namely; food records (or food diaries), 24-hour recall, and Food frequency questionnaire (FFQ) [13].

10.4.5 RISK CHARACTERIZATION

This is in the combination of the preceding steps to acquire a risk estimation (harmful health effects which will occur in a given population, with related lack of certainty) [15]. These adverse effects have been obtained from toxicological studies and study various terms, such as acceptable daily intake (ADI), provisional maximum tolerable daily intake (PMTDI), provisional tolerable weekly intake (PTWI) and provisional tolerable monthly intake (PTMI) [3, 35]. The risk characterization manifests the process that the ingestion of whole food or ingredient is safe for consumers or not. This is worked on various hypothesis for the evidence of safety for the use of bioactive component utilized in the production of nutraceutical and synchronize by the following postulations [16, 26]:

- Bioactive components are physiologically active compounds having a varied range of results in the body, from physiologic effect to therapeutic potency to toxicity. Comprehending the system for pharmacology and toxicology prospects is significant to anticipate the outcomes of exposure at different levels of dosage.
- Bioactive components are diverse. The distinctive safety issues are linked with each and every component.
- The drug-food interactions must be worked upon for safety's sake.

The characterization of components and pre-existing data serve as the background for the testing of toxicological studies, and they can predict the potency of perilous health effects at the distinct levels of dosage as mentioned below [26]:

- Conventional/extracted/synthetic single component ingredients;
- Extracts with the specific ingredients; and
- New product development with recent technologies [26].

10.5 TOXICOLOGICAL STUDIES

Naturally food-derived components are usually considered as safe as per their long history of human usage. Despite the fact that there is insufficient data on the toxicological studies on the biological products as a consequence of their natural structure. Therefore, it is not correct to assume that these compounds are biologically safe and are exempted from any health risk. Nowadays, studies manifest those common bioactive compounds do have advantageous effects nutritionally, but they can also deploy harmful effects at pharmacological doses. There are new methodologies formulated for the assessment of toxicity in foods, such as – predictive toxicology and toxicomics, which reduces the inconsistency in decision-making in regard to natural products [18].

Toxicity in nutraceuticals or food ingredients through the process of production will definitely modify the standard and safety. The risk on health due to the consumption of these products significantly lies on the existence of infrequent high amounts of chemicals which tend to result in severe toxicity or even fatality many times. This is an urgent issue for children, pregnant women, or elderly adults if contaminated nutraceuticals are consumed above the tolerable limit as compared to adults. Some of the toxic chemicals have been investigated with different types of toxic contaminants in functional foods and nutraceuticals [18].

10.6 RISK ASSESSMENT OF TOXIC CONTAMINANTS

Provisional tolerable intake (PTI) values are based upon the maximum quantity of toxic non-metals and metals present in medicinal plant materials. The permissible limits of toxic metals in all the herbal presentations vary from product to product [47, 48]. With the recommendations of JMPR, the limits have been determined, and they also have the supplements of ADI and analytical methodology to determine the various residues. Different world organizations such as WHO, FAO, and EU decide the maximum residual levels (MRL) for animal feed and human food [18].

10.7 REGULATORY BODIES FOR THE ASSESSMENT OF FUNCTIONAL FOODS AND NUTRACEUTICALS

The regulatory bodies have a very significant role in the safety analysis and toxicological studies of functional and nutraceutical part of the food. All countries have their own regulatory framework for the formulations and safety evaluation of functional foods. Table 10.2 indicates the regulatory bodies in different countries [46].

10.8 REGULATION OF CLAIMS FOR NUTRACEUTICALS

The claims for the nutraceuticals and functional foods have not been validated with their technical requirements but still three very important claims have been streamlined as follows [25]:

1. **Nutrient-Content Claims:** To promulgate the content of nutrients on the product.
2. **Nutrient-Comparative Claims:** Estimate the content of various nutrients with the corresponding foods.
3. **Disease-Risk Reduction Claim:** Established by Codex Alimentarius which are commonly not permitted in Asia.

TABLE 10.2 Regulatory Bodies for Functional Foods and Nutraceuticals

Country	Regulation
Australia and New Zealand	Australia New Zealand Therapeutic Products Agency (ANZTPA)
Canada	Bureau of Nutritional Sciences of the Food Directorate of Health Canada
China	State Food and Drug Administration (SFDA)
European Union	European Food Safety Authority (EFSA)
India	Food Safety and Standards Authority of India (FSSAI)
Japan	Food for Specialized Health Use (FoSHU)
Russia	Biologically Active Food Supplements (BAFS)
South Korea	Health/Functional Food (HFF)
Taiwan	Health Food Control Act (HFCA)
United States	Dietary Supplement Health and Education Act of 1994 (DSHEA)

10.9 SUMMARY

The industry of nutraceutical is flourishing at a faster pace. Even though, there is evidence that these bioactive components play a focal and determining role in the prevention of the disease and health enhancement, but the safety considerations should be of utmost importance. Safety concerns have latterly been lifted, specially focusing on random addition of bioactive components to the foods. Moreover, toxicity in nutraceuticals or food constituents with the process of production has an effect on the safety concerns and quality parameters. Therefore, the chapter aims to review the health and safety certitudes of functional foods and nutraceutical constituents, in regard to make the researchers proficiently develop a potent system for catalog and scrutinize the reports of health problems correlated with functional and nutraceuticals foods.

KEYWORDS

- **bioactive components**
- **eicosapentaenoic acid**
- **functional foods**
- **nutraceuticals**
- **pharmaceuticals**
- **safety aspects**

REFERENCES

1. Al-sheraji, S. H., Ismail, A., Manap, M. Y., Mustafa, S., Yusof, R. M., & Hassan, F. A., (2013). Prebiotics as functional foods: A review. *Journal of Functional Foods, 5*(4), 1542–1553.
2. Bhaskarachary, K., Vemula, S. R., Gavaravarapu, S. R. M., & Joshi, A. K. R., (2016). Traditional foods, functional foods and nutraceuticals. *Proceedings of the Indian National Science Academy, 82*(5), 1565–1577.
3. Bradford, B., (2014). Benefits claims of functional food. In: Hock, F., (ed.), *Drug Discovery and Evaluation: Pharmacological Assays* (pp. 1–6). Berlin–Heidelberg: Springer.
4. Cencic, A., & Chingwaru, W., (2010). The role of functional foods, nutraceuticals, and food supplements in intestinal health. *Nutrients, 2*(6), 611–625.
5. Constable, A., Jonas, D., Cockburn, A., Davi, A., Edwards, G., Hepburn, P., Herouet-Guicheney, C., et al., (2007). History of safe use as applied to the safety assessment of novel foods and foods derived from genetically modified organisms. *Food and Chemical Toxicology, 45*(12), 2513–2525.
6. Crutchfield, S. R., & Roberts, T., (2000). Food safety efforts accelerated in the 1990's. *Food Review, 23*(3), 44–49.
7. Crutchfield, S. R., & Weimer, (2000). Nutrition policy in the 1990s. *Food Review, 23*(3), 38–43.
8. Daliu, P., Santini, A., & Novellino, E., (2019). From pharmaceuticals to nutraceuticals: Bridging disease prevention and management. *Expert Review of Clinical Pharmacology, 12*(1), 1–7.
9. Dillard, C. J., & Bruce, G. J., (2000). Phytochemicals: Nutraceuticals and human health. *Journal of the Science of Food and Agriculture, 80*(12), 1744–1756.
10. EC, (1997). Regulation (EC) No. 258/97 of the European Parliament and of the council of 27 January concerning novels foods and novel foods ingredients. *Official Journal of European Commission,* 43, 1–7.
11. Edwards, G., (2005). Safety assessment of novel foods and strategies to determine their safety in use. *Toxicology and Applied Pharmacology, 207*(2 Suppl), S623–S627.
12. EFSA, (2009). General principles for the collection of national food consumption data in the view of a pan-European dietary survey. *EFSA Journal, 7*(12), 1435.

13. Ellwood, K., Balentine, D. A., Dwyer, J. T., Erdman, J. W., Cortney, G. P., & Kwik, U. C. L., (2014). Considerations on an approach for establishing a framework for bioactive food components. *Advances in Nutrition, 5*(6), 693–701.
14. Ellong, E. N., Billard, C., Adenet, S., & Rochefort, K., (2015). Polyphenols, carotenoids, vitamin C content in tropical fruits and vegetables and impact of processing methods. *Food and Nutrition Sciences, 6*(3), 299–313.
15. FAO/WHO, (2003). *Hazard Characterization for Pathogens in Food and Water: Microbiological Risk Assessment Series 3, Guidelines* (p. 76). World Health Organization.
16. GAO, (2000). *Food Safety: Improvements Needed in Overseeing the Safety of Dietary Supplements and Functional Foods* (p. 43). Washington–DC: US General Accounting Office.
17. García, J. L., De Vicente, M., & Galán, B., (2017). Micro-algae, old sustainable food and fashion nutraceuticals. *Microbial Biotechnology, 10*(5), 1017–1024.
18. Gil, F., Hernández, A. F., & Martín-Domingo, M. C., (2016). Toxic contamination of nutraceuticals and food ingredients; Chapter 58. In: Gupta, R. C., (ed.), *Nutraceuticals: Efficacy, Safety and Toxicity* (pp. 825–837). Amsterdam: Elsevier Inc.
19. Gul, K., Singh, A. K., & Jabeen, R., (2016). Nutraceuticals and functional foods: The foods for the future world. *Critical Reviews in Food Science and Nutrition, 56*(16), 2617–2627.
20. Hasler, C., Moag-Stahlberg, A., Webb, D., & Hudnall, M., (2001). How to evaluate the safety, efficacy, and quality of functional foods and their ingredients. *Journal of the American Dietetic Association, 101*(7), 733–736.
21. Hernández, A. F., & Gil, F., (2016). Nutraceuticals and adverse outcome pathways; Chapter 59. In: Gupta, R. C., (ed.), *Nutraceuticals-Efficacy, Safety and Toxicity* (pp. 839–853). Amsterdam: Elsevier Inc.
22. Hugenholtz, J., & Smid, E. J., (2002). Nutraceutical production with food-grade microorganisms. *Current Opinion in Biotechnology, 13*(5), 497–507.
23. Hugenholtz, J., Hunik, J., Santos, H., & Smid, E., (2002). Nutraceutical production by propionibacteria. *Le Lait, 82*(1), 103–112.
24. Kaferstein, F., & Abdussalam, M., (1999). Food safety in the 21st century. *Bulletin of World Health Organization, 77*(4), 47–51.
25. Keservani, R. K., Sharma, A. K., Ahmad, F., & Baig, M. E., (2014). Nutraceutical and functional food regulations in India; Chapter 19. In: Bagchi, D., (ed.), *Nutraceutical and Functional Food Regulations in the United States and Around the World* (2nd edn., pp. 327–342) London–UK: Elsevier.
26. Kruger, C. L., & Mann, S. W., (2003). Safety evaluation of functional ingredients. *Food Chemistry and Toxicology, 41*(6), 793–805.
27. Ku, S., Park, M. S., Ji, G. E., & You, H. J., (2016). Review on *Bifidobacterium bifidum* BGN4: Functionality and nutraceutical applications as a probiotic microorganism. *International Journal of Molecular Sciences, 17*(9), 1544.
28. Lakhanpal, T. N., & Rana, M., (2005). Medicinal and nutraceutical genetic resources of mushrooms. *Plant Genetic Resources, 3*(2), 288–303.
29. Lakhanpal, T. N., Sai, I., & Thakur, M., (2016). Biodiversity of wild edible, mycorrhizal and nutraceutical important mushrooms. In: Sharma, S., Singh, M., Prasad, S., & Rathore, H., (eds.), *Mushroom of India* (pp. 8–37). Delhi: IIT.
30. Malago, J. J., & Koninkx, J. F. J. G., (2011). Modulation of immune system by probiotics to protect against enteric disorders; Chapter 12. In: Malago, J. J., Koninkx, J. F. J. G., &

Marinsek-Logar, R., (eds.), *Probiotic Bacteria and Enteric Infections: Cytoprotection by Probiotic Bacteria* (pp. 263–286). Netherlands: Springer.

31. McElhatton, A., & Sobral, P. J. A., (2012). *Novel Technologies in Food Science: Their Impact on Products, Consumer Trends and the Environment* (p. 424). Verlag–New York: Springer.
32. Molino, A., Iovine, A., Casella, P., Mehariya, S., Chianese, S., Cerbone, A., Rimauro, J., & Musmarra, D., (2018). Microalgae characterization for consolidated and new application in human food, animal feed and nutraceuticals. *International Journal of Environmental Research and Public Health, 15*(11), 2436.
33. Nicoletti, M., (2016). Microalgae nutraceuticals. *Foods, 5*(4), 54.
34. Ozen, A. E., Pons, A., & Tur, J. A., (2012). Worldwide consumption of functional foods: A systematic review. *Nutrition Reviews, 70*(8), 472–481.
35. Pascal, G., (2009). Safety impact–the risk/benefits of functional foods. *European Journal of Nutrition, 48*(Suppl 1), S33–S39.
36. Prakash, D., Gupta, C., & Sharma, G., (2012). Importance of phytochemicals in nutraceuticals. *Journal of Chinese Medicine Research and Development, 1*(3), 70–78.
37. Prakash, V., & Van, B. M. A. J. S., (2010). Chapter 19; nutraceuticals: Possible future ingredients and food safety aspects. In: Boisrobert, C. E., Stjepanovic, A., Oh, S., & Lelieveld, H. L. M., (eds.), *Ensuring Global Food Safety: Exploring Global Harmonization* (pp. 333–338). London–UK: Academic Press.
38. Reyes, R. G., Lou, L., Lopez, M. A., Kumakura, K., & Kalaw, S. P., (2009). *Coprinus comatus*, a newly domesticated wild nutraceutical mushroom in the Philippines. *Journal of Agricultural Technology, 5*(2), 299–316.
39. Shahidi, F., (2009). Nutraceuticals and functional foods: Whole versus processed foods. *Trends in Food Science and Technology, 20*(9), 376–387.
40. Shinde, N., Bangar, B., Deshmukh, S., & Kumbhar, P., (2014). Nutraceuticals: A review on current status. *Research Journal of Pharmacy and Technology, 7*(1), 110–113.
41. Singh, K., & Thakur, M., (2020). Marine genomics: An emerging novel technology; Chapter 14. In: Goyal, M. R., Suleria, H. A. R., & Kirubanandan, S., (eds.), *Technological Processes for Marine Foods, from Water to Fork: Bioactive Compounds, Industrial Applications and Genomics* (pp. 357–366). Oakville–ON: Apple Academic Press Inc.
42. Thakur, M., (2015). Wild mushrooms as natural untapped treasures. In: Chauhan, A. K., Pushpangadan, P., & George, V., (eds.), *Natural Products: Recent Advances* (pp. 214–226). New Delhi–India: Write and Print Publications.
43. Thakur, M., (2016). Significance of prebiotics, probiotics and synbiotics as health potentiators. *Food Science Research Journal, 7*(2), 327–334.
44. Thakur, M., (2020). Marine bioactive components: Sources, health perspective and future prospects; Chapter 3. In: Goyal, M. R., Suleria, H. A. R., & Kirubanandan, S., (eds.), *Technological Processes for Marine Foods, from Water to Fork: Bioactive Compounds, Industrial Applications and Genomics* (pp. 61–72). Oakville–ON: Apple Academic Press Inc.
45. Thakur, M., & Modi, V. K., (2020). *Emerging Technologies in Food Science* (1st edn., p. 287). Singapore: Springer.
46. Thakur, M., Singh, K., & Khedkar, R., (2020). Phytochemicals: Extraction process, safety assessment, toxicological evaluations and regulatory issues; Chapter 11. In: Prakash, B., (ed.), *Functional and Preservative Properties of Phytochemicals* (pp. 341–361). London–UK: Elsevier Inc.

47. WHO, (2007). *WHO Guidelines for Assessing Quality of Herbal Medicines with Reference to Contaminants and Residues* (p. 105). Geneva–Switzerland: World Health Organization.
48. Woteki, C. E., Facinoli, S. L., & Schor, D., (2001). Keep food safe to eat: Healthful food must be safe as well as nutritious. *The Journal of Nutrition, 131*(2), 502S–509S.
49. Zhang, X., Chen, F., & Wang, M., (2015). Bioactive substances of animal origin. In: Cheung, P., & Mehta, B., (eds.), *Handbook of Food Chemistry* (pp. 1009–1033). Berlin–Heidelberg: Springer.

CHAPTER 11

EDIBLE MUSHROOM PHENOLICS: HEALTH AND FUTURE PERSPECTIVES

SANUSI S. NASSARAWA, AHMAD GAMBO, MUNIR A. DANDAGO, and BALARABE B. ISMAIL

ABSTRACT

As secondary metabolites, phenolic compounds are found in many mushroom (MSR) species. These compounds have sparked a lot of interest as common bioactive compounds in the functional foods and pharmaceutical industries for health promotion and disease control because of their possible health benefits to humans, such as antioxidants and other biological activities. In this chapter, a MSR will be introduced; particular focus on classification and distribution of MSR phenols, biological activities of MSR (antioxidant, antitumor, anti-inflammatory, antihyperglycemic, anti-tyrosinase, and antimicrobial activity), MSR phenol as a nutraceutical, pharmaceutical, and cosmeceutical agents, and future perspectives of MSR phenol will be availed.

11.1 INTRODUCTION

MSR is a mega-organism with a typical plant organ that can look either higher or downstairs, prominent adequate to be envisioned [18]. MSR is estimated to be 0.14 million worldwide. More than 2,000 species are protected, with about 700 species exhibiting various biological activities [80]. Since the old era, MSR species are consumed as food and medicinal drug. Apart from taste buds and fragrance, MSR has significant nutritional value.

Novel Processing Methods for Plant-Based Health Foods: Extraction, Encapsulation, and Health Benefits of Bioactive Compounds. Megh R. Goyal, N. Veena & Ritesh B. Watharkar (Eds.)

MSR is rich in proteins, vitamins, carbohydrates, minerals, and minor fat content. Moreover, MSR has contained a few essential amino acids (EAA) and fibers [62]. MSR also contains secondary metabolites [48] and phenolic compounds, all of which have antiviral, antioxidant, antibacterial, and anti-tumor properties [39].

Among various secondary metabolites reported from MSR, polyphenols are widely explored and effective against multiple health benefits [35]. The presence of phenolic compounds as well as flavonoid compounds were reported from mushroom extracts [10].

This chapter discusses recent developments in mushroom phenolic compounds, such as new sources, structural features, biological activities, and possible mechanisms of action, trends in their use as a nutraceutical, medicinal, and cosmeceutical agents, and prospective industrial applications.

11.2 CATEGORIES OF MSR PHENOLS

In phenolic compounds, one or more hydroxyl groups are bound to at least one aromatic ring (C6). Simple molecules to complex polymers can be found in their composition [47]. Various phenolic acids have been discovered in extracts from various MSR species (Figure 11.1).

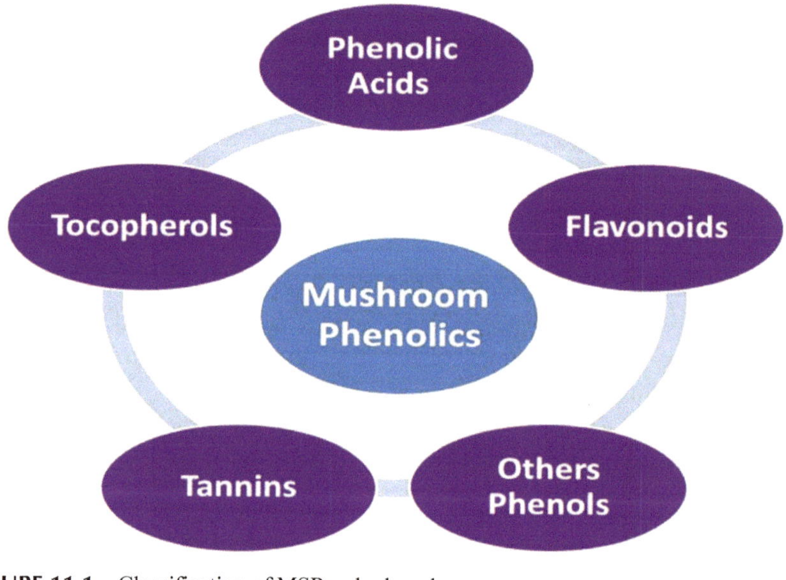

FIGURE 11.1 Classification of MSR polyphenols.

Non-flavonoid acids (e.g., gallic, *p*-coumaric, and caffeic acids) are classified into two categories: C1–C6 and C3–C6 backbone derivatives of cinnamic acid and benzoic acid. According to many studies, phenolic acids are the largest polyphenols responsible for many MSR species' biological activities [50, 53].

Flavonoids are phenolic compounds with two C6 units (A and B ring) and a general structural backbone of C6–C3–C6. Flavonoids are divided into many subgroups, including flavones, flavonols, and anthocyanins. The existence of flavonoid compounds in several MSR species extracts has been demonstrated in many studies [10].

Tannins are phenolic compounds found in plants that can precipitate proteins from aqueous solutions. Tannins may form complexes with various other substances, including polysaccharides, alkaloids, and nucleic acids. Tannins have been found in *Pleurotus tuber-regium (fries)* [3], *Agaricus silvaticus, Hydnum rufescens, and Meripilus giganteus* [30].

Another class of MSR polyphenols is tocopherols. These compounds are classified into four groups (α, β, γ, and δ) based on the number and position of methyl groups. Some research on the content of tocopherol in MSRs has been published [31]. Other phenolic compounds have been obtained from some MSR species are:

- Benzophenone derivatives, daldinals A-C from *Daldinia childiae* [61, 81];
- Grifolin derivatives from *Albatrellus ovinus* [43, 55];
- Grifolin derivatives from *Boletus pseudocalopus* [68];
- Hispidin from *Phellinus linteus* [59].

11.3 BIOLOGICAL ACTIVITIES OF MSR PHENOLIC COMPOUNDS

11.3.1 ANTIOXIDANT PROPERTIES

Oxidative stress is the primary cause of excessive reactive oxygen species (ROS) formation in cellular organisms. Too much (ROS) production can trigger oxidative mutilation to biological macromolecules; including proteins, lipids, and nucleic acids, leading to tissue injury or death [78]. ROS such as peroxide, hydroxyl radicals, and superoxide radicals can damage DNA and cause enzyme and protein structure disruption. These molecules are also involved in the pathogenesis of several chronic diseases, including cancer, diabetes, cataracts, aging, neurological disorders, cardiovascular disease, and rheumatoid arthritis [36].

Free radical scavenging (FRS) and reducing capacities were high in some MSR extracts, as well as excellent DNA protection against H_2O_2-induced damage [5]. Ahmad et al. [2] demonstrated that the observed effects are primarily related to phenolic content's antioxidant properties owing to redox reactions, which enable them to function as reducing agents or hydrogen molecule donors. Also, Bahadori et al. [10] illustrated that phenolic water extracts in MSR species (*Melanoleuca cognata* and *Melanoleuca stridula*) have the strongest *in vitro* antiradical activity followed by methanolic extracts; however, the lowest was found in ethyl acetate extracts.

Antioxidant properties of ethyl acetate extracts and methanolic extracts of various MSR species were investigated in another study (*Meripilus giganteus, Agaricus silvaticus, and Hydnum rufescens* flowed from high flavonoid content and polyphenol content). These effects promote the consumption of these extracts for improving human wellness, equally in food industries and pharmaceuticals [30]. Antioxidant activity was found in *Polyporus fomentarius, Polyporus volvatus, Polyporus badius, Cantharellus cibarius, Polyporus stevenii, Lactarius deliciosus, Trametes Versicolor* MSRs from Turkey and the highest *in vitro* scavenging activity was observed in *P. voluatus* [57]. Similarly, researchers studied the antioxidant activities of some commonly consumed MSR species in China and illustrated that all MSR species showed antioxidant activity; and the highest *in vitro* antioxidant activities were found in *Boletus aereus* (porcino nero), *Hellinus igniarius* (mulberry yellow), *Umbilicaria esculenta* (stone ear), and *Griflola frondosa* (maitake) [35].

11.3.2 ANTITUMOR ACTIVITY

The importance of MSR species as anti-cancer agents has been recognized in recent years, and its use as a biological agent in cancer treatment has been suggested [22]. They exert antitumor activity by inhibiting the increase in cancer cells and tumor growth, as seen both *in vitro* and *in vivo* investigations [54]. Several research showed the potential anticancer activity of phenolic compounds from MSR species such as *Albatrellus confluence* against human ovarian cancer [86] and human gastric cancer cell [85], *Auricularia polytricha* species against the colon, breast, and kidney cancer [7], *Clitocybe alexandri* against Nonsmall cell lung cancer [77], *Coprinopsis atramentaria* against human colon cancer [38], *Ganoderma lucidum* against human lung cancer [20], and *Inonotus obliquus* against leukemia [52].

On the other hand, flavonoid compounds such as rutin, myricetin, naringenin, quercetin, hesperetin, and morin make up most of the polyphenolic content of *Ganoderma lucidum* ethanol extract. Antiproliferative activity and the occurrence of these compounds in MSR extracts are highly correlated [64].

11.3.3 ANTI-INFLAMMATORY ACTIVITY

Inflammation is the immune system's biological reaction to harm caused by chemical, physical, and pathogenic influences. Acute inflammation is a short-term inflammatory response that usually resolves on its own [6]. In some cases, however, uncontrolled acute inflammation can lead to chronic stage [21]. Due to its ability to reduce the synthesis of inflammatory mediators, MSR has demonstrated anti-inflammatory properties [24].

Moro and coworkers stated that different compounds in mushrooms had been pointed out as potential anti-inflammatory agents, including phenolic compounds [49]. They also discovered that pyrogallol inhibits the synthesis of nitric oxide (NO), interleukin-6 (IL-6) mRNAs, and inducible nitric oxide synthase (iNOS), IL-1β in RAW 364.7 macrophages in response to lipopolysaccharide (LPS) stimulation [32, 49].

Using LPS-activated RAW 264.7 macrophages, the anti-inflammatory effects of phenolics from edible MSR species were also investigated *in vitro*. The effects of extracts on the expression of inflammation marker, including IL-6 and IL-1, and the synthesis of NO were investigated. MSR species, *Lactarius deliciosus, Cantharellus cibarius, and Agaricus bisporus* had the most prominent anti-inflammatory effects of the studied model [58]. In addition, grifolins were found that LPS stimulated NO output in RAW 264.7 cells was significantly inhibited, with IC_{50} values varying from 22.9–29 μ M [61].

Pleurotus ostreatus (PO), *Macrolepiota procera* (MP), *Boletus impolitus* (BI), and *Agaricus bisporus* (AB) phenolic extracts displayed the most impregnable anti-inflammatory budding, showing the most eminent inhibition of NO development [74]. These species also had the highest concentration of cinnamic acid, which had the greatest anti-inflammatory activity.

11.3.4 ANTIHYPERGLYCAEMIC ACTIVITY

Diabetes mellitus metabolic changes are associated with lipid metabolism, carbohydrate, and protein [66]. The most powerful method for reducing the risk of diabetes type 2 is to inhibit the polysaccharide hydrolysis enzymes pancreatic amylase and intestinal glucosidase [40].

Many MSR species have been claimed to be active against this disease and can regulate it and its problems such as hypertension, cardiovascular, and renal failure diseases [29]. As such, antioxidants in MSRs, such as polyphenols and flavonoids, have shown important anti-diabetic properties by suppressing the activity of some enzymes, including α-glucosidase and α-amylase [44].

Some edible MSR species collected in Thailand, such as *Russula emetica, Phlebopus portentous*, and *Rugiboletus extremiorientalis* showed anti-hyperglycemic activity. *R. extremiorientalis* had higher inhibitory activities against α-glucosidase than other mushroom species, which may be attributed to its high polyphenol and flavonoid content. *R. extremiorientalis* water extracts and methanolic extracts had higher α-glucosidase repressing activities (54.4 ± 1.2% and 55.5 ± 3.9 reserves) than its ethanol and acid extracts [37].

In another study, Stojkovic, and coworkers investigated the *in vitro* anti-diabetic properties for six edible and medicinal MSR classes, such as *Cordyceps militaris, Inonotus obliquus, Morchella conica, Agaricus blazei, Coprinus comatus, and Phellinus linteus* are only a few of the species. They discovered that all methanolic extracts of selected MSRs inhibited α-glucosidase enzyme activity, while all methanolic extracts except *M. conica and C. militaris* inhibited α-amylase activity. According to the findings, *Inonotus obliquus* was the most promising potential anti-diabetic species [71].

Tremella fuciformis, Agrocybe aegerita, Auricularia auriculajudae, Hericium Erinaceus, Grifola frondosa, Ganoderma lucidum, Lentinus edodes, and Russula sanguinea were among the eight MSR species studied for their anti-diabetic properties. For some fungi, such as *G. lucidum,* because of their high content of phenolic and flavonoid content, as well as high inhibition of α-glycosidase and aldose reductase, *G. lucidum* has anti-diabetic properties. Furthermore, this research discovered a connection between antioxidant activity and anti-diabetic results [83]. In diabetic rats, oral syringic acid administration reduced plasma glucose levels as well as liver and kidney glycoproteins, as evidenced by increased plasma insulin and C-peptide levels [51].

11.3.5 ANTI-OSTEOPOROTIC ACTIVITY

Bone loss and skeletal tissue microarchitectural degradation characterize osteoporosis, follow-on in bone fragility and an amplified risk of fracture. Fast bone turnover causes osteoporotic bone loss, in which bone resorption outpaces bone deposition [75]. In menopausal women, a lack of estrogen (E2) induces an imbalance in bone turnover. The number of osteoporosis patients has risen in recent years because of an aging population, and bone health has become a serious concern. Fitness and dietary habits have a significant stimulus on bone health. Consequently, improved nutrition is likely to be part of the solution to this serious issue. Bioavailable estrogens, such as selective estrogen receptor (ER) modulators, can help to correct the difference in bone turnover [46].

Nutritional compounds extracted from MSR species, including phenolic compounds and vanillic acids, show promise in preventing postmenopausal osteoporosis. Nevertheless, more research is required to fully comprehend the molecular mechanisms that underpin the anti-osteoporotic behavior of syringic and vanillic acids [73].

Furthermore, dietary syringic acid from the mycelium of the shiitake MSR *Lentinula edodes* can protect ovariectomized (OVX) mice from bone loss and microarchitectural deterioration without affecting the uterus [72]. Syringic acid can affect both bone resorption and bone growth, making it beneficial for osteoporosis disease prevention. Syringic and vanillic acids prevent nuclear factor-kB (NFkB), an oxidative stress-responsive factor, from binding to DNA in human colorectal cells, resulting in anti-osteoporotic activity [1].

11.3.6 ANTI-TYROSINASE ACTIVITY

Tyrosinase is a di-nuclear copper active site polyphenol oxidase that is involved in the first stage of melanin formation in humans, animals, and plants. An overactive form of the tyrosinase enzyme is also involved in the enzymatic browning of fresh produce, causes hyperpigmentation (melanin increase) of human skin. As a result, tyrosinase inhibitors have piqued interest as skin whitening and anti-browning agents in the food and beverage industries [17].

Gallic acid, hydroxycinnamic acid, catechin, sinapic acid, protocatechuic acid, rutin, vanillic acid, syringic acid, ferulic acid, and apigenin are large phenolic acids that have significant anti-tyrosinase activity and can be found in many wild MSR species. The antioxidant activity of these compounds may be one of the most effective ways to inhibit tyrosinase [37]. Furthermore, flavonoid compounds contained in mushroom extracts have the ability to act as natural inhibitors of tyrosinase [19].

Furthermore, the *Lentinus lepideus* methanol extract inhibited tyrosinase more effectively than the hot water extract. The phenolic and flavonoid content of both mushroom species had an average relationship with their tyrosinase inhibitory activity [87]. In contrast, the amount of hydroxyl groups in phenolic compounds can affect inhibition of tyrosinase, which forms hydrogen bonds with the low enzyme's active site, resulting in a decrease in enzyme activity [4].

11.4 NUTRACEUTICAL, PHARMACEUTICAL, AND COSMECEUTICAL AGENTS OF MUSHROOM PHENOL

11.4.1 MUSHROOM PHENOL AS PHARMACEUTICAL AND NUTRACEUTICAL AGENTS

Many ancient traditional medicine systems have recorded the use of mushroom species in folk medicine, including traditional Chinese medicine, Indian medicine (*Ayurveda*), Korean medicine (*Hanyak*), and traditional Japanese medicine) (*Kampo*) [60]. For a long time, ancient civilizations such as Egypt, early Greek, Roman, and Mexican cultures used MSRs as delicacies and even medicine [16, 28].

Recently, using MSR bioactive compounds in functional foods to promote its health effects and control of many chronic diseases is on the rise [76]. In this context, fortifying grains such as brown rice, canjica corn, and wheat by some mushroom species with high phenolic components can improve antioxidant, anti-diabetic, anti-obesity, and other biological activities for these products, which present a potential for functional foods [70].

Furthermore, several mushroom species have been used in muffins to upsurge the phenolic content and improve the dietary value and quality characteristics of these foods, according to numerous reports [56]. Furthermore, adding the MSR powder to extruded snacks increases the snack samples' essential volume, water solubility index, total phenolic content (TPC), and antioxidant activity [45].

11.4.2 MUSHROOM PHENOLS AS COSMECEUTICAL AGENTS

The growing number of commercial cosmeceutical formulations that claim to help with fine lines, wrinkles, aging, skin texture, photoprotection, and pigmentation indicate that mushroom cosmeceutical ingredients are making their way into the cosmetic industry [84]. The findings showed strong markers of extract suitability for skincare formulations, as well as the absence of toxicity in keratinocytes and fibroblasts, suggesting that they could provide some protection.

Natural polyphenols have also been shown to have anti-oxidant efficacy, making them promising candidates for use in anti-aging lotions and creams [69]. Several previous studies have discovered anti-collagenase activity in mushroom phenolic acids include ellagic acid [9] and p-coumaric acid [65].

11.5 CHALLENGES AND FUTURE PERSPECTIVES OF MUSHROOM PHENOLS

11.5.1 USING MUSHROOM PHENOLICS IN INDUSTRIAL APPLICATIONS

Processing method and storage influence the stability, solubility, bioactivity, and bioavailability of phenolic compounds, limiting their uses in food and medicines [15]. Food processing, storage conditions (including oxidation, light, and high temperature), and gastrointestinal (GI) tract conditions all have an impact on the effects of phenolic compounds. These conditions can hinder their incorporation into foods or medicines. In addition, the unpleasant taste of phenolic compounds is another challenge for adding them in food products [12].

The use of micro/nano-sized particles/fibers to encapsulate phenolic compounds improved their stability, protected them from the atmosphere, and regulated their release under explicit conditions. Two of the most promising methods for encapsulating and shielding these critical compounds from degradation are electrospinning and electrospraying [26, 39].

11.5.2 NOVEL DRUG DELIVERY SYSTEM FOR MUSHROOM PHENOLICS

Nano-formulations and other novel drug delivery systems may increase the efficacy of therapeutic compounds while lowering their toxicity, doses, and side effects. Nanomaterial structures of sizes between 1 and 100 nm in at least one dimension can be used to increase the efficiency of drug delivery and the solubility of drugs with poor water solubility as well as decreasing side effects of medicines due to their ability to cross cell barriers and to present an increased reaction zone [39, 67]. For this purpose, mushroom bioactive compounds are used as nanoparticle materials added in food industries as active functional food ingredients [79].

11.5.3 USE OF MSR PHENOLICS FOR ACTIVE PACKAGING

Using active packaging technology to extend the shelf life of foods is an innovative concept. Antioxidant, antimicrobial, and carbon dioxide (CO_2) emitting/producing agents contained in packaging materials were used to establish this technology [27]. Active packaging can work by slowly releasing active agents into the environment or absorbing food-degrading compounds [82, 88]. Chitosan is considered a perfect component for edible films due to its peculiar characteristics [23]. Phenolic compounds are widely used as active film or coating additives in food packaging to extend the shelf life of food [16]. Gallic acid grafted chitosan film may be used as a new active packaging material for MSR and other food postharvest storage.

11.5.4 EXTRACTION OF MSR PHENOLIC COMPOUNDS

Phytochemicals were extracted from natural resources using a variety of non-traditional extraction methods [13, 14]. The extraction of MSR polyphenols by green extraction methods has been increased to avoid the problems associated with using conventional extraction methods (CEM). Microwave hydrodiffusion and gravity, autohydrolysis, and supercritical CO_2 extraction were used to extract polyphenols from the *Pleurotus eryngii* [63].

11.6 SUMMARY

MSRs contain secondary metabolites, including polysaccharides [48] and phenolic compounds, which have antioxidant, antibacterial, antiviral, antitumor, and anti-inflammatory properties; they are even good for cardiovascular health. Due to these properties, using MSR bioactive compounds in functional foods for promoting its health effects and control of many chronic diseases has been increased. Several studies have found that MSR fruiting bodies and extracts comprise high levels of phenolic acids that are more effective. Cinnamic, p-hydroxybenzoic, P-coumaric, protocatechuic, and caffeic acids are only a few of the phenolic acids commonly contained

MSR extracts [50]. The efficiency of phenolic compounds is contingent on retaining their stability, bioavailability, and bioactivity during processing, storage, and consumption. So, the most promising future perspectives of phenolic compounds are active packaging by grafted chitosan and novel drug delivery systems by nano-formulations or encapsulation to enhance its biological activities and stability.

KEYWORDS

- **biological activities**
- **caffeic acid**
- **drug delivery**
- **encapsulation**
- **lipopolysaccharide**
- **mushroom**
- **phenolics**

REFERENCES

1. Abaza, M. S., Al-Attiyah, R., Bhardwaj, R., Abbadi, G., Koyippally, M., & Afzal, M., (2013). Syringic acid from *Tamarix aucheriana* possesses antimitogenic and chemo-sensitizing activities in human colorectal cancer cells. *Pharmaceutical Biology, 51*(9), 1110–1124.
2. Ahmad, N., Mahmood, F., Khalil, S. A., Zamir, R., Fazal, H., & Abbasi, B. H., (2014). Antioxidant activity via DPPH, gram-positive and gram-negative antimicrobial potential in edible mushrooms. *Toxicology and Industrial Health, 30*(9), 826–834.
3. Akindahunsi, A. A., & Oyetayo, F. L., (2006). Nutrient and antinutrient distribution of edible mushroom, Pleurotus tuber-regium (Fries) singer. *LWT Food Science and Technology, 39*(5), 548–553.
4. Alam, N., Yoon, K. N., & Lee, T., (2011). Evaluation of the antioxidant and antityrosinase activities of three extracts from *Pleurotus nebrodensis* fruiting bodies. *African Journal of Biotechnology, 10*(15), 2978–2986.
5. Aprotosoaie, A. C., Zavastin, D. E., Mihai, C., Voichita, G., Gherghel, D., Silion, M., Trifan, A., & Miron, A., (2017). Antioxidant and antigenotoxic potential of *Ramaria largentii* Marr & D. E. Stuntz, a wild edible mushroom collected from Northeast Romania. *Food and Chemical Toxicology, 108*, 429–437.
6. Araujo, J. R., Goncalves, P., & Martel, F., (2011). Chemopreventive effect of dietary polyphenols in colorectal cancer cell lines. *Nutrition Research, 31*(2), 77–87.

7. Arora, S., Goyal, S., Balani, J., & Tandon, S., (2013). Enhanced antiproliferative effects of aqueous extracts of some medicinal mushrooms on colon cancer cells. *International Journal of Medicinal Mushrooms, 15*(3), 301–314.
8. Bach, F., Zielinski, A. A. F., Helm, C. V., Maciel, G. M., Pedro, A. C., Stafussa, A. P., Ávila, S., & Haminiuk, C. W. I., (2019). Bio compounds of edible mushrooms: *In vitro* antioxidant and antimicrobial activities. *LWT-Food Science and Technology, 107*, 214–220.
9. Bae, J. Y., Choi, J. S., Kang, S. W., Lee, Y. J., Park, J., & Kang, Y. H., (2010). Dietary compound ellagic acid alleviates skin wrinkle and inflammation induced by UV-B irradiation. *Experimental Dermatology, 19*, 182–190.
10. Bahadori, M. B., Sarikurkcu, C., Yalcin, O. U., Cengiz, M., & Gungor, H., (2019). Metal concentration, phenolics profiling and antioxidant activity of two wild edible melanoleuca mushrooms (*M. cognata and M. stridula*). *Microchemical Journal, 150*, 104–172.
11. Barros, L., Baptista, P., Estevinho, L. M., & Ferreira, I. C. F. R., (2007). Effect of fruiting body maturity stage, on chemical composition and antimicrobial activity of *Lactarius Sp.* mushrooms. *Journal of Agricultural and Food Chemistry, 55*(21), 8766–8771.
12. Bell, L. N., (2002). Stability testing of nutraceuticals and functional foods; Chapter 30. In: Robert, E. C., (ed.), *Handbook of Nutraceuticals and Functional Foods* (pp. 504–519). Boca Raton: CRC Press.
13. Belwal, T., Ezzat, S. M., Rastrelli, L., Bhatt, I. D., Daglia, M., Baldi, A., Devkota, H. A., et al., (2018). A critical analysis of extraction techniques used for botanicals: Trends, priorities, industrial uses and optimization strategies. *Trends in Analytical Chemistry, 100*, 82–102.
14. Belwal, T., Huang, H., Li, L., Duan, Z. H., Zhang, X. B., Aalim, H., & Luo, Z., (2019). Optimization model for ultrasonic-assisted and scale-up extraction of anthocyanins from Pyrus communis 'starkrimson' fruit peel. *Food Chemistry*, 297, 124993.
15. Bhushani, J. A., & Anandharamakrishnan, C., (2014). Electrospinning and electrospraying techniques: Potential food based applications. *Trends in Food Science and Technology, 38*(1), 21–33.
16. Bladé, C., Aragonès, G., Arola-Arnal, A., Muguerza, B., Bravo, F. I., Salvadó, M. J., Arola, L., & Suárez, M., (2016). Proanthocyanidins in health and disease. *Biofactors, 42*(1), 5–12.
17. Carcelli, M., Rogolino, D., Bartoli, J., Pala, N., Compari, C., Ronda, N., Bacciottini, F., et al., (2020). Hydroxyphenyl thiosemicarbazones as inhibitors of mushroom tyrosinase and antibrowning agents. *Food Chemistry, 303*, 125–310.
18. Chang, S. T., & Miles, P. G., (2004). *Mushrooms: Cultivation, Nutritional Value, Medicinal Effect, and Environmental Impact* (2nd edn., p. 480). Boca Raton, FL: CRC Press.
19. Chang, T. S., (2009). An updated review of tyrosinase inhibitors. *International Journal of Molecular Sciences, 10*(6), 2440–2475.
20. Chen, Y., Lv, J., Li, K., Xu, J., Li, M., Zhang, W., & Pang, X., (2016). Sporoderm-broken spores of *Ganoderma lucidum* inhibit the growth of lung cancer: Involvement of the Akt/mTOR signaling pathway. *Nutrition and Cancer, 68*(7), 1151–1160.
21. Dennis, E. A., & Norris, B. C., (2015). Eicosanoid storm in infection and inflammation. *Nature Reviews Immunology, 15*(8), 511–523.
22. Ding, X., Hou, Y., & Hou, W., (2012). Structure feature and antitumor activity of a novel polysaccharide isolated from *Lactarius deliciosus* gray. *Carbohydrate Polymers, 89*(2), 397–402.

23. Elsabee, M. Z., & Abdou, E. S., (2013). Chitosan based edible films and coatings: A review. *Materials Science & Engineering: C, 33*(4), 1819–1841.
24. Elsayed, A. E., Hesham, E. L. E., Mohammad, A. M. W., & Ramlan, A., (2014). Mushrooms: A potential natural source of anti-inflammatory compounds for medical applications. *Mediators of Inflammation, 2014*, 805841.
25. Erjavec, J., Ravnikar, M., Brzin, J., Grebenc, T., Gosak, M. Z., Sabotic, J., Kos, J., & Dreo, T., (2016). Antibacterial activity of wild mushroom extracts on bacterial wilt pathogen *Ralstonia solanacearum*. *Plant Disease, 100*(2), 453–464.
26. Esfanjani, F. A., & Jafari, S. M., (2016). Biopolymer nanoparticles and natural nano-carriers for nano-encapsulation of phenolic compounds. *Colloids and Surfaces B: Biointerfaces, 146*, 532–543.
27. Fang, Z., Zhao, Y., Warner, R. D., & Johnson, S. K., (2017). Active and intelligent packaging in the meat industry. *Trends in Food Science & Technology, 61*, 60–71.
28. Feeney, M. J., Dwyer, J., Hasler-Lewis, C. M., Milner, J. A., Noakes, M., Rowe, S., Wach, M., et al., (2014). Mushrooms and health summit proceedings. *The Journal of Nutrition, 144*(7), 1128S–1136S.
29. Garduno-Diaz, S. D., & Khokhar, S., (2012). Prevalence, risk factors and complications associated with type 2 diabetes in migrant South Asians. *Diabetes / Metabolism Research and Reviews, 28*(1), 6–24.
30. Garrab, M., Edziri, H., ElMokni, R., Mastouri, M., Mabrouk, H., & Douki, W., (2019). Phenolic composition, antioxidant and anticholinesterase properties of the three mushrooms *Agaricus silvaticus Schaeff., Hydnum rufescens Pers.* and *Meripilus giganteus (Pers.) Karst.* in Tunisia. *South African Journal of Botany, 124*, 359–363.
31. Heleno, S. A., Barros, L., Sousa, M. J., Martins, A., & Ferreira, I. C. F. R., (2010). Tocopherols composition of Portuguese wild mushrooms with antioxidant capacity. *Food Chemistry, 119*(4), 1443–1450.
32. Hu, Q., Yuan, B., Xiao, H., Zhao, L., Wu, X., Rakariyatham, K., Zhong, L., et al., (2018). Polyphenols-rich extract from *Pleurotus eryngii* with growth inhibitory of HCT116 colon cancer cells and anti-inflammatory function in RAW264.7 cells. *Food and Function, 9*(3), 1601–1611.
33. Hugo, W. B., & Russell, A. D., (1983). *Pharmaceutical Microbiology* (3rd edn., p. 470). London: Blackwell Scientific Publications.
34. Hwang, M. S., Lee, I., Choi, H. J., & Yun, B., (2015). Anti-influenza activities of polyphenols from the medicinal mushroom *Phellinus baumii*. *Bioorganic and Medicinal Chemistry Letters, 25*(16), 3256–3260.
35. Islam, T., Yu, X., & Xu, B., (2016). Phenolic profiles, antioxidant capacities, and metal chelating ability of edible mushrooms commonly consumed in China. *LWT-Food Science and Technology, 72*, 423–431.
36. Jeong, E. M., Liu, M., Sturdy, M., Gao, G., Varghese, S. T., Sovari, A. A., & Dudley, Jr. S. C., (2012). Metabolic stress, reactive oxygen species and arrhythmia. *Journal of Molecular and Cellular Cardiology, 52*(2), 454–463.
37. Kaewnarin, K., Suwannarach, N., Kumla, J., & Lumyong, S., (2016). Phenolic profile of various wild edible mushroom extracts from Thailand and their antioxidant properties, anti-tyrosinase and hyperglycaemic inhibitory activities. *Journal of Functional Foods, 27*, 352–364.
38. Khan, A. A. K., Gani, A., Ahmad, M., Masoodi, F. A., Amin, F., & Kousar, S., (2016). Mushroom varieties found in the Himalayan regions of India: Antioxidant, antimicrobial, and antiproliferative activities. *Food Science and Biotechnology, 25*(4), 1095–1100.

39. Khoshnoudi-Nia, S., Sharif, N., & Jafari, S. M., (2020). Loading of phenolic compounds into electrospun nanofibers and electrosprayed nanoparticles. *Trends in Food Science & Technology, 95*, 59–74.
40. Koike, D., (2005). Effect of a wheat amylase inhibitor on canine carbohydrate digestion, gastrointestinal function and pancreatic growth. *Gastroenterology, 108*(4), 1221–1229.
41. Kosanic, M., Rankovic, B., & Dasic, M., (2012). Mushrooms as possible antioxidant and antimicrobial agents. *Iranian Journal of Pharmaceutical Research, 11*(4), 1095–1102.
42. Kosanic, M., Rankovic, B., Rancic, A., & Stanojkovi, T., (2016). Evaluation of mental concentration and antioxidant, antimicrobial and anticancer potentials of two edible mushrooms *Lactarius deliciosus* and *Macrolepiota procera. Journal of Food and Drug Analysis, 24*(3), 477–484.
43. Lee, I. K., Seok, S. J., Kim, W. G., & Yun, B. S., (2006). Diaporthin and orthosporin from the fruiting body of *Daldinia concentrica. Mycobiology, 34*(1), 38–40.
44. Liu, Y. T., Sun, J., Luo, Z. Y., Rao, S. Q., Su, Y. J., Xu, R. R., & Yang, Y. J., (2012). Chemical composition of five wild edible mushrooms collected from southwest China and their antihyperglycemic and antioxidant activity. *Food and Chemical Toxicology, 50*(5), 1238–1244.
45. Lu, X., Brennan, M. A., Narciso, J., Guan, W., Zhang, J., Yuan, L., Serventi, L., & Brennan, C. S., (2020). Correlations between the phenolic and fiber composition of mushrooms and the glycaemic and textural characteristics of mushroom enriched extruded products. *LWT-Food Science and Technology, 118*, 108730.
46. Martinkovich, S., Shah, D., Planey, S. L., & Arnott, J. A., (2014). Selective estrogen receptor modulators: Tissue specificity and clinical utility. *Clinical Interventions in Aging, 9*, 1437–1452.
47. Michalak, A., (2006). Phenolic compounds and their antioxidant activity in plants growing under heavy metal stress. *Polish Journal of Environmental Studies, 15*(4), 523–530.
48. Mingyi, Y., Belwal, T., Devkota, H. P., Li, L., & Luo, Z., (2019). Trends of utilizing mushroom polysaccharides (MPs) as potent nutraceutical components in food and medicine: A comprehensive review. *Trends in Food Science & Technology, 92*, 94–110.
49. Moro, C., Palacios, I., Lozano, M., D'Arrigo, M., Guillamon, E., Villares, A., Martínez, J. A., & García-Lafuente, A., (2012). Anti-inflammatory activity of methanolic extracts from edible mushrooms in LPS activated RAW 264.7 macrophages. *Food Chemistry, 130*(2), 350–355.
50. Muszyńska, B., Sułkowska-Ziaja, K., & Ekiert, H., (2013). Phenolic acids in selected edible Basidiomycota species: *Armillaria mellea, Boletus badius, Boletus edulis, Cantharellus cibarius, Lactarius deliciosus and Pleurotus ostreatus. Acta Scientiarum Polonorum-Hortorum Cultus, 12*(4), 107–116.
51. Muthukumaran, J., Srinivasan, S., Venkatesan, R. S., Ramachandran, V., & Muruganathan, U., (2013). Syringic acid, a novel natural phenolic acid, normalizes hyperglycemia with special reference to glycoprotein components in experimental diabetic rats. *Journal of Acute Disease, 2*(4), 304–309.
52. Nakajima, Y., Nishida, H., Matsugo, S., & Konishi, T., (2009). Cancer cell cytotoxicity of extracts and small phenolic compounds from Chaga [*Inonotus obliquus* (persoon) pilat]. *Journal of Medicinal Food, 12*(3), 501–507.
53. Nowacka, N., Nowak, R., Drozd, M., Olech, M., Los, R., & Malm, A., (2014). Analysis of phenolic constituents, antiradical and antimicrobial activity of edible mushrooms growing wild in Poland. *LWT-Food Science and Technology, 59*(2), 689–694.

54. Nowacka-Jechalkev, M., Olech, M., & Nowak, R., (2018). Mushroom polyphenols as chemopreventive agents; Chapter 11. In: Watson, R. R., Preedy, V. R., & Zibadi, S., (eds.), *Polyphenols: Prevention and Treatment of Human Disease* (2nd edn., Vol. 2; pp. 137–150). London–UK: Academic Press.
55. Nukata, M., Hashimoto, T., Yamamoto, I., Iwasaki, N., Tanaka, M., & Asakawa, Y., (2002). Neogrifolin derivatives possessing anti-oxidative activity from the mushroom *Albatrellus ovinus*. *Phytochemistry, 59*(7), 731–737.
56. Olawuyi, I. F., & Lee, W. Y., (2019). Quality and antioxidant properties of functional rice muffins enriched with shiitake mushroom and carrot pomace. *International Journal of Food Science & Technology, 54*(7), 2321–2328.
57. Orhan, I., & Üstün, O., (2011). Determination of total phenol content, antioxidant activity and acetylcholinesterase inhibition in selected mushrooms from Turkey. *Journal of Food Composition and Analysis, 24*(3), 386–390.
58. Palacios, I., Lozano, M., Moro, C., D'Arrigo, M., Rostagno, M. A., Martínez, J. A., García-Lafuente, A., et al., (2011). Antioxidant properties of phenolic compounds occurring in edible mushrooms. *Food Chemistry, 128*(3), 674–678.
59. Park, I. H., Chung, S. K., Lee, K. B., Yoo, Y. C., Kim, S. K., Kim, G. S., & Song, K. S., (2004). An antioxidant hispidin from the mycelial cultures of *Phellinus linteus*. *Archives of Pharmacal Research, 27*(6), 615–618.
60. Paul, N., Slathia, P. S., Vaid, A., & Kumar, R., (2018). Traditional knowledge of *Gucchi Morchella esculenta* (Ascomycetes) in Doda District, Jammu and Kashmir, India. *International Journal of Medicinal Mushrooms, 20*(5), 445–450.
61. Quang, D. N., Harinantenaina, L., Nishizawa, T., Hashimoto, T., Kohchi, C., Soma, G. I., & Asakawa, Y., (2006). Inhibitory activity of nitric oxide production in raw 264.7 cells of daldinals A–C from the fungus *Daldinia childiae* and other metabolites isolated from inedible mushrooms. *Journal of Natural Medicines, 60*, 303.
62. Rathore, H., Prasad, S., & Sharma, S., (2017). Mushroom nutraceuticals for improved nutrition and better human health: A review. *Pharma Nutrition, 5*(2), 35–46.
63. Rodríguez-Seoane, P., Díaz-Reinoso, B., González-Muñoz, M. J., Portela, C. F. A., & Domínguez, H., (2019). Innovative technologies for the extraction of saccharidic and phenolic fractions from *pleurotus eryngii*. *LWT-Food Science and Technology, 101*, 774–782.
64. Saltarelli, R., Palma, F., Gioacchini, A. M., Calcabrini, C., Mancini, U., Bellis, R. D., Stocchi, V., & Potenza, L., (2019). Phytochemical composition, antioxidant and antiproliferative activities and effects on nuclear DNA of ethanolic extract from an Italian mycelial isolate of *Ganoderma lucidum*. *Journal of Ethnopharmacology, 231*(1), 464–473.
65. Seok, J. K., & Boo, Y. C., (2015). *p*-coumaric acid attenuates UVB-induced release of stratifin from keratinocytes and indirectly regulates matrix metalloproteinase 1 release from fibroblasts. *The Korean Journal of Physiology & Pharmacology, 19*(3), 241–247.
66. Shobana, S., Sreerama, Y. N., & Malleshi, N. G., (2009). Composition and enzyme inhibitory properties of finger millet (*Eleusine coracana L.*) seed coat phenolics: Mode of inhibition of α-glucosidase and pancreatic amylase. *Food Chemistry, 115*(4), 1268–1273.
67. Soleymani, S., Iranpanah, A., Najafi, F., Belwal, T., Ramola, S., Abbasabadi, Z., Momtaz, S., & Farzaei, M. H., (2019). Implications of grape extract and its nano formulated bioactive agent resveratrol against skin disorders. *Archives of Dermatological Research, 311*(8), 577–588.

68. Song, J., Manir, M., & Soon, S. S., (2009). Cytotoxic grifolin derivatives isolated from the wild mushroom *Boletus pseudocalopus* (Basidiomycetes). *Chemistry and Biodiversity, 6*(9), 1435–1442.
69. Soto, M. L., Falqué, E., & Domínguez, H., (2015). Relevance of natural phenolics from grape and derivative products in the formulation of cosmetics: Review. *Cosmetics, 2*(3), 259–276.
70. Stoffel, F., Santana, W. D. O., Fontana, R. C., Gregolon, J. G. N., Kist, T. B. L., Siqueira, F. G. D., Mendonça, S., & Camassola, M., (2019). Chemical features and bioactivity of grain flours colonized by macrofungi as a strategy for nutritional enrichment. *Food Chemistry, 297*, 124988.
71. Stojkovic, D., Smiljkovic, M., Ciric, A., Glamoclija, J., Griensven, L. V., Ferreira, I. C. F. R., & Sokovic, M., (2019). An insight into anti-diabetic properties of six medicinal and edible mushrooms: Inhibition of α-amylase and α-glucosidase linked to type-2 diabetes. *South African Journal of Botany, 120*, 100–103.
72. Tanaka, T., Kawaguchi, N., Zaima, N., Moriyama, T., Fukuta, Y., & Shirasaka, N., (2017). Anti-osteoporotic activity of a syringic acid diet in ovariectomized mice. *Journal of Natural Medicines, 71*(4), 632–641.
73. Tanaka, T., Onuma, H., Shigihara, T., Kimura, E., Fukuta, Y., Shirasaka, N., Moriyama, T., & Homma, Y., (2019). Anti-osteoporotic effects of syringic acid and vanillic acid in the extracts of waste beds after mushroom cultivation. *Journal of Bioscience and Bioengineering, 128*(5), 622–629.
74. Taofiq, O., Calhelha, R. C., Heleno, S., Barros, L., Martins, A., Santos-Buelga, C., Queiroz, M. J. R. P., & Ferreira, I. C. F. R., (2015). The contribution of phenolic acids to the anti-inflammatory activity of mushrooms: Screening in phenolic extracts, individual parent molecules and synthesized glucuronate and methylated derivatives. *Food Research International, 76*, 821–827.
75. Tella, S. H., & Gallagher, J. C., (2014). Prevention and treatment of postmenopausal osteoporosis. *The Journal of Steroid Biochemistry and Molecular Biology, 142*, 155–170.
76. Tepsongkroh, B., Jangchud, K., Jangchud, A., Chonpracha, P., Ardoin, R., & Prinyawiwatkul, W., (2019). Consumer perception of extruded snacks containing brown rice and dried mushroom. *International Journal of Food Science & Technology, 55*(1), 46–54.
77. Vaz, J. A., Almeida, G. M., Ferreira, I. C. F. R., Martins, A., & Vasconcelos, M. H., (2012). *Clitocybe alexandri* extract induces cell cycle arrest and apoptosis in a lung cancer cell line: Identification of phenolic acids with cytotoxic potential. *Food Chemistry, 132*(1), 482–486.
78. Vaz, J. A., Barros, L., Martins, A., Santos-Buelga, C., Vasconcelos, M. H., & Ferreira, I. C. F. R., (2011). Chemical composition of wild edible mushrooms and antioxidant properties of their water-soluble polysaccharide and ethanolic fractions. *Food Chemistry, 126*(2), 610–616.
79. Wang, W., Wu, Z., Wang, X., Li, C., Zhang, K., Zhou, J., Cheng, S., & Lu, F., (2018). Enzymatic hydrolysis combined with high-pressure homogenization for the preparation of polysaccharide-based nanoparticles from the by-product of *Flammulina velutipes*. *International Journal of Food Science & Technology, 53*(10), 2422–2429.
80. Wasser, S. P., & Weis, A. L., (1999). Therapeutic effects of substances occurring in higher basidiomycetes mushrooms: A modern perspective. *Critical Reviews in Immunology, 19*(1), 65–96.

81. Witkowska, A. M., Zujko, M. E., & Mirończuk-Chodakowska, I., (2011). Comparative study of wild edible mushrooms as sources of antioxidants. *International Journal of Medicinal Mushrooms, 13*(4), 335–341.
82. Wrona, M., Bentayeb, K., & Nerín, C., (2015). A novel active packaging for extending the shelf-life of fresh mushrooms (*Agaricus bisporus*). *Food Control, 54*, 200–207.
83. Wu, T., & Xu, B., (2015). Anti-diabetic and antioxidant activities of eight medicinal mushroom species from China. *International Journal of Medicinal Mushrooms, 17*(2), 129–140.
84. Wu, Y., Choi, M. H., Li, J., Yang, H., & Shin, H. J., (2016). Mushroom cosmetics: The present and future. *Cosmetics, 3*(3), 22.
85. Wu, Z., & Li, Y., (2017). Grifolin exhibits anti-cancer activity by inhibiting the development and invasion of gastric tumor cells. *Oncotarget, 8*(13), 21454–21460.
86. Yan, H., Che, X., Lv, Q., Zhang, L., Dongol, S., Wang, Y., et al., (2017). Grifolin induces apoptosis and promotes cell cycle arrest in the A2780 human ovarian cancer cell line via inactivation of the ERK1/2 and Akt pathways. *Oncology Letters, 13*(6), 4806–4812.
87. Yoon, K. N., Alam, N., Lee, K. R., Shin, P. G., Cheong, J. C., Yoo, Y. B., & Lee, T. S., (2011). Antioxidant and antityrosinase activities of various extracts from fruiting bodies of *Lentinus lepideus*. *Molecules, 16*(3), 2334–2347.
88. Zarandona, I., Puertas, A. I., Dueñas, M. T., Guerrero, P., & De La Caba, K., (2020). Assessment of active chitosan films incorporated with gallic acid. *Food Hydrocolloids, 101*, 105486.

CHAPTER 12

NUTRITIONAL CONSTITUENTS AND *IN VITRO* ANTIOXIDANT ACTIVITIES OF SELECTED WILD EDIBLE FRUITS

SANGITA MUCHAHARY, ASHOK DAIMARY, DEBRAJ PRADHAN, SANTOSH MANDAL, FUNGJA BASUMATARY, and ANUCK ISLARY

ABSTRACT

This chapter investigates the phytochemical and antioxidant activities of fruit extracts from locally available edible wild fruits, such as *Ficus carica* (figs), *Melastoma malabathricum* (tinku), *Ziziphus mauritiana* (plum). Flavonoid and total phenolic content (TPC) were investigated by using Folin-Ciocalteu and aluminum chloride method, respectively. Moisture content, ash content, and protein content in the selected fruits were determined for the study on nutritional composition. The assessment of the antioxidant activities were measured using 2,2-diphenyl-1-picrylhydrazyl (DPPH) and ferric reducing antioxidant power (FRAP) method. Results revealed that the nutritional compositions in studied wild fruits (fig, tinku, and plum) were in significant amounts. Total phenolic compound both in methanol and aqueous extract of tinku was recorded highest 13212.40 ± 0.06 µg gallic acid equivalent (GAE)/g and 12818.13 ± 0.08 µg GAE/g, respectively. Flavonoid content both in methanol and aqueous extract of plum was found highest 2.04 ± 0.12 mM quercetin equivalent (QE)/100 g and 10.18 ± 0.22 mM QE/100 g, respectively. In DPPH assay, fig both in aqueous and methanol extracts were recorded highest activity 56.49 ± 2.01% and 60.40 ± 0.18%, respectively whereas tinku showed the highest activity of 4350.20 ± 0.18 mM Fe^{2+}/g in FRAP method. Results revealed that edible wild fruits may be beneficial for human health as the fruits exhibited potent source of healthy compounds.

Novel Processing Methods for Plant-Based Health Foods: Extraction, Encapsulation, and Health Benefits of Bioactive Compounds. Megh R. Goyal, N. Veena & Ritesh B. Watharkar (Eds.)

12.1 INTRODUCTION

12.1.1 BACKGROUND AND BIOLOGY OF FICUS CARICA (FIG FRUIT), MELASTOMA MALABATHRICUM (TINKU) AND ZIZIPHUS MAURITIANA (PLUM)

Fruits are rich in many essential nutrients, phytochemicals, and many other bioactive compounds that exhibit antioxidant activities, antimicrobial properties, and various other health benefits. In some places, many wild fruits are still not known properly which are abundant in numerous bioactive molecules such as polyphenols, alkaloids, vitamins, β-carotene, and many others which have an important role in antioxidant and antimicrobial activity. Generally, wild fruits contain higher nutritive values than cultivated fruits. There are some wild fruits in the rural area, which have been investigated for various nutrients but still proper study or report on its nutrition and phytochemicals are not available. Therefore, such plants/fruits should be investigated for a better understanding of their properties, safety, and efficiency, which will help in various fields like medical, food processing, cosmetic industries, and so on.

Some edible wild fruits (Adumbra (*Ficus carica*), Bwigri (*Ziziphus mauritiana*), and Tinku (*Melastoma malabathricum*)) in the district of Kokrajhar (Assam) are very rich in nutrients and many other phyto-chemicals. Adumbra is also known as fig or common fig, which is said to be one of the traditional fruits in Asia [14]. As an agricultural product, it has global economic importance. Fig belongs to the genus *Ficus* and according to Stover et al. [32], Ficus ranges from 600–1,900 species, where fig plant only with the bunch of fruits in the stem is considered as edible variety.

Tinku from *Melastoma* is also a wild fruit that is rarely known worldwide as a vegetable. Among 22 species in Southeast Asia and Malaysia, tinku is one of them that is believed as folk medicine by some population in India, China, and Indonesia [15].

Plum belongs to the genus *Ziziphus* is also locally known as bwigri in Kokrajhar, Assam, which has a sweet and sour taste when it is ripened. It contains about 40 different species and since ancient times, it has been widely used as a traditional medicinal purpose [26]. It is consumed as in many ways, raw, ripe, raw chutney, jam, jelly, traditional loaf, etc., as reported by Muchuweti et al. [21].

12.1.2 PRODUCTION

According to FAOSTAT [9], worldwide fig production in the year 2018 was reported at 1.14 million tons, whereas India is contributing very less production of fig. Production of tinku in India is still unknown because, mostly tinku plants are grown wild and in India, there is no proper record of its cultivation.

In India, the production of plum is quite minor, because it is mostly considered as an underutilized crop, due to which cultivation is scanty. As per FAOSTAT [9], the production of plum in combination with sloes was 0.25 million tons in 2018. The majority of the plum is grown wild and very little quantity is cultivated by some states in India like Uttar Pradesh, Madhya Pradesh, Rajasthan, Punjab, Bihar, and Tamil Nadu [12].

12.1.3 NUTRITIONAL CONTENTS AND PHYTOCHEMICALS

According to Shamin-Shazwan et al. [29], figs are rich in protein, sugar, fiber, minerals, polyphenols, antioxidants, and amino acids. Tinku was reported for its richness in anthocyanin content along with other beneficial secondary metabolites [2]. They are also rich in some minerals like iron, calcium, manganese, and copper as reported by Nayak and Basak [22]. Plum is a good source of nutrition, alkaloids, terpenoids, flavonoids, pectin, and saponin [12]. A good amount of sugar content of plum is also playing a major role in imparting its sweet flavor along with its sour taste due to the presence of ascorbic acid as reported by Pareek [25]. The majority of the wild fruits are very nutritive and rich in phytochemicals due to their wild habitat and grown wild in the fertile area.

12.1.4 ANTIOXIDANTS

As fruit contains high phytochemicals, it turns to dark purple while getting ripe [2]. According to Joffry et al. [15]; and Omar et al. [24], almost every part of the plant shows medicinal activities like antimicrobial activity, antioxidant activity, anti-inflammatory, antinociceptive, etc., and reported for its ability to treat stomach pain, toothache, dysentery, diarrhea, scar prevention, etc. Researchers have studied the etiologies and shown that many phytonutrients in fruits exhibits potential medicinal effect to treat cardiovascular disease, cancerous cells, diabetes, and some diseases caused by oxygen radicals. Every fruit and vegetable has different phytochemistry and different compositions and content that impart its specialty towards pharmaceutical use. Variations in nutrition and phytochemical compositions may have different tendencies and medicinal properties against various prevalent diseases and health problems caused by oxidative damages and many others.

Free radical is an ion, atom, or molecule having an unpaired valence electron with hanging covalent bonds [8], which makes it very reactive and unstable. Although they have many beneficial functions in the origin of life, evolution, gene transcription, signal transduction, etc., mostly free radicals damage some cells in the human body, as they have a highly reactive and oxidative nature causing cell disruption, DNA damage [8, 19] and lipid oxidation. In the case of food processing and preservation, lipid oxidation is undesirable leading spoilage of food causing microbial populations and giving foul odor in the food. Besides promoting aging, it causes many diseases such as rheumatoid arthritis, cancer [27, 35], and atherosclerosis [7]. This oxidative stress is due to the imbalance situation with increased oxidants or decreased antioxidants in the medium.

12.1.5 MEDICINAL DEMAND OF WILD FRUITS

Wild fruits mostly grow in a place where there is a lack of human interaction or without any disturbances in their surroundings. They mainly need good climatic conditions for survival that may not be suitable for a human being. Because of habitat in rich climatic conditions, those wild fruits are highly nutritious, i.e., good supplementary food and medicinal food. Wild fruits are very important for the rural population existing in the area as a great source of income. They produce byproducts from wild fruits as highly nutritional foods [30]. Figs are reported for its various health beneficial properties like antidiabetic, antimicrobial, antioxidant, antispasmodic, etc., which is also studied for its ability to treat cardiovascular disorders, gastrointestinal (GI) disorders, inflammation [31], etc.

Besides the fruit part, other parts of plum are well reported for its medicinal uses, such as to enhance digestion, blood purification [17], as antidiabetic, anticancer, etc., and seeds are also reported for its sedative property [26], preventive property against insomnia, anxiety [20], and many others. According to Schippmann et al. [28], the upsurge in the market and demand for wild fruits is increasing as their nutritional importance are getting noticed, due to which, many cultivars got some requests for the cultivation of the wild fruits.

This chapter focuses on the nutritional contents, antioxidant activities and some polyphenols of three edible wild fruits *viz*., fig, tinku, and plum. Present investigation will help to spread awareness of its importance among the society and to the researchers for further study on its medicinal value and food processing.

12.2 EXPERIMENTAL SETUP

12.2.1 MATERIALS AND METHODS

Mature fruits of fig (*Ficus carica*), plum (*Ziziphus maritiana*), and tinku (*Melastoma malabathricum*) were gathered in the month of January from the local market near Central Institute of Technology (CIT), Kokrajhar, India (26.5136°N, 90.2245°E). Analytical grade chemicals such as bovine serum albumin (BSA), quercetin, 2, 2-diphenyl-1-picrylhydrazyl (DPPH) and gallic acid were procured from Sigma Aldrich, USA. Each chemical and reagent was of analytical grade.

12.2.2 SAMPLE PREPARATION

Collected wild fruits of *Ficus carica* (fig fruit), *Melastoma malabathricum* (tinku) and *Ziziphus mauritiana* (plum) (Figure 12.1) were washed thoroughly and then rinsed with distilled water. Washed fruits were sliced (~ 5 mm thick) separately to separate the seed and unwanted parts like dust, stem, and any infected area. The fruits were smashed to paste and then kept at 4°C until required for the experiment. For aqueous extract, the ground wild fruits were diluted using double distilled water at 2:30 just before the experiment, and similarly, for methanol extract, the ground fruits were diluted using methanol at 2:30 just before use. The mixture was agitated frequently up to 5 min for better separation of supernatant. Then, the supernatant was collected in a beaker after filtration with Whatman No. 1 filter paper and directly subjected to experiments or stored at 4°C for further analysis.

12.2.3 NUTRITIONAL ANALYSIS

Moisture content and ash content was determined according to the protocol by Garcia et al. [10]. The moisture content was expressed in percentage at wet basis as follows:

$$Moisture\,content(\%) = \frac{(A_i - A_f)}{A_i} \times 100 \tag{1}$$

where; A_i is initial sample weight; and A_f is the sample weight after drying.

Ash content was also expressed in percentage at wet basis by following equation:

$$Ash\,content(\%) = \frac{A_f}{A_i} \times 100 \tag{2}$$

where; A_i is the initial sample weight; and A_f is the sample weight after ashing.

FIGURE 12.1 Wild fruits: (a) figs; (b) tinku; and (c) plum.

Determination of protein content was assessed by the following method as described by Lowry [18]. At beginning, reagent A, B, and C were prepared. Regent A is the mixture of 7 mM NaK tartrate, 0.8 M sodium carbonate and 0.5 N sodium hydroxide solution. Reagent B was prepared by addition of 70 mM NaK tartrate and 40 mM $CuSO_4$. Then, reagent C was Folin-Ciocalteau reagent (1:15) in distilled water.

The standard BSA linear equation obtained from the calibration with different concentrations of BSA ranged from 1 to 10 μg/ml was prepared. Aliquot of 0.9 ml reagent A was added in each 1 ml of standard solutions, sample, and blank separately in test tubes. All the test tubes containing the mixture were then incubated in a dark chamber for 10 min. Similarly, again 3 ml of reagent B was poured in each mixture and allowed to rest for 10 min. Rapidly, reagent C was mixed and then after 10 min incubation. Each test tube containing the mixture was measured for its absorbance and recorded at 660 nm, using a UV-Vis spectrophotometer (Lambda 35, Perkin Elmer, USA). The content of protein was expressed in percentage.

12.2.4 PHYTOCHEMICAL ANALYSIS

Flavonoid content and total phenolic content (TPC) were investigated in both methanol and aqueous extract of fruit samples. Determination of flavonoid content was done by following Susanti [34]. Aliquoted 1 ml of extract or five concentrations of quercetin solution (20 to 100 mg/l) and then 4 ml distilled water or methanol was added. $NaNO_2$ (5%) of 0.3 ml was poured into the mixture, and then 0.3 ml of $AlCl_3$ (10%) was added after five minutes. After five minutes, 2 ml aliquots of NaOH (1N) was mixed and then diluted with distilled water/methanol up to 10 ml total volume. After mixing the solution thoroughly, the absorbance was observed against blank solution prepared without sample and standard at 570 nm by using a spectrophotometer. The flavonoid content in each fruit sample was expressed in milligrams of quercetin equivalent per 100 g sample (mg QE/100 g).

In the determination of TPC, aliquoted 1 ml extract or five concentrations of gallic acid standard solution (20 to 100 mg/l) was prepared and a blank solution using distilled water without sample and standard was also prepared. Folin-Ciocalteau reagent of 1 ml was mixed to the aliquots and then shaken vigorously. Solution mixtures were kept for rest for a while and then 10 ml of sodium carbonate (7%) was added. The whole mixture was brought up to 25 ml with de-ionized water and incubated for 1.5 h at 35°C. Absorbance for each solution mixture was noted at 784 nm. Whole experiments were conducted in triplicate to minimize the error in the result.

12.2.5 ASSESSMENT OF ANTIOXIDANT ACTIVITY BY DPPH METHOD

Antioxidant activities were assessed by following Hemalatha et al. [13]; and Sumazian et al. [33] with slight modification. Proceeding the protocol where 2 mg of DPPH reagent was dissolved in 40 ml of methanol/distilled water and then absorbance was noted carefully. Three sample extracts of fig, plum, and tinku at 1 mg/ml were made ready separately, and 0.1 ml of each mixture along with 4 ml of DPPH reagent solution was allowed to complete reaction in separate test tubes at 35°C. After the interaction for half-hour duration, each mixture was recorded for its absorbance at 517 nm. The inhibition (%) for each sample was estimated by using the equation given below:

$$\% Inhibition = \left[\frac{(A_C - A_S)}{A_C}\right] \times 100 \quad (3)$$

where; A_C is the absorbance of DPPH control; and A_S is the absorbance of a mixture of sample and DPPH.

12.2.6 ASSESSMENT OF ANTIOXIDANT ACTIVITY BY FRAP METHOD

Antioxidant activity was assessed by following the protocol used by Goulas and Manganaris [11]; and Elfalleh et al. [5]. A solution of sodium acetate buffer of 300 mM (pH 3.6), a mixture of 3.1 g $C_2H_9NaO_5$ and 15 ml of glacial CH_3COOH was made ready and the solution was poured with distilled water up to 1 L. FRAP reagent comprising 2.5 ml of 2,4,6-tripyridyl-s-triazine (TPTZ) of 10 mM in hydrochloric acid (40 mM), 2.5 ml of $FeCl_3.6H_2O$ (20 mM) and 20 ml of acetate buffer was prepared. After that, fruit sample (0.3 ml) and FRAP reagent (2 ml) were allowed to rest for five minutes at 35°C. Then, the solution was observed for its absorbance at 593 nm by a spectrophotometer. For the preparation of the calibration curve, five concentrations of the ferrous standard (Fe^{2+}) (0.1 to 1.5 mM/g) were used, and all the absorbance was measured as for sample, and the antioxidant activity was expressed in mM Fe^{2+}/g.

12.3 DISCUSSION AND RESULTS

12.3.1 NUTRITIONAL CONTENT

Table 12.1 shows some of the nutritional compositions in the three studied sample extracts; fig (*Ficus carica*), tinku (*Melastoma malabathricum*), and plum (*Ziziphus mauritiana*). Major ash (%) and moisture (%) content were found in fig at 3.20 ± 0.00 and 82.33 ± 3.15, respectively whereas, higher protein content (%) was seen in tinku at 5.20 ± 0.02. In determination of protein content, the standard calibration curve of BSA was prepared as concentration of protein against absorbance. Linear plot of the data is given by "y = (0.0015x + 0.1668)" equation with 0.99 correlation coefficient (R^2) (Figure 12.2). In preparation of standard calibration curve, as a higher R^2 (more than 0.85) value is very important to obtain the higher precision of results.

TABLE 12.1 Nutritional Content in Figs, Tinku, and Plum

Wild Fruits	Ash Content (%)	Moisture Content (%)	Protein Content (%)
Fig	3.20 ± 0.00	82.33 ± 3.15	4.01 ± 1.15
Tinku	3.00 ± 1.00	61 ± 1.82	5.20 ± 0.02
Plum	3.00 ± 1.00	77 ± 2.080	1.18 ± 0.13

Note: Data shown in mean ± standard deviation (at n = 3).

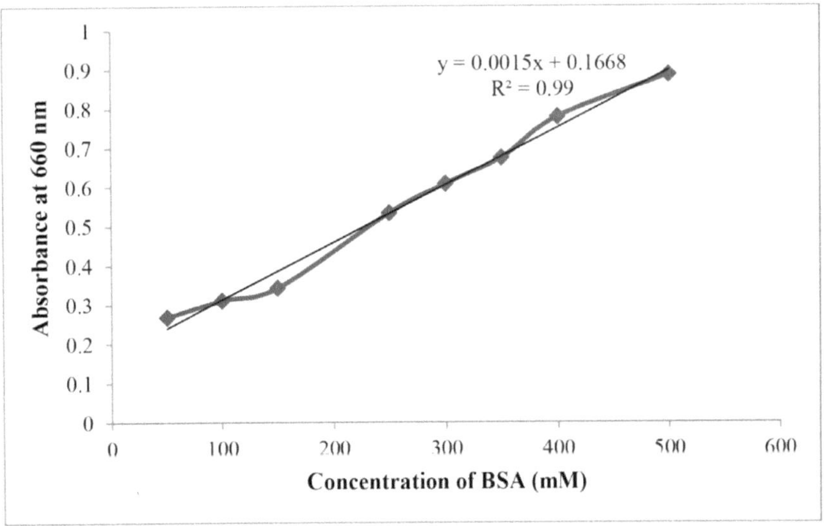

FIGURE 12.2 BSA standard calibration curve at 660 nm.

Higher content of ash indicates the presence of a good amount of minerals in the fruit, while the presence of higher moisture in the fruit indicates the juiciness of the fresh produce that is indirectly relatable with the perishability of the fruit. Results in this study are quite similar with some previous research reports where moisture content and protein content of tinku was found to be 56.60 ± 0.71% and 5.48 ± 0.58%, respectively as reported by Nayak and Basak [22]. Ash content and moisture content in fig were found to be at 4.65% and 4.67% [31]. According to Pareek [25], plum contain 0.3–0.59 g/100 g ash content, 0.8 g/100 g protein content, and 81–83% moisture content. As nutrient bulk foods are highly in demand for maintaining a healthy diet to avoid various diseases and disorders, therefore, fig, plum, and tinku are encouragingly nutritious wild fruits and good sources of nutrients that need to be explored in vast.

12.3.2 PHYTOCHEMICALS

12.3.2.1 FLAVONOIDS CONTENT

Table 12.2 shows the flavonoid content in aqueous (aq) and methanol (meth) extract of fig, tinku, and plum. In aqueous and methanol extract of fruits, flavonoid content in plum was seen higher, i.e., 2.04 ± 0.12 and 10.18 ± 0.22 expressed in millimolar of quercetin equivalent (mM QE) per 100 g sample, respectively. Quercetin calibration curve obtained at five concentrations (20 to 100 mg/L) was "y = (1.2969x + 0.4999)" with R^2 = 0.99 (Figure 12.3).

TABLE 12.2 Flavonoid Content in Fig, Tinku, and Plum

Wild Fruits	Flavonoid Content (aq) (mM QE/100 g)	Flavonoid Content (meth) (mM QE/100 g)
Fig	1.16 ± 1.01	1.75 ± 0.34
Tinku	0.05 ± 0.08	10.18 ± 0.22
Plum	2.04 ± 0.12	3.18 ± 0.05

Note: aq: aqueous; meth: methanol; QE: quercetin equivalent.
Data shown in mean ± standard deviation (at n = 3).

In comparison with phytochemical contents reported by other researchers, results obtained in this study were quite different, which may be because of climatic conditions and soil quality of the plant grown.

Flavonoid content was found to be 2.75 mg CE/g and 8.36 to 21.97 mg CE/100 g in plum as revealed by Soni et al. [31]; and Koley et al. [16], respectively. Similarly, Dureja, and Dhiman [4] also reported that flavonoid content of plum was 2528.00 ± 6.55 mg QE/100 g. Flavonoid contents in methanol extracts were seen at a little higher content than in the aqueous extract, which might be due to the polarity of the extraction solvent. Flavonoids have a class of sub-flavonoids which are soluble in both organic solvents (*viz.* methanol, ethanol, acetone, etc.), and water, but there are some flavonoids with aglycone class are only soluble in methanol, ethanol, etc. [33]. Therefore, the solubility of constituents in fruit samples was more in methanol than the water.

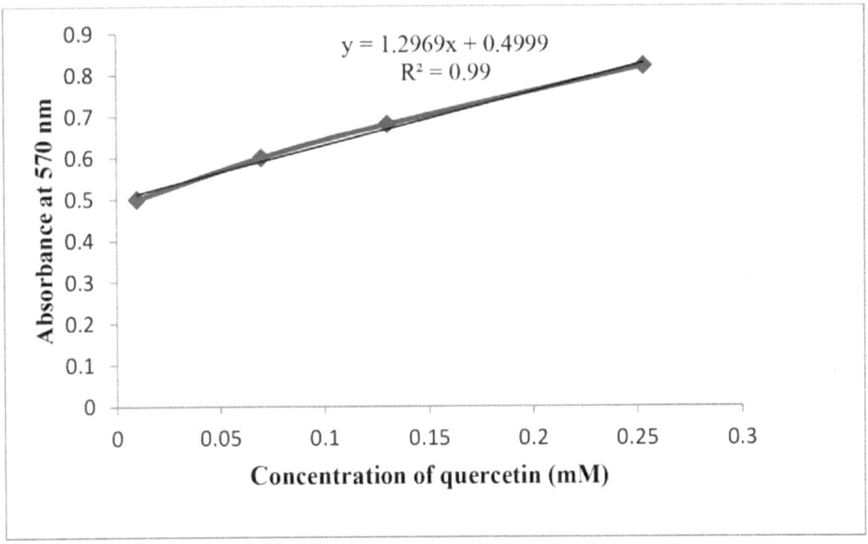

FIGURE 12.3 Quercetin standard calibration curve at 570 nm.

12.3.3 *TOTAL PHENOLIC CONTENT (TPC)*

Total phenolic content (TPC) in the studied wild fruits *viz.* fig, tinku, and plum were seen in a good amount. Phenolic compounds were reported for their strong potential antioxidant activities [33] that exhibit various health benefits. In this study, the total phenolic compound was found to be 13212.40 ± 0.06 and 12818.13 ± 0.08 μg GAE/g in methanol and aqueous extract of tinku, respectively. Secondly, plum contains 10801.70 ± 0.00 μg GAE/g of phenolic content in methanol extract. The calibration curve of gallic acid standard in the five concentrations (20 to 100 mg/L) with equation " y = (0.0014 x + 0.1699)" with $R^2 = 0.98$ (Figure 12.4).

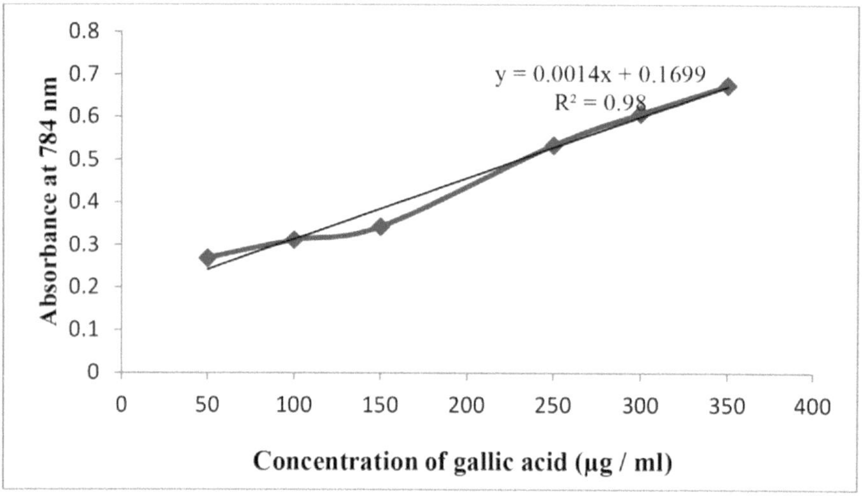

FIGURE 12.4 Standard calibration curve of gallic acid at 784 nm.

This study revealed a superior content of TPC in fig, tinku, and plum in comparison with the results reported by Soni et al. [31], the total phenolic compound was recorded 10.90 μg GAE/mg in methanol extract of fig (Table 12.3). Tinku contains 19.20 mg GAE/g reported by Nayak and Basak [23] and plum contains 14.20 ± 0.00 mg GAE/g for aqueous extract and 5.30 ± 1.10 mg GAE/g for methanol extract by Chalise et al. [3].

TABLE 12.3 Total Phenolic Content of Fig, Tinku, and Plum

Wild Fruits	Total Phenolic Content (aq) (μg GAE/g)	Total Phenolic Content (meth) (μg GAE/g)
Fig	2100.00 ± 1.12	2163.30 ± 0.02
Tinku	12818.13 ± 0.08	13212.40 ± 0.06
Plum	10660.50 ± 0.56	10801.70 ± 0.00

Note: aq: aqueous; meth: methanol; GAE: gallic acid equivalent.

Data shown in mean ± standard deviation (at n = 3).

12.3.4 ANTIOXIDANT ACTIVITIES

12.3.4.1 ANTIOXIDANT DETERMINATION BY DPPH METHOD

In the DPPH method, free radicals are inhibited by transferring one hydrogen atom to the free radicals present in the wild fruits. DPPH inhibitions (%) recorded in the methanol extract and aqueous extract of 0.05 g/ml fig were recorded highest, i.e., 60.40 ± 0.18% and 56.49 ± 2.01%, respectively. Phytochemicals present in the fruits showed various inhibitions against DPPH as given in Table 12.4. Reason of inhibition might be the presence of the secondary metabolites, polyphenols, and other phytochemicals that act as natural antioxidants that inhibit the oxidation reaction takes place inside the medium and as described in 1.4. However, those antioxidants inhibit the oxidation by donation of a hydrogen atom to the free radicals in the complex and lead to neutralization of the oxidation reaction.

TABLE 12.4 DPPH Inhibition Activity of Fig, Tinku, and Plum

Sample Name	Inhibition % (aq)	Inhibition % (meth)
Fig	11.50 ± 1.15	12.15 ± 1.87
Tinku	56.49 ± 2.01	60.40 ± 0.18
Plum	10.137 ± 0.07	58.52 ± 0.03

Note: aq: aqueous; meth: methanol.
Data shown in mean ± standard deviation (at n = 3).

Therefore, during the experiment a DPPH solution was allowed to reduce by different sample extracts (*viz.* fig, tinku, and plum) and then lesser absorbance of the mixture was recorded. More the antioxidant capacity of the sample, lighter will be the color of the DPPH and sample mixture showing lesser absorbance value than the DPPH control solution at 517 nm. That defines spectrophotometric absorbance of the DPPH and sample mixture will be always lesser than the DPPH control solution if the sample has any antioxidant activity.

Results obtained in current investigation were in accordance with the DPPH free radical inhibition (%) was reported to be 21.20 ± 0.91% in fig (50 μg/ml) by Ali et al. [1] and according to results of Chalise et al. [3], plum was reported to be 45% at 50 μg/ml methanol extract. An investigation of full phytochemical profiling exhibiting antioxidant activities in the wild fruits are encouraged.

12.3.4.2 ANTIOXIDANT ACTIVITY DETERMINATION BY FRAP METHOD

In assessing antioxidant activities by FRAP method, a linear ferrous standard (Fe^{2+}) calibration curve at 593 nm was drawn and standard equation such as: "y = (1.0632x – 0.0533)" with R^2 = 0.95 (Figure 12.5 and Table 12.5).

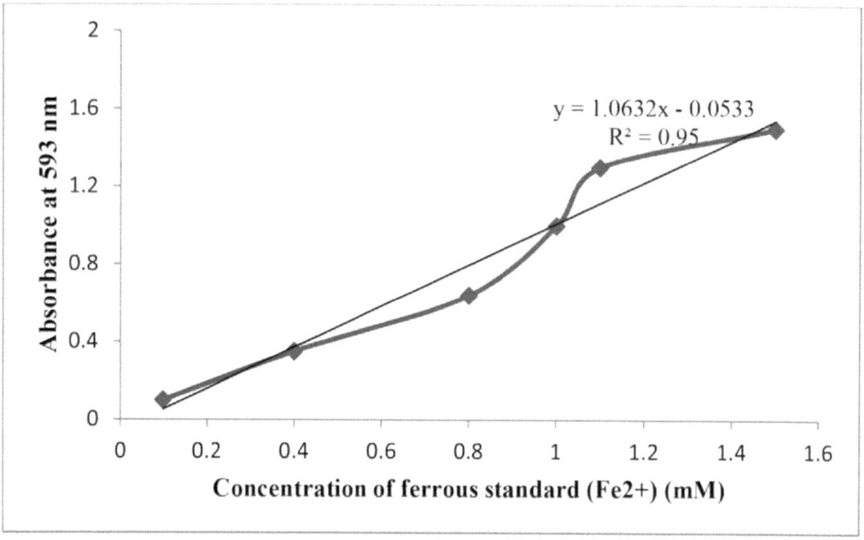

FIGURE 12.5 Ferrous standard (Fe^{2+}) calibration curve at 593 nm.

TABLE 12.5 FRAP Activity of Fig, Tinku, and Plum

Sample Name	Antioxidant Activity (Methanolic Extract) (mM Fe^{2+}/g)
Fig	142.51 ± 1.33
Tinku	4350.20 ± 0.18
Plum	957.00 ± 0.02

Note: Data shown in mean ± standard deviation (at n = 3).

In comparison with Ferrous standard (Fe^{2+}) calibration curve, the antioxidant activity of tinku was 4350.20 ± 0.18 mM Fe^{2+}/g. Tinku showed higher ferric reducing capacity than the antioxidant activity shown by fig and plum, which is in accordance with the result reported by the Nayak and Basak [23] showed high antioxidant activity by tinku, i.e., 5878.35 ± 0.05 μM Ascorbic acid equivalent antioxidant capacity (AEAC)/g.

According to Ercisli et al. [6], the antioxidant activity of plum was 1.87 mM Fe^{2+}/g. In the experiment, the antioxidants present in the samples reduce the ferrous ion (free radical compound) by donating a hydrogen ion. Reducing mixtures formed blue colored solutions. Variations in the results obtained may be due to the environmental factor and extraction parameters used during analysis.

In this study, phytochemicals in the form of polyphenols and many others might have major role in the reduction of the ferrous and oxidizing agents in the system and therefore, phytochemicals present in the wild fruits (fig, tinku, and plum) were inhibiting the antiradical or oxidizing agents. This indicates the extract of the wild fruit possesses antioxidant activity that has a potent ability to prevent numerous health problems such as cell damage, inflammation, and many other oxidative damages caused by free radicals in the human body.

A comparative evaluation of antioxidant activities and inhibitions for wild fruits from the comparison chart (Figure 12.6) was done. By the observation of antioxidant activities (FRAP method) and inhibition (%) of DPPH shown by fig, tinku, and plum were in the increasing order as fig < plum < tinku. Similarly, in the case of phytochemical contents, total phenolic and flavonoid contents of the wild fruits also increased as fig < plum < tinku. This indicates antioxidant activities were directly dependent on the phenolic compounds present in the fruits.

Antioxidants present in the three studied fruits; fig, tinku, and plum might be in the form of phenolic compounds and other secondary metabolites or active compounds. On the other side, in the determination of phytochemical contents, the extraction medium used plays a great role. Different phytochemical compounds exhibit various properties, whereas an extracting solvent used must be the medium where a compound dissolves partially or fully. Also, the extraction efficiency can be achieved by a certain limit of extracting parameters; temperature, agitation, force, extracting solvent used. In this study, as methanol has more polar behavior than the distilled water, phytochemicals obtained from the methanol extracts were comparatively higher than the aqueous extract.

In this study, tinku revealed the content of a good amount of nutrition and phytochemicals with higher antioxidant activity, which defines that it might contribute various medicinal properties. As per the literature, there are no such products processed from tinku fruit. There is a great scope to study and rise its demand toward food processing sector. Further study (*in vivo*/*in vitro*) on specific compounds present in the studied fruits for their antioxidant role is needful.

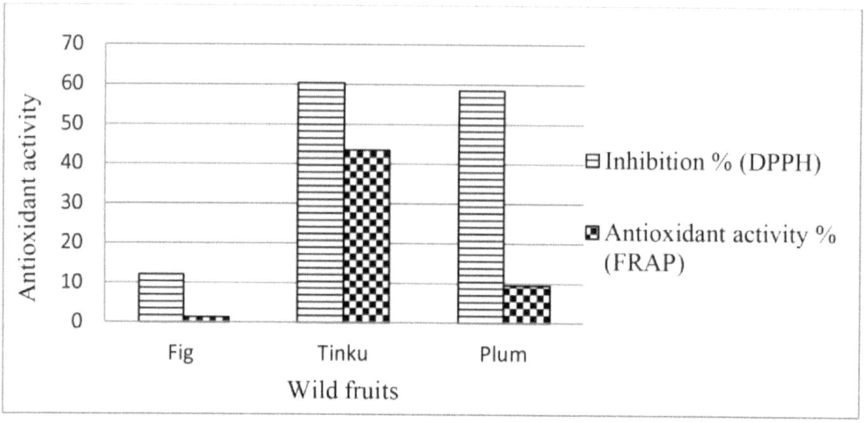

(a)

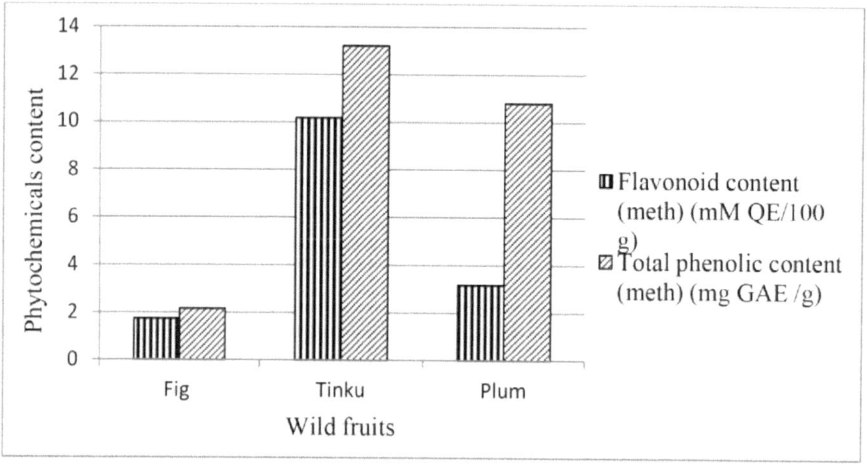

(b)

FIGURE 12.6 Comparative chart of antioxidant activities and phytochemicals of wild fruits viz. fig, tinku, and plum; (a) Antioxidant activities by DPPH and FRAP; and (b) phytochemicals content (flavanoid and total phenolic content).

12.4 SUMMARY

Wild fruits are a rich source of nutrients, beneficial phytochemicals, and some antioxidants which can inhibit the oxidation reaction and can remove the free radical compounds existed in the system. In the current study, the TPC, flavonoid content, and other nutritional compositions in wild fruits of Assam; fig, tinku, and plum were found in a good amount. Among three studied wild fruits, tinku was recorded for the highest antioxidant and free radical inhibition activity that was estimated by DPPH inhibition assay and FRAP method. Therefore, these wild fruits may be a potent source to exhibit various health-beneficial properties in human beings and animals. Therefore, the intake of these fruits may prevent free radical-mediated diseases. Underutilized wild fruits with very low cultivation rates need to be focused more to highlight their importance in their potential activities towards various diseases in human beings. This study might help to study pharmaceutical products in the future. The study of bioactive compounds and mechanisms responsible for antioxidants is needed. The present study will help in spreading awareness that is necessary among the population as an important source of the well-being of the indigenous population in the region.

ACKNOWLEDGMENTS

The authors are sincerely thankful to the laboratory of Food Engineering and Technology, Central Institute of Technology (CIT), Assam, for the research facilities.

KEYWORDS

- **antioxidant activity**
- **bovine serum albumin**
- **flavonoid**
- **nutritional value**
- **phytochemicals**
- **wild fruit**

REFERENCES

1. Ali, H., Monga, J., Gupta, L., Manigauha, A., Trivedi, V. B., & Ahi, J., (2009). Antioxidant potential of *Ficus carica* by the DPPH free radical method: *In vitro* analysis. *Oriental Journal of Chemistry, 25*(1), 257–265.

2. Anuar, N., Mohd, A. A. F., Saat, N., Aziz, N., & Mat Taha, R., (2013). Optimization of extraction parameters by using response surface methodology, purification, and identification of anthocyanin pigments in *Melastoma malabathricum* fruit. *The Scientific World Journal, 2013,* 810547.
3. Chalise, J. P., Acharya, K., Gurung, N., Bhusal, R. P., Gurung, R., Skalko-Basnet, N., & Basnet, P., (2010). Antioxidant activity and polyphenol content in edible wild fruits from Nepal. *International Journal of Food Sciences and Nutrition, 61*(4), 425–432.
4. Dureja, A. G., & Dhiman, K., (2012). Free radical scavenging potential and total phenolic and flavonoid content of *Ziziphus mauritiana* and *Ziziphus nummularia* fruit extracts. *International Journal of Green Pharmacy, 6*(3), 187–192.
5. Elfalleh, W., Tlili, N., Nasri, N., Yahia, Y., Hannachi, H., Chaira, N., Ying, M., & Ferchichi, A., (2011). Antioxidant capacities of phenolic compounds and tocopherols from tunisian pomegranate (*Punica granatum*) fruits. *Journal of Food Science, 76*(5), C707–C713.
6. Ercisli, S., Tosun, M., Karlidag, H., Dzubur, A., Hadziabulic, S., & Aliman, Y., (2012). Color and antioxidant characteristics of some fresh fig (*Ficus carica* L.) genotypes from Northeastern Turkey. *Plant Foods for Human Nutrition, 67*(3), 271–276.
7. Esterbauer, H., Wäg, G., & Puhl, H., (1993). Lipid peroxidation and its role in atherosclerosis. *British Medical Bulletin, 49*(3), 566–576.
8. Fang, Y. Z., Yang, S., & Wu, G., (2002). Free radicals, antioxidants, and nutrition. *Nutrition, 18*(10), 872–879.
9. FAOSTAT, (2019). *UN Food and Agriculture Organization Corporate Statistical Database*. http://www.fao.org/faostat/en/#data/QC (accessed on 30 May 2022).
10. Garcia, L. G. C., Silva, F. A. D., Asquieri, E. R., Boas, E. V. D. B. V., Ogando, F. I. B., Aguiar, C. L. D., & Damiani, C., (2018). Proximate composition, minerals profile, and predominant sugars by ion chromatography along the physiological development of jabuticaba var. *Pingo De Mel. Food Science and Technology, 38,* 16–21.
11. Goulas, V., & Manganaris, G. A., (2012). Exploring the phytochemical content and the antioxidant potential of citrus fruits grown in Cyprus. *Food Chemistry, 131*(1), 39–47.
12. Goyal, M., Nagori, B. P., & Sasmal, D., (2012). Review on ethnomedicinal uses, pharmacological activity and phytochemical constituents of *Ziziphus mauritiana* (Z. jujuba Lam., Non Mill). *Spatula DD, 2*(2), 107–116.
13. Hemalatha, R., Nivetha, P., Mohanapriya, C., Sharmila, G., Muthukumaran, C., & Gopinath, M., (2016). Phytochemical composition, GC-MS analysis, *in vitro* antioxidant and antibacterial potential of clove flower bud (*Eugenia caryophyllus*) methanolic extract. *Journal of Food Science and Technology, 53*(2), 1189–1198.
14. Ikegami, H., Nogata, H., Hirashima, K., Awamura, M., & Nakahara, T., (2009). Analysis of genetic diversity among European and Asian fig varieties (*Ficus carica* L.) using ISSR, RAPD, and SSR markers. *Genetic Resources and Crop Evolution, 56*(2), 201–209.
15. Joffry, S. M., Yob, N. J., Rofiee, M. S., Affandi, M. M. R., Suhaili, Z., Othman, F., Akim, A. Md., Desa, M. N. M., & Zakaria, Z. A., (2012). *Melastoma malabathricum* (L.) smith ethnomedicinal uses, chemical constituents, and pharmacological properties: A review. *Evidence-Based Complementary and Alternative Medicine, 2012,* 258434.
16. Koley, T. K., Kaur, C., Nagal, S., Walia, S., & Jaggi, S., (2016). Antioxidant activity and phenolic content in genotypes of Indian jujube (*Zizyphus mauritiana* Lamk.). *Arabian Journal of Chemistry, 9*(2), S1044–S1052.

17. Krishna, H., & Parashar, A., (2013). Phytochemical constituents and antioxidant activities of some Indian jujube (*Ziziphus mauritiana* Lamk.) cultivars. *Journal of Food Biochemistry, 37*(5), 571–577.
18. Lowry, O. H., Rosebrough, N. J., Farr, A. L., & Randall, R. J., (1951). Protein measurement with the folin phenol reagent. *The Journal of Biological Chemistry, 193*(1), 265–275.
19. McCord, J. M., (2000). The evolution of free radicals and oxidative stress. *The American Journal of Medicine, 108*(8), 652–659.
20. Mishra, T., Paice, A. G., & Bhatia, A., (2011). Use of seeds of Malay apple (*Ziziphus mauritiana*) and related species in health and disease. In: Preedy, V. V., Watson, R. R., & Vinood, P., (eds.), *Nuts and Seeds in Health and Disease Prevention* (pp. 733–739). London : Elsevier Inc.
21. Muchuweti, M., Zenda, G., Ndhlala, A. R., & Kasiyamhuru, A., (2005). Sugars, organic acid and phenolic compounds of *Ziziphus mauritiana* fruit. European *Food Research and Technology, 221*(3, 4), 570–574.
22. Nayak, J., & Basak, U. C., (2015). Analysis of some nutritional properties in eight wild edible fruits of Odisha, India. *International Journal Current Science, 14*, 55–62.
23. Nayak, J., & Basak, U. C., (2015). Antioxidant potential of some lesser-known wild edible fruits of Odisha. *European Journal of Experimental Biology, 5*(8), 60–70.
24. Omar, S. N. C., Abdullah, J. O., Khairoji, K. A., Chin, S. C., & Hamid, M., (2012). Potentials of *Melastoma malabathricum* Linn. flower and fruit extracts as antimicrobial infusions. *American Journal of Plant Sciences, 3*(8), 1127–1134.
25. Pareek, S., (2013). Nutritional composition of jujube fruit. *Emirates Journal of Food and Agriculture, 25*(6), 463–470.
26. San, A. M. M., Thongpraditchote, S., Sithisarn, P., & Gritsanapan, W., (2013). Total phenolics and total flavonoids contents and hypnotic effect in mice of *Ziziphus mauritiana* lam. seed extract. *Evidence-Based Complementary and Alternative Medicine, 2013*, 835854.
27. Sarban, S., Kocyigit, A., Yazar, M., & Isikan, U. E., (2005). Plasma total antioxidant capacity, lipid peroxidation, and erythrocyte antioxidant enzyme activities in patients with rheumatoid arthritis and osteoarthritis. *Clinical Biochemistry, 38*(11), 981–986.
28. Schippmann, U. W. E., Leaman, D., & Cunningham, A. B., (2006). A comparison of cultivation and wild collection of medicinal and aromatic plants under sustainability aspects. In: Bogers, R. J., Craker, L. E., & Lange, D., (eds.), *Medicinal and Aromatic Plants: Agricultural, Commercial, Ecological, Legal, Pharmacological and Social Aspects* (pp. 75–95). Netherlands : Springer.
29. Shamin-Shazwan, K., Shahari, R., Amri, C. N. A. C., & Tajuddin, N. S. M., (2019). Figs (*Ficus Carica* L.): Cultivation method and production based in Malaysia. *Engineering Heritage Journal (GWK), 3*(2), 06–08.
30. Sharma, I. P., Kanta, C., Semwal, S. C., & Goswami, N., (2017). Wild fruits of Uttarakhand (India): Ethnobotanical and medicinal uses. *International Journal of Complementary & Alternative Medicine, 8*(3), 00260.
31. Soni, N., Mehta, S., Satpathy, G., & Gupta, R. K., (2014). Estimation of nutritional, phytochemical, antioxidant and antibacterial activity of dried fig (*Ficus carica*). *Journal of Pharmacognosy and Phytochemistry, 3*(2), 158–165.
32. Stover, E., Aradhya, M., Ferguson, L., & Crisosto, C. H., (2007). The fig: Overview of an ancient fruit. *HortScience, 42*(5), 1083–1087.

33. Sumazian, Y., Syahida, A., Hakiman, M., & Maziah, M., (2010). Antioxidant activities, flavonoids, ascorbic acid and phenolic contents of Malaysian vegetables. *Journal of Medicinal Plants Research, 4*(10), 881–890.
34. Susanti, D., Sirat, H. M., Ahmad, F., Ali, R. M., Aimi, N., & Kitajima, M., (2007). Antioxidant and cytotoxic flavonoids from the flowers of *Melastoma malabathricum* L. *Food Chemistry, 103*(3), 710–716.
35. Vieira, S. A., Zhang, G., & Decker, E. A., (2017). Biological implications of lipid oxidation products. *Journal of the American Oil Chemists' Society, 94*(3), 339–351.

CHAPTER 13

DELIVERY OF PROBIOTICS IN THE FOOD INDUSTRIES

B. RENUKA and B. B. BORSE

ABSTRACT

Probiotics are a source of promising health benefits to the host. The change in dietary habits, today's modern lifestyle and urbanization has compelled the consumers who are conscious about health to search for a systematic treatment alternative to various illness caused due to the lifestyle-related diseases. Hence, the significance of consumption of probiotics to enhance the quality of life is clearly demonstrated by the number of probiotic-enriched products. Any microorganism to be considered as probiotic should resist the harsh gut environment and should be stable as well to be used in food application. This is important in the food industries to emphasis on the varied application of probiotics in the preparation of food products to produce a new generation of 'probiotic health' foods.

13.1 INTRODUCTION

Probiotic is the term derived from a Latin word, meaning "for life." In the history of humankind, the use of probiotic go back to 2000 BC, as it was ascertained with the concept of preserving milk for longer periods and using fermented milk products in our diet [76]. Recent scientific papers have elucidated that much earlier than 2000 BC, many fermented beverages were produced using bacteria and yeasts, though their existence was not aware [88, 133]. Till date, the fermented dairy products of the Middle East, such

Novel Processing Methods for Plant-Based Health Foods: Extraction, Encapsulation, and Health Benefits of Bioactive Compounds. Megh R. Goyal, N. Veena & Ritesh B. Watharkar (Eds.)

as 'Laban Rayad' and 'Laban Khed' are widespread and they were used in early as 3500 BC [1].

Over the period of time, wine, and beer, the fermented beverages were produced using diverse cultures, fruits, and grains [46] by leaving them in covered containers for an extended time period. Though the process was not understood, it was termed as 'fermentation.' Temperature and the exposure to air are the key factors for fermentation was learnt later [5]. In 1856, with the help of Pasteur, a man got a sour milk kind substance as a substitute for alcohol when he was fermenting sugar beetroot. Pasteur distilled the sour milk and sediment of the sour milk was analyzed chemically as lactic acid (LA) and the same was observed as a smaller substance than yeast under microscope. He also noticed a large number of yeasts in the substance where the alcoholic fermentation had occurred. According to Pasteur, two types of fermentation occurs through yeast and bacteria called 'alcoholic fermentation' and 'lactic acid (LA) fermentation,' respectively [78].

In 1857, Pasteur was the first to discover bacteria producing LA and after 21 years, these bacteria were isolated by Lister in 1878 from rancid milk. The isolation of lactic acid bacteria (LAB) from intestinal tract was carried out at the same institute from 1880 to 1888 [90]. *Bifidobacterium* spp. was found as the dominant microorganism in the breast-fed infant's gut, which was discovered by Henry Tissier in 1889. He is the first scientist to focus on friendly bacterium, i.e., *Bifidobacterium* spp. used to treat acute gastroenteritis and/or intestinal diseases [61, 141]. However, in many scientific papers the probiotics benefits on the host health was given by Elie Metchnikoff hence, he is been regarded as the 'father of probiotics.' To understand the general and intestinal health of the host with probiotics in a better way, it was defined.

In 1965, "Lilly and Stillwell" described probiotics as "Growth promoting factors produced by microorganisms" [70]. In 1974, Parker proposed it as that "Organisms with beneficial effects for host by influencing the intestinal microflora." Fuller [42] defined it as "A feed supplement with live microorganisms that affects the host beneficially by improving the intestinal microbial balance." In 1992, "Havenaar and Huis Int Veld" (Probiotics was assumed) assumed probiotics as "Live microorganisms either mixed or a single culture improves the properties of microflora which are indigenous and hence the host is affected beneficially" [51]. In 1998, the "International Life Sciences Institute" believed it as "A viable microbial food supplement which beneficially influences the health of the host." Whereas, Diplock et al. [32]; and Naidu et al. [86] said that "Foods based on probiotic are functional when they affect the physiology of host beneficially through modulating systemic and mucosal immunity and reducing the risk of diseases." The FAO/WHO defined it as "Live microorganisms which when administered in adequate amounts confer health benefits on the host" [33]. The microorganism to be considered as probiotic has to fulfill the following criteria.

- Should resist in the gut environment – "< pH and > bile salt conditions" and adherence to the epithelial cells of the intestine.
- Ability to produce antimicrobial substances and to incur the immunomodulatory responses.

Having the above criteria, the strains of *Bifidobacterium, Streptococcus, Lactobacillus,* and other microorganisms are considered as probiotic microorganisms (Table 13.1).

TABLE 13.1 Microorganisms as Probiotics

Lactobacillus spp.	*Lb. acidophilus, Lb. plantarum, Lb. rhamnosus, Lb. casei, Lb. paracasei, Lb. fermentum, Lb. reuteri, Lb. johnsonii, Lb. brevis, Lb. lactis, Lb. gasseri, Lb. crispatus, Lb. helveticus, Lb. sporogenes, Lb. gallinarum, Lb. amylovorus, Lb. salivarius, Lb. delbrueckii* subsp. *Bulgaricus*
Bifidobacterium spp.	*B. Breve, B. infantis, B. longum, B. bifidum, B. thermophilum, B. adolescentis, B. animalis, B. lactis, B. essensis, B. laterosporus*
Others	*Bacillus cereus, Bacillus coagulans, Enterococcus faecium, Enterococcus faecalis, Escherichia coli* Nissle, *P. freudenreichii, P. freudenreichii* subsp. *shermanii, P. Jensenii, Kluyveromyces lactis, Leu. lactis* subsp. *cremoris, L. lactis, Clostridium butyricum, Pediococcus acidilactic, S. thermophilus, S. cremoris, S. diacetylactis, S. intermedius, S. salivarius* subsp. *thermophilus, Saccharomyces cerevisiae, S. boulardii, Leu. mesenteroides, Sporolactobacillus inulinus.*

The probiotics mechanism of action is the result of *in vitro* experiment and results were further validated by clinical research as *in vivo* studies to explore its health benefits. Probiotics are suggested to be associated with a wide range of therapeutic effects such as mitigation of lactose intolerance [69], immunomodulation [40], alleviation of bacterial, viral, and antibiotic associated or radiotherapy induced diarrheas [48, 100, 138], anti-mutagenic [21], anti-carcinogenic [72], lowering blood cholesterol [96], inflammatory bowel disease (IBD), anti-pathogenicity, anti-diabetic, anti-obesity, treating rheumatoid arthritis, preventing or reducing the effects of atopic dermatitis as well as urinary tract infections (UTI) [114]. A little attention has received for their beneficial health effects as the exact probiotic mechanisms of action are not well known because of their diverse, heterogeneous, and strain-specific characteristics.

This chapter explores probiotics mechanism of action, probiotic dairy and non-dairy products, specifications, safety criteria, and probiotics market.

13.2 MECHANISM OF ACTION

Probiotics are considered to be safe microorganisms because of the administration of adequate doses at appropriate periods affects beneficially on the host. Figure 13.1 depicts the mechanism of action, which includes:

- Epithelial barrier enhancement;
- Attachment to intestinal mucosa and inhibition of pathogen;
- Production of antimicrobial substances and immunomodulation.

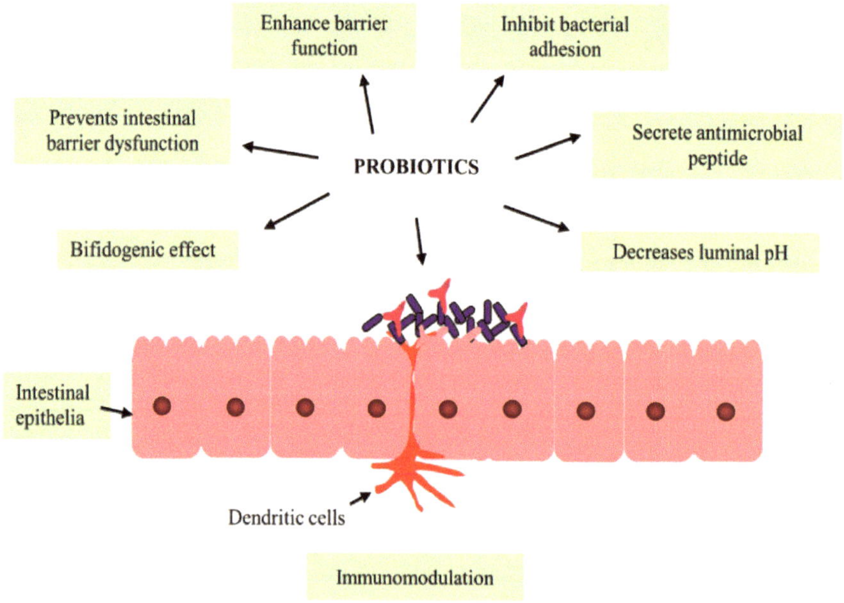

FIGURE 13.1 Probiotic mechanism of action.

13.2.1 EPITHELIAL BARRIER ENHANCEMENT

The contact between the luminal content and the enteric flora is the epithelium of the intestine. The integrity of epithelium is maintained by the intestinal barrier, which is a most important defense mechanism. The complex of "epithelial junction adhesion, mucous layer, antimicrobial peptides, and secretory IgA" are the defenses of the intestinal barrier [94]. If the function of this barrier is disrupted, antigens of the bacteria reach submucosa and hence inflammatory responses are formed, which results in IBD, the disorder of the intestine [58, 59, 125]. Many research findings have shown that the probiotics used to be very significant in enhancing the epithelial barrier function after damage [43, 67, 97, 99, 136, 157]. From the studies, it resulted that the probiotics form a barrier in the gastrointestinal (GI) epithelium lining, which reduces the entry of commensal organisms.

13.2.2 ATTACHMENT TO INTESTINAL MUCOSA AND INHIBITION OF PATHOGEN

The primary characteristics of probiotic are colonization [12, 63], antagonistic action towards pathogens [56] and immunomodulation [103, 127] by adhering to intestinal mucosa. Hence adhesion is considered as important criteria for selecting any microorganism as probiotics, leads to its positive effects on the host [27]. LAB produces surface determinants which interact with the epithelial cells of the intestinal mucosa. These cells produce and secrete small peptides which are active against bacteria, fungi, and viruses called defensin and mucin, a glycoprotein respectively which stabilizes gut barrier function by preventing pathogen binding. The adhering property of the pathogenic bacteria to the surfaces of the epithelial mucosa of the intestine is considered to be the first step of infection [12, 37]. Once the pathogens interact with the secretory proteins of probiotic LAB, the pathogens interaction and competitive exclusion occurs [148]. Many researchers have studied the attachment of probiotics to intestinal epithelium, and its inhibition to the adhesion of pathogens was investigated using Caco-2 cell line (from a human colon carcinoma) [15]. Several research studies have reported that the colonization of pathogens such as *E. coli, Salmonella, Listeria monocytogenes, H. pylori* and *Rotavirus* on intestinal epithelium was inhibited by *Lactobacilli* and *Bifidobacteria* [24, 26, 83, 87, 142, 143].

13.2.3 PRODUCTION OF ANTIMICROBIAL SUBSTANCES

Another probiotics mechanism of action is the synthesis of antimicrobial compounds (bacteriocins, mycocins) and organic acids (LA and other acids), which is a low molecular weight compound that inhibits the activity of gram-negative bacteria [4, 29, 75, 113]. The organic acids in the ionized form enter the bacterial cell, leads to lowering the intracellular pH followed by death of the pathogenic microorganisms [98, 116]. Antibacterial peptides produced by probiotic LAB such as lactacin B, plantaricin, *and* nisin from *Lb. acidophilus, Lb. plantarum* and *L. lactis*, respectively, acts against food-borne pathogens [91]. Mechanisms involved include the inhibition of cell wall synthesis or destructive pore formation of target cells [50].

13.2.4 IMMUNOMODULATION

The modulation of the immune system occurs when probiotics act together with the epithelial and dendritic cells, macrophages, and lymphocytes by producing immunoglobulin A (IgA), enhancing the quantity of natural killer cells, or improving macrophages phagocytic activity [52, 71, 102, 128]. The results of many *in vitro* and *in vivo* experiments have shown to reduce not only the intestinal disorders but also other diseases such as diabetes, obesity, food allergy, etc., due to its immunomodulating effects [101, 137, 149].

13.3 PROBIOTIC DAIRY PRODUCTS

Dairy products are considered as natural healthy products, benefiting the consumer by preventing certain diseases when consumed daily. Some of the factors like viability of probiotics [105, 130], the physicochemical properties [3, 11], the therapeutic effects of probiotics [100, 122] and the regulatory and labeling matters of the final products [33, 121] need to be addressed when the probiotics are used in dairy products. Few dairy products based on probiotic fermentation are discussed below.

13.3.1 YOGURT

Yogurt is considered as a novel probiotics source for very long time and till date it remained as an accepted product of probiotic. Yogurt, a fermented milk product traditionally known as "Dahi" in the Indian subcontinent, is obtained from the milk by the action of *Lb. delbrueckii* subsp. *bulgaricus* and *S. salivarius* subsp. *thermophilus*. LA is produced when lactose is fermented by probiotic LAB, which contributes to the typical taste and texture of yogurt

[156]. The origin of yogurt dates back to the 6000 BC and is originated in the Central Asia. The first industrial production of yogurt started in 1919 at Danone, a company in Barcelona, Spain. Yogurt is a nutritious drink, rich in milk proteins, carbohydrates, minerals (calcium and phosphorous) [38] and vitamins (A, B_1, B_2, B_3, B_{12}, and folate), hence digested easily in the body. Based on the combination of starter culture used and the milk obtained and also the duration of the fermentation process, the nutritional composition of yogurt varies. Yogurt strain and the culture viability are depending on the specific health benefits [81]. Several studies have recommended that the probiotics used in the preparation of yogurt helps in maintaining a healthy gut along with preventing some of the GI disorders like constipation, colon cancer, diarrheal diseases, lactose intolerance, IBD, infection by *Helicobacter pylori* and allergies [2]. Some of the commercially available probiotic yogurts are Danone Activia with *Bifidobacterium lactis* CNCM I-2494, Activia Strawberry with *B. regularis* and *B. lactis* DN 173010, and Probiotic La yogurt, and Siggis yogurt for kids.

13.3.2 CURD (DAHI)

Curd is traditionally prepared from milk by the fermentation of bacteria [19]. The species involved in the fermentation include single or mixed cultures in a combination of acid and flavor producing bacteria such as *L. lactis, L. cremoris, Lb. delbrueckii* subsp. *bulgaricus, S. thermophilus, S. diacetylactis, S. cremoris, L. diacetylactis*, along with *Leuconostoc* species. The therapeutic and health promoting properties such as improving gut health, immune response, mineral absorption, lipid profile, managing hypertension and reduce the risk of hypertension are nurtured by probiotic curd. Some of the commercially available probiotic curds are Nestlé Actiplus with *Lb. acidophilus*, Mother Dairy Advanced Dahi, Amul probiotic Dahi and Nilgiris Dahi.

13.3.3 MISHTI DOI

Mishti Doi (Payodhi or Lal Dahi) is a popular fermented sweet Doi (yogurt) in the eastern parts of India. Boiled milk with slightly thick consistency added with brown sugar or date molasses is used to prepare Mishti Doi. Starter culture added as in the case of Dahi and allowed to set and packaged in earthen pots [74, 139]. A little success was found with the commercialization of Mishti Doi, though the earlier prevention of food adulteration (PFA) standards are not obtained for the product [47, 108] but the FSSAI approval is mandatory now. The probiotic starters with a combination *S. salivarius* ssp. *thermophilus*, *Lb. acidophilus* and *Lb. delbrueckii* ssp. *bulgaricus* and the other combination with *Lb. acidophilus*, *L. lactis* ssp. *lactis* and *S. cerevisiae* was used in the preparation of Mishti Doi. It also holds great promise for those suffering from IBD, anti-inflammatory, and pro-regenerative roles in cases of IBD that cause prolonged inflammation of the digestive tract. Some of the commercially available Mishti Doi is with different fruits (cherry, mango, and bananas), flavors (butterscotch, cardamom, and vanilla), grains, and nuts (almond and pistachio).

13.3.4 SHRIKHAND

It is a sweetish-sour fermented milk dessert with semi-solid consistency, made out of strained yogurt or dahi. Though Shrikhand is considered as an Indian staple dessert, it forms the part of meal on festival season in Maharashtra and Gujarat [62, 66]. During the summer season, it is very refreshing with distinguishing odor, taste, more palatable, high nutrition, and promising therapeutic properties. Shrikhand is prepared traditionally from chakka, a solid mass obtained from cultured milk (curd) by separating the whey using muslin cloth. The sugar powder was blended to chakka to smooth and homogenous consistency of Shrikhand. Since olden days Shrikhand is found to have therapeutic properties such as anti-carcigenic and anti-cholesterolemic [17]. The Shrikhand is found to be stable for more than 30 days, which is long when compared with other cultured milk products. The shelf life of symbiotic (prebiotic fructooligosaccharides and probiotics) Shrikhand is much longer (60 days) [129]. Shrikhand is available with different flavors, such as amrakhand (mango), badam (almond), pista (pistachio), butterscotch, elaichi (cardamom), Kesar (saffron), rajbhog, and strawberry.

13.3.5 BUTTERMILK (LASSI)

In India, lassi is the local name of buttermilk, prepared by churning the curd. At the time of butter manufacturing, cream is churned and the aqueous phase obtained after the extraction of butter from churned curd or yogurt is called buttermilk [28, 82]. The globule of milk-fat membrane forms proteins and polar lipids composition in lassi, hence it is considered as an important foodstuff. Pasteurized milk cultured with microorganisms producing flavor is used in the preparation of lassi [6]. Some of the commonly associated LAB with lassi is *L. lactis* ssp. *lactis, L. lactis* ssp. *cremoris, L. lactis* ssp. *lactis* biovar. *diacetyl lactis*, *Lb. delbrucki, Lb. acidophilus, Lb. helveticus, S. thermophilus* and *Leu. mesenteroides* ssp. *Cremoris* [41]. In India, lassi is also appreciated in various flavored forms as a thirst-quenching beverage for its palatability apart from its rich therapeutic properties and is now commercially available in many places.

13.3.6 KEFIR

Kefir is a popular fermented beverage traditionally prepared using kefir grains. In Central Asia, kefir is consumed from the time of the Middle Ages. Now, in many regions of the world it is considered as an important consumer commodity. The geographic regions affect the proportion of dense populations of LAB, yeast, and acetobacter in kefir and it is composed of a complex of proteins and polysaccharides [30]. The studies on animals were conducted for kefir by *in vitro,* which confirmed its positive effects on lactose intolerance [55], immunomodulation, antimicrobial activity, and harmonization of intestinal microflora [147]. Kefir has essential vitamins, minerals, amino acids, and enzymes. The Commercially available Kefir are Babushka, Liberte, and Lifeway.

13.3.7 CHEESE

Cheese is a fermented product of dairy, manufactured in varieties of flavors and textures (soft and hard cheese) using cultured milk. There are 2,000 varieties of cheeses present worldwide, in which Cheddar, Cream, Feta, Blue, and goat are the most important ones. In acid coagulated cheeses probiotics have been incorporated, as they show similarity to fermented milks [115, 150]. However, research was carried out on the preparation of different cheeses like Cheddar [79, 95], Edam [146], brined cheeses [64, 118, 152], semi-hard [13] cheese by using probiotics. The most important strains of *Streptococci* and *Lactobacilli* present in Domiati cheese are *Lb. casei, Lb. plantarum*, *Lb. brevis*, *Lb. fermenti*, *S. faecalis* and *L. lactis.* Commercially available probiotic rich cheeses include aged, traditional Cheddar, Gouda, and alpine cheeses like Gruyere and other cheeses available with different brands are Vintage Cheddar, Mozzarella, Amul, Queso Asadero, Colby, Nancy Cottage cheese, etc.

13.3.8 SOUR CREAM

Sour cream is soured by LAB, especially *S. lactis* and/or any yogurt starter culture can be used to ferment cream as they are known to produce LA. Fayed et al. [35] developed a sour cream with low-calorie probiotic-containing *S. thermophilus*, *Lb. acidophilus* and *Bifidobacterium spp*. Cultured cream will be savory or sweet with high fat content (10–40%), significantly fewer calories than mayonnaise. The sour cream has a mild, tangy flavor and thick, smooth texture. In baking products such as cookies, cakes, breads, and pies, sour cream is considered to be appropriate.

13.3.9 ACIDOPHILUS MILK

Acidophilus milk is the milk either fermented or unfermented with *Lb. acidophilus* and *Bacillus acidophilus* possessing desirable sensory properties [25]. However, Myers stated that acidophilus milk is having an unpleasant taste and LA was not important in transforming the intestinal flora. Hence, he explained that, by incorporating *Lb. acidophilus* directly to the unfermented milk, retention of the taste and physical properties of milk is possible with the viable *Lb. acidophilus* for 7 days at 2 to 5°C [85]. Acidophilus milk is rich in free amino acids compared to normal milk.

13.3.10 ICE CREAM

Ice cream is a potential Haulier for probiotics, as it holds better viability of probiotics at the time of production as well as storage period when compared with fermented milks. Preparation methods like oxygen-resistant strain selection and its application, molecular oxygen elimination, heat treatment application, microencapsulation techniques utilization and product formulation by fortifying the milk with prebiotics and other nutrients could improve the viable probiotic cells in the final formulation. *B. lactis* (Bb12) and *Lb. casei* (Lc01) exhibited the maximum resistance to simulated gastric conditions in comparison to other bacterial strains, making them appropriate strains for use in ice cream [57]. Commercially available ice creams are Culture Republick, Probiotic Wellness and Foxy's ice cream and much other variety of brands and flavors to choose world-over.

13.4 NON-DAIRY PRODUCTS

Dairy products are exhibiting promising features of probiotics by maintaining the bacterial viability in the finished foods [120]. However, the consumption of these products is not possible for those who are sensitive to milk proteins, lactose intolerance and high cholesterol levels. Therefore, an alternative product free of milk/dairy is used to deliver probiotics which includes fruits and vegetable products (Table 13.2), cereals products, dessert products, and meat products [44, 49, 76, 107]. Probiotic non-dairy foods developed throughout the world have been reviewed by Rivera-Espinoza and Gallardo-Navarro [111].

TABLE 13.2 List of Fruit and Vegetable-Based Non-Dairy Products

Fruit and Vegetable Juices/Beverages	References
Vegetable-based drinks	[68]
Beetroot	[153]
Tomato	[154]
Cabbage	[155]
Carrot	[89]
Onion	[112]
Ginger	[23]
Fermented banana pulp	[145]
Fermented banana and Banana puree	[144]
Pineapple, orange	[131]
Grape, passion fruit	[117]
Noni	[151]
Non fermented fruit juice beverages	[110]
Black currant	[73]
Plum	[132]
Cashew apple	[104]

Fruit and vegetable-based products are gaining importance as they contain nutrients unlike in dairy products and also probiotics growth is enhanced due to the presence of sugars in it [31]. Some of the fruits and vegetables exploited for this purpose are apple, banana, blueberry, black currant, cashew apple, orange, pineapple, raspberry, pomegranate, and cantaloupe melon juice [7, 93, 126, 153] and carrot, beetroot, and mixed vegetable juice [92]. The juices do not stay for a longer period in the stomach and hence the probiotics used in the juices withstand the harsh condition of the stomach. The stability in terms of viable counts of probiotic *Lb. casei, Lb. plantarum, Lb. acidophilus* in cashew apple juice, melon juice, pineapple juice, blackcurrant juices was observed to be more than 8.00 log CFU/ml after the storage period of 6 weeks, which indicates that the product is as competent as dairy products [39, 73, 104].

Cereals are considered as one of the main foods consumed in our daily diet, and are carriers of probiotics. They contain protein, vitamins, carbohydrates, minerals, and fiber along with non-digestible carbohydrates (NDC) and/or dietary fiber through which *Lactobacilli* and *Bifidobacteria* gets stimulated in the colon [22]. The minerals bioavailability is increased when the microbial enzymes act on cereals or by the organic acids production during fermentation [60, 140] of phosphorous, iron, and zinc [123]. In Asia and African countries, cereals fermentation by LAB is considered as the ancient practice which resulted in the manufacture of beverages, gruels, and porridge. Cereal grains used for this purpose are wheat, millet, maize, oats, sorghum, barley, and rye. *Lb. plantarum*, *Lb. acidophilus* and *Lb. rhamnosus* GG are used in the preparation of barley and malt, Sorghum flour-based yogurt and pan bread slices, respectively [109, 124, 135]. Some of the products based on cereal include puddings [54], rice-based yogurt [19], oat-based drink and milk [8, 14], oat, barley, and malt-based products [119], "Yosa," an oat-bran pudding [16] and "Mahewu," a maize-based fermented beverage [80].

Dessert foods have an enormous potential for market because probiotic strains such as *Lb. paracasei* subsp. *paracasei* LBC 82 along with inulin, a prebiotic are incorporated in chocolate mousse preparation [9, 106]. The retention of significant viability of *Lb. paracasei, B. lactis, Lb. rhamnosus* and *S. boulardii* was found in soy dessert throughout the storage period [53]. Fermented acerola (*Malpighia emarginata*) ice cream and coconut flan have exhibited probiotic property as they contain *B. longum* and *B. lactis* [34] and *B. lactis* and *Lb. paracasei* [65] as good carrier with high viable cell counts.

Meat in the form of sausages, as an alternate probiotic-based dairy foods is being used. It is reported that the alginate encapsulated with *Lb. reuteri* and *B. longum* is used to prepare sausages based on meat [84]. In the preparation of meat-based probiotic products, the use of probiotic cultures has been reviewed by Rivera Espinoza and Gallardo-Navarro [111]. The food industry is looking for an improvement in the meat products quality through maintaining significant viable probiotic cell counts to extend its shelf life [134]. An excellent survivability of probiotics was found in the dry-cured sausage with orange fiber, a Spanish non-fermented food, hence it is regarded as an exceptional medium for these bacteria. Some of the probiotics LAB suitable to use in the fermentation of meat are *Lb. acidophilus, Lb. gallinarum, Lb. gasseri, Lb. amylovorus, Lb. crispatus,* and *Lb. johnsonii*, which enhance the safety of product [10].

13.5 PRODUCT SPECIFICATIONS

Product specifications are blueprints that describe exactly what the product will be, what it should look like and what function it will perform. Once the product enriched with probiotic is developed, the product should be specified with its outline information on packaging. The information about the safety of the product containing the probiotic microorganisms should be provided to the concerned authority by the manufacturer before marketing. The probiotic microorganisms used in the product should be documented as follows:

- Antimicrobial and antibiotic resistant genes of the strain should be reported;
- Antimicrobial property of the strain should be evaluated through *in vivo* and *in vitro*;
- Documentation must be in the form of published peer-reviewed articles;
- Information related to the extrachromosomal DNA and lack of virulence factor must be provided;
- The accession number should be submitted for the obtained sequence data of the microorganisms;
- The deposition of the identified microorganism must be in a recognized international culture collection center after obtaining its accession and strain number;
- The names of the identified microorganism must be in accordance with the international code for nomenclature of microorganisms;
- The phenotypic and genotypic characteristic of the strains should be carried out.

Once the information regarding the probiotic microorganisms is documented, it can be used in the manufacture of dairy and non-dairy foods. The product must furnish the following specifications.

- Viable probiotics cells in the product should be expressed as log CFU per gram;
- Number of times the product to be used per day and the serving size;
- The necessary conditions such as time, temperature, exposure to air and relative humidity required for storage;
- Stability of the product after opening the packaging.

It is the accountability of the producer/manufacturer to provide information on the safety assessments carried out on the products to the consumers and/or clinicians.

13.6 SAFETY CRITERIA

As the markets for probiotic are rapidly growing worldwide, the evaluation of probiotic bacteria for its safety and efficacy are gaining importance for the management of national and international regulations and guidelines. There is high significance for the deliberation of safety of probiotic bacteria, as they are souk through food and feed supplements. The different safety policies for probiotics-based products have been maintained in different nations. It is believed that for the general population, all the species of probiotics to be safe (as stated by the European Food Safety Authority) and thus, probiotics have grabbed a significant place in healthcare [20, 36]. Most probiotics used in products for decades are derived either from fermented foods or from the microbes colonizing a healthy human. The generally regarded as safe (GRAS) status of the LAB is encouraging the consumption of the processed foods containing these bacteria in the past have not resulted in any harmful effects [45]. Ecology of breast-fed infant's intestinal tract has Bifidobacteria as predominant bacteria; as a result, the infant is healthy. There is no question on the use of Bifidobacteria in any formulations, as it is safe and till now there is no research report on its harmful effect on the host.

To measure the strain of probiotic to be safe, following approaches could be studied:

1. **The Fundamental Characteristics of the strain:** *In vitro* characterization on the extreme bile salts deconjugation.
2. **The Pharmacokinetics of the Strain:** The difference in the survivability of probiotics ingested at various stages of the GI tract varies with the strains, which is determined by *in vivo* fecal samples and biopsies of the intestinal mucosa. Outcome of the ingested strains could be predicted by a number of *in vitro* models.

3. **The Strain-Host Interactions:** The adverse interactions of microbiological agents in food can lead to host illness which is hard to envisage than infection caused from the chemical means. The quantitative risk assessment is necessary to understand through the food containing pathogens which reveals that there is no existence of zero risk. However, it is very difficult to understand the concept of minimal infective dose since the involvement of a greater amount of microbial and host factors and the differences of the individuals at high potential.

13.7 TESTING METHODS

The guidelines deal with the use of probiotic microorganism, the prerequisite to evaluate the probiotic strain for its safety efficacy, its health claims, and the products labeling with probiotics. According to FAO recommendation, *in vitro* and *in vivo* studies are necessary to document the safety of these strains which includes the assessment of metabolic activities, determining the profiles of antibiotic resistance, antimicrobial drug resistance patterns, toxin production and hemolytic activity, administration of dose and method (oral or otherwise), absences of allergenic material, determination of genetic and pathological side effects in humans during clinical trials. The guidelines also deal with post-market surveillance of epidemiology, the consuming population's physiological status, special consideration to the vulnerable populations, new-born infants and to those who are critically ill. The history on the isolation of probiotic microorganisms and its taxonomic classification should be recorded. During the manufacturing of products, if probiotics are cross-contaminated with other microbes or substances between batches, such batches must be eliminated.

13.8 PROBIOTICS MARKET

Before 2022, the probiotics market is anticipated to acquire \$57.4 billion with the registered compound annual growth rate (CAGR) of 7.7% throughout 2016–2022 which is the estimated period. Nowadays, probiotic products are employed to diagnose many diseases as they are known to support the immune system of the human. The positive effects of probiotics are on host health including the gut microbiome apart from providing basic nutrition, therefore consumption of functional foods is rising which leads to the demand for probiotics. By increasing the preference of consumers for natural products, the major market growth factor can be enhanced. Capsules, gel, and powders are the forms of probiotic ingredients placed on the market.

The probiotic market in India is driving the market growth tremendously due to the therapeutic properties of probiotics and consumers awareness on preventive healthcare. As per the Research study by TechSci on "India Probiotic Market Forecast and Opportunities," in 2019, the revenue projected during 2014–2019 was 19.80% which is an average growth on the probiotic market. The Indian Probiotic market is dominated by Northern India, followed by Southern and Central regions in terms of sales revenue. In India, Amul, Danone, Epigamia, Mother Dairy, Milky Mist, Nestle India, and Yakult are the foremost producers of probiotic beverages and functional foods. As the need for women and pediatric nutrition are rising, there is a dominancy on the probiotic drugs segment and dietary supplements by various firms such as Tablets India, Dr. Reddy's laboratories and USV India, Sun Pharma, Mankind, Lupin, Cipla, Himalaya Drugs, Curatio, Corona Remedies, Glenmark, Torrent, Fourrts India, Pharmed, Unichem, etc. In India, "Prolife (Amul), Yakult Yogurt, Yakult Acidophilus Plus, Healthvit, ViBact, inLife, Organic Low Fat Kefir, b-Active, NesVita, Neo, Bio-K Plus, Doctor's Best, Probiotic Tea, Coffee, Vista Nutritions, etc., are the major probiotic products.

Global scenario of probiotics products growth has been quite amazing. Knowledge in probiotic continues to expand during the last decades due to their protective role in the gut to keep our gut healthy and fit. In 2018, the size of the market is exceeded to USD 2 billion and is estimated to grow at over 8.15% CAGR between 2020 and 2025. The highly competitive market with key players includes Nestle SA, Danone SA, PepsiCo Inc., and Yakult Honsha Co. Ltd. Several companies like Sanofi, Abbott, Euro Lifecare, Blis technology, Sunwave Pharma, Aceto, etc., and also manufacture products of international probiotic giants like CHR Hansen, DSM, DuPont Nutrition, Lesaffre, TOA (Tennessee orthopedic alliance) pharmaceuticals, UAS Labs, Morinaga, etc., are involved in research using Japanese technology and rigid process engineering and controls. Probiotics are manufactured in the form of capsules, sachets, tablets, dry syrup lozenges and the latest development is probiotic drops.

13.9 SUMMARY

The promising physiological and therapeutic effects of probiotics is growing and intense research efforts are underway in developing value added food products with the incorporation of *Lactobacillus* and *Bifidobacterium* species. The benefits of probiotics have helped food industries in recognizing the probiotics role on human health and exploiting different formulations. In vulnerable populations, the comprehensive post-marketing surveillance of therapeutic properties of probiotic-based dairy and non-dairy products by *in vivo* method is necessary to monitor the incidence of any unintended consequences.

ACKNOWLEDGMENT

The authors are thankful to the Director, CFTRI, Mysore, India.

KEYWORDS

- **bifidobacteria**
- **dairy products**
- **health benefits**
- **lactobacillus**
- **non-dairy products**
- **probiotics**
- **safety concern**

REFERENCES

1. Abou-Donia, S. A., (2008). Origin, history and manufacturing process of Egyptian dairy products: A overview. *Alexandria Journal of Food Science and Technology, 5*(1), 51–62.
2. Adolfsson, O., (2004). Yogurt and gut function. *American Journal of Clinical Nutrition, 80*(2), 245–256.
3. Akın, M., Akın, M., & Kırmacı, Z., (2007). Effects of Inulin and sugar levels on the viability of yogurt and probiotic bacteria and the physical and sensory characteristics in probiotic ice-cream. *Food Chemistry, 104*(1), 93–99.
4. Alakomi, H. L., Skytta, E., Saarela, M., Mattila-Sandholm, T., Latva-Kala, K., & Helander, I. M., (2000). Lactic acid permeabilizes gram-negative bacteria by disrupting the outer membrane. *Applied Environmental Microbiology, 66*(5), 2001–2005.
5. Alba-Lois, L., & Segal-Kischinevzky, C., (2010). Beer and winemakers. *Nature Education, 3*(9), 17.

6. Aneja, R. P., Vyas, M. N., Sharma, D., & Samal, S. K., (1989). *A Method for Manufacturing Lassi*. Indian Patent no. 17374.
7. Anekella, K., & Orsat, V., (2013). Optimization of microencapsulation of probiotics in raspberry juice by spray drying. *LWT Food Science and Technology, 50*(1), 17–24.
8. Angelov, A., Gotcheva, V., Kuncheva, R., & Hrstozova, T., (2006). Development of a new oat-based probiotic drink. *International Journal of Food Microbiology, 112*(1), 75–80.
9. Aragon-Alegro, L. C., Alarcon-Alegro, J. H., Carderelli, H. R., Chiu, M. C., & Saad, S. M. I., (2007). Potentially probiotic and synthetic chocolate mousse. *LWT-Food Science and Technology, 40*(4), 669–675.
10. Arihara, K., Ota, H., Itoh, M., Kondo, Y., Sameshim, T., Yamanaka, H., Akimoto, M., et al., (1998). *Lactobacillus acidophilus* group lactic acid bacteria applied to meat fermentation. *Journal of Food Science, 63*(3), 544–547.
11. Bais, H. P., Weir, T. L., Perry, L. G., Gilroy, S., & Vivanco, J. M., (2006). The role of root exudates in rhizosphere interactions with plants and other organisms. *Annual Review of Plant Biology, 57*, 233–266.
12. Beachy, E. H., (1981). Bacterial adherence: Adhesin-receptor interactions mediating the attachment of bacteria to mucosal surfaces. *Journal of the Infectious Diseases, 143*(3), 325–345.
13. Bergamini, C. V., Hynes, E. R., & Zalazar, C. A., (2006). Influence of probiotic bacteria on the proteolysis profile of a semi hard cheese. *International Dairy Journal, 16*(8), 856–866.
14. Bernat, N., Cháfer, M., González-Martínez, C., Rodríguez-García, J., & Chiralt, A., (2015). Optimization of oat milk formulation to obtain fermented derivatives by using probiotic *Lactobacillus reuteri* microorganisms. *Food Science and Technology International, 21*(2), 145–157.
15. Bernet, M. F., Brassart, D., Neeser, J. R., & Servin, A. L., (1994). *Lactobacillus acidophilus* LA1 binds to cultured human intestinal cell lines and inhibits cell attachment and cell invasion by enterovirulent bacteria. *Gut, 35*(4), 483–489.
16. Blandino, A., Al-Aseeri, M. E., Pandiella, S. S., Cantero, D., & Webb, C., (2003). Cereal-based fermented foods and beverages. *Food Research International, 36*(6), 527–543.
17. Boghra, V. R., & Mathur, O. N., (2000). Physico-chemical status of major milk constituents and minerals at various stages of shrikhand preparation. *Journal of Food Science and Technology, 37*(2), 111–115.
18. Boonyaratanakornkit, M., & Wongkhalaung, C., (2000). Development of a yogurt-type product from saccharified rice. *Kasetsart Journal, 34*(1), 107–116.
19. Caballero, B., Finglas, P., & Trugo, L., (2003). *Encyclopedia of Food Sciences and Nutrition* (2nd edn., p. 6000). London–UK: Academic Press.
20. Carabotti, M., Scirocco, A., Maselli, M. A., & Severi, C., (2015). The gut-brain axis: Interactions between enteric microbiota, central and enteric nervous systems. *Annals of Gastroenterology, 28*(2), 203–209.
21. Chalova, V. I., Lingbeck, J. M., Kwon, Y. M., & Ricke, S. C., (2008). Extracellular antimutagenic activities of selected probiotic *Bifidobacterium* and *Lactobacillus* spp. as a function of growth phase. *Journal of Environmental Science and Health Part B, 43*(2), 193–198.
22. Charalampopoulos, D., Wang, R., Pandiella, S. S., & Webb, C., (2002). Application of cereals and cereal components in functional foods: A review. *International Journal of Food Microbiology, 79*(1–2), 131–141.

23. Chen, I. N., Ng, C. C., Wang, C. Y., & Chang, T. L., (2009). Lactic fermentation and antioxidant activity of *Zingiberaceae* plants in Taiwan. *International Journal of Food Science and Nutrition, 60*(suppl. 2), 57–66.
24. Chenoll, E., Casinos, B., Bataller, E., Astals, P., Echevarría, J., Iglesias, J. R., Balbarie, P., et al., (2011). Novel probiotic *Bifidobacterium bifidum* CECT 7366 strain active against the pathogenic bacterium *Helicobacter pylori. Applied Environmental Microbiology, 77*(4), 1335–1343.
25. Cheplin, H. A., & Rettger, L. F., (1920). Studies on the transformation of the intestinal flora, with special reference to the implantation of *Bacillus acidophilus*, II. feeding experiments on man. *Proceedings of the National Academy of Sciences of the United States of America, 6*(12), 704–705.
26. Chu, H., Kang, S., Ha, S., Cho, K., Park, S. M., Han, K. H., Kang, S. K., et al., (2005). *Lactobacillus acidophilus* expressing recombinant K99 adhesive fimbriae has an inhibitory effect on adhesion of enterotoxigenic *Escherichia coli. Microbiology and Immunology, 49*(11), 941–948.
27. Collado, M. C., Gueimonde, M., Hernández, M., Sanz, Y., & Salminen, S., (2005). Adhesion of selected *Bifidobacterium* strains to human intestinal mucus and the role of adhesion in enteropathogen exclusion. *Journal of Food Protection, 68*(12), 2672–2678.
28. Corredig, M., & Dalgleish, D. G. (1997). Isolates from industrial buttermilk: Emulsifying properties of materials derived from the milk at globule membrane. *Journal of Agricultural and Food Chemistry, 45*(12), 4595–4600.
29. De Keersmaecker, S. C., Verhoeven, T. L., Desair, J., Marchal, K., Vanderleyden, J., & Nagy, I., (2006). Strong antimicrobial activity of *Lactobacillus rhamnosus* GG against *Salmonella typhimurium* is due to accumulation of lactic acid. *FEMS Microbiology Letters, 259*(1), 89–96.
30. De Oliveira, L. A. M., Miguel, M. A., Peixoto, R. S., Rosado, A. S., Silva, J. T., & Paschoalin, V. M., (2013). Microbiological, technological and therapeutic properties of kefir: A natural probiotic beverage. *Brazilian Journal of Microbiology, 44*(2), 341–349.
31. Ding, W. K., & Shah, N. P., (2008). Survival of free and microencapsulated probiotic bacteria in orange and apple juices. *International Food Research Journal, 15*(2), 219–232.
32. Diplock, A. T., Aggett, P., & Ashwell, M., (1999). Scientific concepts of functional food science in Europe: Consensus document. *British Journal of Nutrition, 81*(1), S1–S27.
33. FAO/WHO, (2001). Health and nutritional properties of probiotics in food including powder milk with live lactic acid bacteria. In: *Food and Agriculture Organization of the United Nations and World Health Organization Expert Consultation Report* (pp. 1–4). Cordoba, Argentina: FAO/WHO.
34. Favaro-Trindade, C., Bernardi, S., Bodini, R., Balieiro, J. C. C., & DeAlmeida, E., (2006). Sensory acceptability and stability of probiotic microorganisms and vitamin C in fermented acerola (*Malpighia emarginata* DC.) ice cream. *Journal of Food Science, 71*(6), S492–S495.
35. Fayed, A. E., Hussein, G. A. M., & Farahat, A. M., (2006). Production of probiotic low-calorie sour cream. *Arab Universities Journal of Agricultural Sciences, 14*(2), 697–710.
36. Festi, D., Schiumerini, R., Eusebi, L. H., Marasco, G., Taddia, M., & Colecchia, A., (2014). Gut microbiota and metabolic syndrome. *World Journal of Gastroenterology, 20*(43), 16079–16094.
37. Finlay, B. B., & Falkow, S., (1997). Common themes in microbial pathogenicity revisited. *Microbiology and Molecular Biology Reviews, 61*(2), 136–169.

38. Foissy, H., (1983). Nutritional and physiological aspects of yogurt. *Food Science and Technology Abstract, 16*(3), 2404–2408.
39. Fonteles, T. V., Costa, M. G. M., De Jesus, A. L. T., Fontes, C. P. M. L., Fernandes, F. A. N., & Rodrigues, S., (2013). Stability and quality parameters of probiotic cantaloupe melon juice produced with sonicated juice. *Food Bioprocess Technology, 6*(10), 2860–2869.
40. Forsythe, P., & Bienenstock, J., (2010). Immunomodulation by commensal and probiotic bacteria. *Immunological Investigations, 39*(4, 5), 429–448.
41. Frede, E., (2003). Butter-the product and its manufacture. In: Roginski, H., Fuquay, J. W., & Fox, P. F., (eds.), *Encyclopedia of Dairy Sciences* (Vol. 1, pp. 220–227). London–UK: Academic Press.
42. Fuller, R., (1989). Probiotics in man and animals. *Journal of Applied Bacteriology, 66*(5), 365–378.
43. Ganji-Arjenaki, M., & Rafieian-Kopaei, M., (2018). Probiotics are a good choice in remission of inflammatory bowel diseases: A meta-analysis and systematic review. *Journal of Cell Physiology, 233*(3), 2091–2103.
44. Gawkowski, D., & Chikindas, M. L., (2013). Non-dairy probiotic beverages: The next step into human health. *Beneficial Microbes, 4*(2), 127–142.
45. Gilliland, S. E., (1990). Health and nutritional benefits from lactic acid bacteria. *FEMS Microbiology Reviews, 7*(1, 2), 175–188.
46. Godoy, A., Herrera, T., & Ulloa, M., (2003). *Más Allá Del Pulque y el Tepache: Las Bebidas Alcohólicas no Destiladas Indígenas de México* (p. 107). Instituto de Investigaciones Antropológicas, Mexico: Universidad Nacional Autonoma de Mexico.
47. Goel, B. K., (1998). Studies on feasibility of commercial production of *mishtidoi*, a techno-economic aspect. *Indian Dairyman, 50*(3), 11–15.
48. Guandalini, S., (2006). Probiotics for children: Use in diarrhoea. *Journal of Clinical Gastroenterology, 40*(3), 244–248.
49. Gupta, S., & Abu-Ghannam, N., (2012). Probiotic fermentation of plant based products: Possibilities and opportunities. *Critical Reviews in Food Science and Nutrition, 52*(2), 183–199.
50. Hassan, M., Kjos, M., Nes, I. F., Diep, D. B., & Lotfipour, F., (2012). Natural antimicrobial peptides from bacteria: Characteristics and potential applications to fight against antibiotic resistance. *Journal of Applied Microbiology, 113*(4), 723–736.
51. Havenaar, R., & Huis in't Veld, J. H. J., (1992). Probiotics: A general view; Chapter 6. In: Wood, B. J. B., (ed.), *The Lactic Acid Bacteria: Volume 1-The Lactic Acid Bacteria in Health and Disease* (Vol. 1; pp. 151–170). US: Springer.
52. Hawrelak, J., (2013). Probiotics; Chapter 116. In: Pizzorno, J. E., & Murray, M. T., (eds.), *Text Book of Natural Medicine* (4th edn., pp. 979–994). St. Louis–Missouri: Elsevier.
53. Heenan, C. N., Adams, M. C., Hosken, R. W., & Fleet, G. H., (2004). Survival and sensory acceptability of probiotic microorganisms in a nonfermented frozen vegetarian dessert. *LWT-Food Science and Technology, 37*(4), 461–466.
54. Helland, M. H., Wciklund, T., & Narvhus, J. A., (2004). Growth and metabolism of selected strains of probiotic bacteria in milk and water-based cereal puddings. *International Dairy Journal, 14*(11), 957–965.
55. Hertzler, S. R., & Clancy, S. M., (2003). Kefir improves lactose digestion and tolerance in adults with lactose maldigestion. *Journal of the American Dietetic Association, 103*(5), 582–587.

56. Hirano, J., Yoshida, T., Sugiyama, T., Koide, N., Mori, I., & Yokochi, T., (2003). The effect of *Lactobacillus rhamnosus* on enterohemorrhagic *Escherichia coli* infection of human intestinal cells *in vitro*. *Microbiology and Immunology, 47*(6), 405–409.
57. Homayouni, A., Azizi, A., Javadi, M., Mahdipour, S., & Ejtahed, H., (2012). Factors influencing probiotic survival in ice cream: A review. *International Journal of Dairy Science, 7*(1), 1–10.
58. Hooper, L. V., Stappenbeck, T. S., Hong, C. V., & Gordon, J. I., (2003). Angiogenins: A new class of microbicidal proteins involved in innate immunity. *Nature Immunology, 4*(3), 269–273.
59. Hooper, L. V., Wong, M. H., Thelin, A., Hansson, L., Falk, P. G., & Gordon, J. I., (2001). Molecular analysis of commensal host-microbial relationships in the intestine. *Science, 291*(5505), 881–884.
60. Hotz, C., & Gibson, R. S., (2007). Traditional food-processing and preparation practices to enhance the bioavailability of micronutrients in plant based diets. *The Journal of Nutrition, 137*(4), 1097–1100.
61. Hughes, D. B., & Hoover, D. G., (1991). Bifidobacteria: Their potential for use in American dairy products. *Food Technology, 45*(4), 74–83.
62. Jain, A., Desai, H. K., & Upadhyay, K. G., (1998). Chemical quality of market shrikhand sold in Gujarat market. *Indian Journal of Dairy Science, 51*(6), 423–426.
63. Juntunen, M., Kirjavainen, P. V., Ouwehand, A. C., Salminen, S. J., & Isolauri, E., (2001). Adherence of probiotic bacteria to human intestinal mucus in healthy infants and during rotavirus infection. *Clinical Diagnostic Laboratory Immunology, 8*(2), 293–296.
64. Kasımoğlu, A., Göncüoğlu, M., & Akgün, S., (2004). Probiotic white cheese with *Lactobacillus acidophilus*. *International Dairy Journal, 14*(12), 1067–1073.
65. Khalf, M., Dabour, N., Kheadr, E., & Fliss, I., (2010). Viability of probiotic bacteria in maple sap products under storage and gastrointestinal conditions. *Bioresource Technology, 101*(20), 7966–7972.
66. Kulkarni, C., Belsare, N., & Lele, A., (2006). Studies on shrikhand rheology. *Journal of Food Engineering, 74*(2), 169–177.
67. Lallemand Health Solutions, (2018). *Bifidobacterium Animalis spp. Lactis LAFTI B94 Recognized for its Benefits in Children with IBS by Health Canada.* https://lallemand-health-solutions.com/en/*bifidobacterium*-animalis-ssp-lactis-lafti-b94-recognized-for-its-benefits-in-children-with-ibs-by-health-canada/ (accessed on 30 May 2022).
68. Lambo, A. M., Oste, R., & Nyman, M. G. E. L., (2005). Dietary fibre in fermented oat and barley β-glucan rich concentrates. *Food Chemistry, 89*(2), 283–293.
69. Levri, K. M., Ketvertis, K., Deramo, M., Merenstein, J. H., & D'Amico, F., (2005). Do probiotics reduce adult lactose intolerance? a systematic review. *Journal of Family Practice, 54*(7), 613–620.
70. Lilly, D. M., & Stillwell, R. H., (1965). Probiotics: Growth promoting factors produced by microorganisms. *Science, 147*(3659), 747, 748.
71. Link-Amster, H., Rochat, F., Saudan, K. Y., Mignot, O., & Aeschlimann, J. M., (1994). Modulation of a specific humoral immune response and changes in intestinal flora mediated through fermented milk intake. *FEMS Immunology and Medical Microbiology, 10*(1), 55–63.
72. Liong, M. T., (2008). Roles of probiotics and prebiotics in colon cancer prevention: Postulated mechanisms and *in-vivo* evidence. *International Journal of Molecular Sciences, 9*(5), 854–863.

73. Luckow, T., & Delahunty, C., (2004). Which juice is healthier? a consumer study of probiotic non-dairy juice drinks. *Food Quality Preferences, 15*(7, 8), 751–759.
74. Macrae, R., Robinson, R. K., & Sadler, M. J., (1993). *Encyclopedia of Food Science, Food Technology and Nutrition* (1st edn., Vol. 1, pp. 655–664). San Diego: Academic Press.
75. Makras, L., Triantafyllou, V., Fayol-Messaoudi, D., Adriany, T., Zoumpopoulou, G., Tsakalidou, E., Servin, A., & DeVuyst, L., (2006). Kinetic analysis of the antibacterial activity of probiotic lactobacilli towards *Salmonella enterica* serovar typhimurium reveals a role for lactic acid and other inhibitory compounds. *Research Microbiology, 157*(3), 241–247.
76. Martins, E. M. F., Ramos, A. M., Vanzela, E. S. L. B., Stringheta, P. C., Pinto, C. L. O., & Martins, J. M., (2013). Products of Vegetable origin: A new alternative for the consumption of probiotic bacteria. *Food Research International, 51*(2), 764–770.
77. Max, M., (1900). *Sacred Books of the East* (p. 457). New York: The Colonial Press.
78. Mazzarello, P., (2002). Life out of nowhere? *Nature, 417*(6891), 792–793.
79. Mc Brearty, S., Ross, R. P., Fitzgerald, G. F., Collins, J. K., Wallace, J. M., & Stanton, C., (2001). Influence of two commercially available bifidobacteria cultures on cheddar cheese quality. *International Dairy Journal, 11*(8), 599–610.
80. McMaste, L. D., Kokott, S. J., & Abratt, V. R., (2005). Use of traditional African fermented beverages as delivery vehicles for *Bifidobacterium lactis* DSM 10140. *International Journal of Food Microbiology, 102*(2), 231–237.
81. Miller, G. D., DiRienzo, D. D., Reusser, M. E., & McCarron, D. A., (2008). Benefits of dairy product consumption on blood pressure in humans: A summary of the biomedical literature. *Journal of the American College of Nutrition, 19*(suppl. 2), 147S–164S.
82. Morin, P., Jiménez-Flores, R., & Pouliot, Y., (2007). Effect of processing on the composition and microstructure of buttermilk and its milk fat globule membranes. *International Dairy Journal, 17*(10), 1179–1187.
83. Muñoz, J. A., Chenoll, E., Casinos, B., Bataller, E., Ramón, D., Genovés, S., Montava, R., et al., (2011). Novel probiotic *Bifidobacterium longum* subsp. infantis CECT 7210 strain active against rotavirus infections. *Applied Environmental Microbiology, 77*(24), 8775–8783.
84. Muthukumarasamy, P., & Holley, R. A., (2006). Microbiological and sensory quality of dry fermented sausages containing alginate microencapsulated *Lactobacillus reuteri. International Journal of Food Microbiology, 111*(2), 164–169.
85. Myers, R. P., (1931). Transformation of the intestinal flora through the feeding of unfermented acidophilus milk. *American Journal of Public Health and the Nation's Health, 21*(8), 867–872.
86. Naidu, A. S., Biblack, W. R., & Clemens, R. A., (1999). Probiotic spectra of lactic acid bacteria (LAB). *Critical Reviews in Food Science and Nutrition, 39*(1), 13–126.
87. Nakamura, S., Kuda, T., An, C., Kanno, T., Takahashi, H., & Kimura, B., (2012). Inhibitory effects of *Leuconostoc mesenteroides* 1RM3 isolated from narezushi, a fermented fish with rice, on *Listeria monocytogenes* infection to caco-2 cells and A/J mice. *Anaerobe, 18*(1), 19–24.
88. Nakazawa, Y., & Hosono, A., (1992). Fermented milk in the orient. In: Nakazawa, Y., & Hosono, A., (eds.), *Functions of Fermented Milk: Challenges for the Health Sciences* (pp. 61–78). London-UK: Elsevier Applied Science.
89. Nazzaro, F., Fratinni, F., Sada, A., & Orlando, P., (2008). Synbiotic potential of carrot juice supplemented with *Lactobacillus* spp. and inulin or fructooligosaccharides. *Journal of the Science of Food and Agriculture, 88*(13), 2271–2276.

90. Neubauer, H., & Mollet, B., (2002). Biotechnological research and dairy industry. In: Hoffman, M., & Thonart, P., (eds.), *Engineering and Manufacturing for Biotechnology* (pp. 399–412). New York–USA: Kluwer Academic Publishers.
91. Nielsen, D. S., Cho, G. S., Hanak, A., Huch, M., Franz, C. M., & Arneborg, N., (2010). The effect of bacteriocin-producing *Lactobacillus plantarum* strains on the intracellular pH of sessile and planktonic *Listeria monocytogenes* single cells. *International Journal of Food Microbiology, 141*, S53–S59.
92. Nosrati, R., Hashemiravan, M., & Talebi, M., (2014). Fermentation of vegetables juice by probiotic bacteria. *International Journal of Biosciences, 4*(3), 171–180.
93. Nualkaekul, S., Lenton, D., Cook, M. T., Khutoryanskiy, V. V., & Charalampopoulos, D., (2012). Chitosan-coated alginate beads for the survival of microencapsulated *Lactobacillus plantarum* in pomegranate juice. *Carbohydrate Polymer, 90*(3), 1281–1287.
94. Ohland, C. L., & Macnaughton, W. K., (2010). Probiotic bacteria and intestinal epithelial barrier function. *American Journal of Physiology-Gastrointestinal and Liver Physiology, 298*(6), G807–G819.
95. Ong, L., Henriksson, A., & Shah, N. P., (2007). Chemical analysis and sensory evaluation of cheddar cheese produced with *Lactobacillus acidophilus, Lb. casei, Lb. paracasei* or *Bifidobacterium* sp. *International Dairy Journal, 17*(8), 937–945.
96. Ooi, L. G., & Liong, M. T., (2010). Cholesterol-lowering effects of probiotics and prebiotics: A review of *in vivo* and *in vitro* findings. *International Journal of Molecular Sciences, 11*(6), 2499–2522.
97. Otte, J. M., & Podolsky, D. K., (2004). Functional modulation of enterocytes by gram-positive and gram-negative microorganisms. *American Journal of Physiology-Gastrointestinal and Liver Physiology, 286*(4), G613–G626.
98. Ouwehand, A. C., (1998). Antimicrobial components from lactic acid bacteria; Chapter 5. In: Salminen, S., & Von, W. A., (eds.), *Lactic Acid Bacteria: Microbiology and Functional Aspects* (2nd edn., pp. 139–159). New York: Marcel Dekker.
99. Parassol, N., Freitas, M., Thoreux, K., Dalmasso, G., Bourdet-Sicard, R., & Rampal, P., (2005). *Lactobacillus casei* DN-114001 inhibits the increase in paracellular permeability of enteropathogenic *Escherichia coli*-infected T84 cells. *Research Microbiology, 156*(2), 256–262.
100. Parvez, S., Malik, K. A., Ah Kang, S., & Kim, H. Y., (2006). Probiotics and their fermented food products are beneficial for health. *Journal of Applied Microbiology, 100*(6), 1171–1185.
101. Pelto, L., Isolauri, E., Lilius, E. M., Nuutila, J., & Salminen, S., (1998). Probiotic bacteria down-regulate the milk-induced inflammatory response in milk-hypersensitive subjects but have an immunostimulatory effect in healthy subjects. *Clinical and Experimental Allergy, 28*(12), 1474–1479.
102. Perdigon, G., Alvarez, S., Rachid, M., Aguero, G., & Gobbato, N., (1995). Immune system stimulation by probiotics. *Journal of Dairy Science, 78*(7), 1597–1606.
103. Perdigon, G., Maldonado, G. C., Valdez, J. C., & Medici, M., (2002). Interaction of lactic acid bacteria with the gut immune system. *European Journal of Clinical Nutrition, 56*(4), S21–S26.
104. Pereira, A. L. F., Maciel, T. C., & Rodrigues, S., (2011). Probiotic beverage from cashew apple juice fermented with *Lactobacillus casei*. *Food Research International, 44*(5), 1276–1283.

105. Phillips, M., Kailasapathy, K., & Tran, L., (2006). Viability of commercial probiotic cultures (*L. acidophilus, Bifidobacterium* sp., *L. casei, L. paracasei*, and *L. rhamnosus*) in cheddar cheese. *International Journal of Food Microbiology, 108*(2), 276–280.
106. Possemiers, S., Marfzorati, M., Verstrate, W., & Wiele, T. V., (2010). Bacteria and chocolate: A successful combination for probiotic delivery. *International Journal of Food Microbiology, 141*(1–2), 97–103.
107. Prado, F. C., Parada, J. L., Pandey, A., & Soccol, C. R., (2008). Trends in non-dairy probiotic beverages. *Food Research International, 41*(2), 111–123.
108. Raju, P. N., & Pal, D., (2009). The physicochemical, sensory, and textural properties of Misti Dahi prepared from reduced-fat buffalo milk. *Food and Bioprocess Technology, 2*(1), 101–108.
109. Rathore, S., Salmerón, I., & Pandiella, S. S., (2012). Production of potentially probiotic beverages using single and mixed cereal substrates fermented with LAB cultures. *Food Microbiology, 30*(1), 239–244.
110. Renuka, B., Kulkarni, S. G., Vijayanand, P., & Prapulla, S. G., (2009). Fructooligosaccharide fortification of selected fruit juice beverages: Effect on the quality characteristics. *LWT-Food Science and Technology, 42*(5), 1031–1033.
111. Rivera-Espinoza, Y., & Gallardo-Navarro, Y., (2010). Non-dairy probiotic products. *Food Microbiology, 27*(1), 1–11.
112. Roberts, J. S., & Kidd, D. R., (2005). Lactic acid fermentation of onions. *LWT-Food Science and Technology, 38*(2), 185–190.
113. Rolfe, R. D., (2000). The role of probiotic cultures in the control of gastrointestinal health. *The Journal of Nutrition, 130*(2S), 396s–402s.
114. Rout, G. K., Jayanta, K. P., Sushanto, G., Yooheon, P., Han-Seung, S., & Gitishree, D., (2018). Benefaction of probiotics for human health: A review. *Journal of Food and Drug Analysis, 26*(3), 927–939.
115. Roy, D., Mainville, I., & Mondou, F., (1997). Selective enumeration and survival of bifidobacteria in fresh cheese. *International Dairy Journal, 7*(12), 785–793.
116. Russell, J. B., & Diez-Gonzalez, F., (1997). The effects of fermentation acids on bacterial growth. *Advances in Microbial Physiology, 39*, 205–234.
117. Saarela, M., Virkarjarvi, I., Alakomi, H. L., Matilla, P. S., & Matto, J., (2006). Stability and functionality of freeze-dried probiotic *Bifidobacterium* cells during storage in juice and milk. *International Dairy Journal, 16*(12), 1477–1482.
118. Sabbagh, N., Gheisari, H. R., & Aminlari, M., (2010). Monitoring the chemical and microbiological changes during ripening of Iranian probiotic low-fat white cheese. *American Journal of Animal and Veterinary Sciences, 5*(4), 249–257.
119. Salmerón, I., Thomas, K., & Pandiella, S. S., (2014). Effect of substrate composition and inoculum on the fermentation kinetics and flavor compound profiles of potentially non-dairy probiotic formulations. *LWT-Food Science and Technology, 55*(1), 240–247.
120. Samia El-Dieb, Abd Rabo, F. H. R., Badran, S. M., Abd El-Fattah, A. M., & Elshaghabee, F. M. F., (2012). The growth behavior and enhancement of probiotic viability in bioyogurt. *International Dairy Journal, 22*(1), 44–47.
121. Sanders, M. E., & Huis in't, V. J., (1999). Bringing a probiotic-containing functional food to the market: Microbiological, product, regulatory and labeling issues. *Antonie Van Leeuwenhoek, 76*, 293–315.
122. Sanders, M. E., & Klaenhammer, T., (2001). Invited review: The scientific basis of *Lactobacillus acidophilus* NCFM functionality as a probiotic. *Journal Dairy Science, 84*(2), 319–331.

123. Sankara, R. D. S., & Deosthale, Y. G., (1983). Mineral composition, ionizable iron and soluble zinc in malted grains of pearl millet and ragi. *Food Chemistry, 11*(3), 217–223.
124. Sanni, A., Franz, C., Schillinger, U., Huch, M., Guigas, C., & Holzapfel, W., (2013). Characterization and technological properties of LAB in the production of sorghurt, a cereal-based product. *Food Biotechnology, 27*(2), 178–198.
125. Sartor, R. B., (2006). Mechanisms of disease: Pathogenesis of Crohn's disease and ulcerative colitis. *Nature Clinical Practice Gastroenterology and Hepatology, 3*(7), 390–407.
126. Savard, T., Gardner, N., & Champagne, C., (2003). Growth of *Lactobacillus* and *Bifidobacterium* cultures in a vegetable juice medium, and their stability during storage in a fermented vegetable juice. *Food Sciences, 23*, 273–283.
127. Schiffrin, E. J., Brassart, D., Servin, A. L., Rochat, F., & Donnet-Hughes, A., (1997). Immune modulation of blood leukocytes in humans by lactic acid bacteria: Criteria for strain selection. *American Journal of Clinical Nutrition, 66*(2), 515S–520S.
128. Schiffrin, E. J., Rochat, F., Link-Amster, H., Aeschlimann, J. M., & Donnet-Hughes, A., (1995). Immunomodulation of human blood cells following the ingestion of lactic acid bacteria. *Journal of Dairy Science, 78*(3), 491–497.
129. Senapati, A. K., Pandey, A., Ann, A., Raj, A., Gupta, A., Das, A. J., Renuka, B., et al., (2015). Indigenous fermented foods involving acid fermentation; Chapter 8. In: Joshi, V. K., (ed.), *Indigenous Fermented Foods of South Asia* (Vol. 7, pp. 431–500). Boca Raton–FL: CRC Press.
130. Shah, N., (2000). Probiotic bacteria: Selective enumeration and survival in dairy foods. *Journal of Dairy Science, 83*(4), 894–907.
131. Sheehan, V. M., Ross, P., & Fitzgerald, G. F., (2007). Assessing the acid tolerance and the technological robustness of probiotic cultures for fortification in fruit juices. *Innovative Food Science and Emerging Technology, 8*(2), 279–284.
132. Sheela, T., & Suganya, R. S., (2012). Studies on anti-diarrhoeal activity of synbiotic plums juice. *International Journal of Scientific and Research Publication, 2*(2), 1–5.
133. Sicard, D., & Legras, J. L., (2011). Bread, beer and wine: Yeast domestication in the *Saccharomyces sensu stricto* complex. *Comptes Rendus Biologies, 334*(3), 229–236.
134. Sidira, M., Karapetsas, A., Galanis, A., Kanellaki, M., & Kourkoutas, Y., (2014). Effective survival of immobilized *Lactobacillus casei* during ripening and heat treatment of probiotic dry-fermented sausages and investigation of the microbial dynamics. *Meat Science, 96*(2), 948–955.
135. Soukoulis, C., Yonekura, L., Gan, H. H., Behboudi-Jobbehdar, S., Parmenter, C., & Fisk, I., (2014). Probiotic edible films as a new strategy for developing functional bakery products: The case of pan bread. *Food Hydrocolloids, 39*, 231–242.
136. Stetinova, V., Smetanova, L., Kvetina, J., Svoboda, Z., Zidek, Z., & Tlaskalova-Hogenova, H., (2010). Caco-2 cell monolayer integrity and effect of probiotic *Escherichia coli* nissle 1917 components. *Neuro Endocrinology Letter, 31*(2), 51–56.
137. Sutas, Y., Hurme, M., & Isolauri, E., (1996). Down-regulation of anti-CD3 antibody-induced IL-4 production by bovine caseins hydrolyzed with *Lactobacillus* GG-derived enzymes. *Scandinavian Journal of Immunology, 43*(6), 687–689.
138. Szajewska, H., & Mrukowicz, J. Z., (2005). Use of probiotics in children with acute diarrhea. *Pediatric Drugs, 7*(2), 111–122.
139. Tamang, J. P., (2017). Indian dietary culture. *Journal of Ethnic Foods, 3*(4), 243–245.
140. Teucher, B., Olivares, M., & Cori, H., (2004). Enhancers of iron absorption: Ascorbic acid and other organic acids. *International Journal for Vitamin and Nutrition Research, 74*(6), 403–419.

141. Tissier, H., (1900). *Recherches sur la Flore Intestinale des Nourrissons (e'tat Normal et Pathologique) [Research on the Intestinal Flora of Infants (Normal state and Pathological)]* (p. 253). PhD Degree Thesis; G. Carré et C. Naud; Paris, France.
142. Todoriki, K., Mukai, T., Sato, S., & Toba, T., (2001). Inhibition of adhesion of food-borne pathogens to Caco-2 cells by lactobacillus strains. *Journal of Applied Microbiology, 91*(1), 154–159.
143. Tsai, C. C., Lin, P. P., & Hsieh, Y. M., (2008). Three lactobacillus strains from healthy infant stool inhibit enterotoxigenic *Escherichia coli* grown *in vitro*. *Anaerobe, 14*(2), 61–67.
144. Tsen, J. H., Lin, Y. P., & King, A. E., (2009). Response surface methodology optimization of immobilized *Lactobacillus acidophilus* banana puree fermentation. *International Journal of Food Science and Technology, 44*(1), 120–127.
145. Tsen, J. H., Lin, Y. P., & King, V. A., (2004). Fermentation of banana media by using k-carrageenan immobilized *Lactobacillus acidophilus*. *International Journal of Food Microbiology, 91*(2), 215–220.
146. Tucović, N., Radulović, Z., Karić, A., Čuk, M., Ipač, N., & Obradović, D., (2004). Primena probiotika u proizvodnji edamskog sira. *Food Industry–Milk and Dairy Products, 15*(1, 2), 12–15.
147. Urdaneta, E., Barrenetxe, J., Aranguren, P., Irigoyen, A., Marzo, F., & Ibáñez, F. C., (2007). Intestinal beneficial effects of kefir-supplemented diet in rats. *Nutrition Research, 27*(10), 653–658.
148. Van, T. M. L., & Miller, M. J., (2011). Lactobacillus adhesion to mucus. *Nutrients, 3*(5), 613–636.
149. Vanderhoof, J. A., & Young, R. J., (1998). Use of probiotics in childhood gastrointestinal disorders. *Journal of Pediatric Gastroenterology and Nutrition, 27*(3), 323–332.
150. Vinderola, C. G., Prosello, W., Ghiberto, D., & Reinheimer, J. A., (2000). Viability of probiotic (*Bifidobacterium, Lactobacillus acidophilus* and *Lactobacillus casei*) and nonprobiotic microflora in Argentinian fresco cheese. *Journal of Dairy Science, 83*(9), 1905–1911.
151. Wang, J., Guo, Z., Zhang, Q., Yan, L., Chen, W., Liu, X. M., & Zhang, H. P., (2009). Fermentation characteristics and transit tolerance of probiotic *Lactobacillus casei* in soymilk and bovine milk during storage. *Journal of Dairy Science, 92*(6), 2468–2476.
152. Yilmaztekin, M., Ozer, B. H., & Atasoy, F., (2004). Survival of *Lactobacillus acidophilus* la-5 and *Bifidobacterium bifidum* BB-02 in white-brined cheese. *International Journal of Food Science and Nutrition, 55*(1), 53–60.
153. Yoon, K. Y., Woodamns, E. E., & Hang, Y. D., (2005). Fermentation of beet juice by beneficial lactic acid bacteria. *LWT-Food Science and Technology, 38*(1), 73–75.
154. Yoon, K. Y., Woodamns, E. E., & Hang, Y. D., (2004). Probiotication of tomato juice by LAB. *Journal of Microbiology, 42*(4), 315–318.
155. Yoon, K. Y., Woodamns, E. E., & Hang, Y. D., (2006). Production of probiotic cabbage juice by lactic acid bacteria. *Bioresource Technology, 97*(12), 1427–1430.
156. Zourari, A., Accolas, J. P., & Desmazeaud, M. J., (1992). Metabolism and biochemical characteristics of yogurt bacteria: A review. *Lait, 72*(1), 1–34.
157. Zyrek, A. A., Cichon, C., Helms, S., Enders, C., Sonnenborn, U., & Schmidt, M. A., (2007). Molecular mechanisms underlying the probiotic effects of *Escherichia coli* nissle 1917 involve ZO-2 and PKC redistribution resulting in tight junction and epithelial barrier repair. *Cell Microbiology, 9*(3), 804–816.

INDEX

D

F

Q

R

T

U